Theory and Applications
of
Numerical Analysis

Theory and Applications of Numerical Analysis

G. M. PHILLIPS
University of St. Andrews

and

P. J. TAYLOR
University of Stirling

1973

Academic Press: London and New York

A Subsidiary of Harcourt Brace Jovanovich, Publishers

ACADEMIC PRESS INC. (LONDON) LTD.
24/28 Oval Road,
London NW1

United States Edition published by
ACADEMIC PRESS INC.
111 Fifth Avenue
New York, New York 10003

Library of Congress Catalog Card Number: 73-7033
ISBN: 0-12-553556-2

PRINTED IN GREAT BRITAIN BY
WILLIAM CLOWES & SONS, LIMITED
LONDON, BECCLES AND COLCHESTER

Preface

Although the student or teacher of numerical analysis is faced with a bewildering choice of text books, we believe that it is still extremely difficult to find a satisfactory text. We have written this book as an introductory text on numerical analysis for undergraduate mathematicians, computer scientists, engineers and other scientists. The material is equivalent to about sixty lectures, to be taken after a first year calculus course, although some calculus is included in the early chapters both to refresh the reader's memory and to provide a foundation on which we may build. We do not assume that the reader has a knowledge of matrix algebra and so have included a brief introduction to matrices. It would, however, help the understanding of the reader if he has taken a basic course in matrix algebra or is taking one concurrently with any course based on this book.

We have tried to give a logical, self-contained development of our subject *ab initio*, with equal emphasis on practical methods and mathematical theory. Thus we have stated algorithms precisely and have usually given proofs of theorems, since each of these aspects of the subject illuminates the other. We believe that numerical analysis can be invaluable in the teaching of mathematics. Numerical analysis is well motivated and uses many important mathematical concepts. Where possible we have used the theme of approximation to give a unified treatment throughout the text. Thus the different types of approximation introduced are used in the chapters on the solution of non-linear algebraic equations, numerical integration and differential equations.

It is not easy to be a good numerical analyst since, as in other branches of mathematics, it takes considerable skill and experience to be able to leap nimbly to and fro from the general and abstract to the particular and practical. A large number of worked examples has been included to help the reader develop these skills. The problems, which are given at the end of each chapter, are to be regarded as an extension of the text as well as a test of the reader's understanding.

We have taken care not to be unfair to those who do not have easy access to a computer. Accordingly, nearly all calculations required in checking our examples and in solving the problems may be performed with the aid of tables or a desk machine. We have included some flow diagrams to emphasize the algorithmic aspect of the text, but have not included any computer programs as we think that these do not promote efficient communication between humans. However, we hope that our material will encourage those with access to a computer to carry out some more extensive calculations. Much can be learned from numerical experiments.

There are some obvious, but deliberate, omissions from our text. We have included only a little on finite differences as they are not extensively used nowadays and are of little conceptual interest. Due to shortage of space we have not included a special treatment of polynomial equations as this subject does not lead directly on to further topics. Regretfully, we felt unable to include any material on eigenvalues or partial differential equations. For a meaningful study of these topics, one requires more knowledge of matrix algebra than we assume the reader has at this stage.

Some of our colleagues have most kindly read all or part of the manuscript and offered wise and valuable advice. These include Dr J. D. Lambert, Prof. P. Lancaster, Dr J. H. McCabe and Prof. A. R. Mitchell. Our thanks go to them and, of course, any errors or omissions which remain are entirely our responsibility. We are also much indebted to our students and to our colleagues for the stimulus given by their encouragement.

Finally, the largest debt of gratitude is due to our wives, who so heroically typed the manuscript and assisted with the final preparation, checking and proof reading.

G. M. PHILLIPS
Department of Applied Mathematics
University of St. Andrews
Scotland

P. J. TAYLOR
Department of Computing Science
University of Stirling
Scotland

December 1972

Contents

Chapter 1

Introduction

1.1 What is numerical analysis?

Numerical analysis is concerned with the mathematical derivation, description and analysis of methods of obtaining numerical solutions of mathematical problems. We shall be interested in *constructive methods* in mathematics; these are methods which show how to construct solutions of mathematical problems. For example, a constructive proof of the existence of a solution to a problem not only shows that the solution exists but also describes how a solution may be determined. A proof which shows the existence of a solution by *reductio ad absurdum*† is not constructive. Consider the following simple example.

Example 1.1 Prove that the quadratic equation

$$x^2 + 2bx + c = 0 \tag{1.1}$$

with real coefficients b and c satisfying $b^2 > c$ has at least two real roots.

Constructive proof For all x, b and c,

$$x^2 + 2bx + c = x^2 + 2bx + b^2 + c - b^2$$
$$= (x + b)^2 + c - b^2.$$

Thus x is a root of (1.1) if, and only if,

$$(x + b)^2 + c - b^2 = 0$$

that is,

$$(x + b)^2 = b^2 - c.$$

On taking square roots, remembering that $b^2 - c > 0$, we see that x is a root if, and only if,

$$x + b = \pm \sqrt{(b^2 - c)}$$

that is

$$x = -b \pm \sqrt{(b^2 - c)}. \tag{1.2}$$

---

† In such a proof we assume that the solution does not exist and obtain a contradiction.

1

Thus there are two real roots and (1.2) shows how to compute these.

Non-constructive proof Let

$$q(x) = x^2 + 2bx + c.$$

Suppose first that there are no roots of (1.1) and hence no zeros of q. As q is a continuous function (see §2.2), we see that $q(x)$ is either always positive or always negative for all x. Now

$$q(-b) = b^2 - 2b^2 + c$$
$$= c - b^2 < 0$$

by the condition on b and c. Thus $q(x)$ must always be negative. However, for $|x|$ large,

$$x^2 > |2bx + c|$$

and thus $q(x) > 0$ for $|x|$ large. We have a contradiction.

Secondly, suppose that q has only one zero. Then from the continuity of q we must have $q(x) > 0$ for x large and $q(x) < 0$ for $-x$ large or vice versa. This contradicts the fact that $q(x) > 0$ for $|x|$ large, regardless of the sign of x.

This proof does not show how to compute the roots. $\square$

The word "*algorithm*" has for a long time been used as a synonym for "method" in the context of constructing the solution of a mathematical problem. It is only recently that a proper mathematical definition of the term has been devised. Much of this important work is due to A. M. Turing (1912–54) who gave a definition based on an abstract concept of a computer. For the purposes of this book we define an algorithm to be a complete and unambiguous description of a method of constructing the solution of a mathematical problem. One difficulty in making a formal definition is deciding precisely what operations are allowed in the method. In this text we shall take these intuitively to mean the basic arithmetic operations of addition, subtraction, multiplication and division. In general, the solution and method occurring in an algorithm need not be numerical. For example, the early Greek geometers devised algorithms using ruler, pencil and compass. One such algorithm is that for constructing the perpendicular to a straight line at a given point. In this book we shall deal only with numerical algorithms.

It is interesting to note that until the beginning of this century much mathematical analysis was based on constructive methods. Famous early mathematicians, for example Isaac Newton (1642–1727), L. Euler (1707–83) and K. F. Gauss (1777–1855), produced many important algorithms and constructive proofs. It is perhaps ironic that the trend in recent years has been towards non-constructive methods, even though the introduction of calculating devices has made the execution of algorithms much easier.

The rapid advances in the design of digital computers have, of course, had a profound effect on numerical analysis. At the time of writing this book, computers can perform well over a million arithmetic operations in a second and this was quite unthinkable thirty years ago. Even ten years ago, speeds were nearer one thousand operations a second. This means that lengthier and more complex calculations may be tackled but the room for error has greatly increased. Unfortunately, the expectations of users of numerical algorithms have more than matched the improvements in computers. It is also becoming difficult for a user to select the best algorithm for his problem, unless he has some understanding of the theoretical principles involved.

Although most numerical algorithms are designed for use on digital computers, the subject numerical analysis should not be confused with *computer programming* and *information processing* (or *data processing*). Those who design and use numerical algorithms need a knowledge of the efficient use of a digital computer for performing calculations. Computer programming is primarily concerned with the problem of coding algorithms (not necessarily numerical) in a form suitable for a computer. Information processing is concerned with the problems of organizing a computer so that data can be manipulated. The data is not necessarily numerical. An introduction to computers is given by Phillips and Taylor [1969].

1.2 Numerical algorithms

For many problems there are no algorithms for obtaining exact numerical values of solutions and we have to be satisfied with approximations to the values we seek. The simplest example is when the result is not a rational number and our arithmetic is restricted to rational numbers. What we do ask of an algorithm, however, is that the error in the result can be made as small as we please. Usually, the higher the accuracy we demand, the greater is the amount of computation required. This is illustrated by the next example, in which a demand for higher accuracy in the result means completing more basic arithmetical operations, as well as increasing the accuracy with which these are performed.

Example 1.2 One algorithm† for square root finding is based on *bisection*. Given $a > 0$ and $\varepsilon > 0$, we seek an approximation to $\sqrt{a}$ with error not greater than ε. We will assume that $a \geqslant 1$. The process is described by the *flow diagram*, Fig. 1.1. At each stage we have two numbers x_0 and x_1. In the calculation, we change x_0 and x_1, keeping $x_0 \leqslant \sqrt{a} \leqslant x_1$ and at the same time repeatedly halving the interval $[x_0, x_1]$.

At the start we take $x_0 = 1$ and $x_1 = a$, as $1 \leqslant \sqrt{a} \leqslant a$. We compute the

† This algorithm is not recommended; it is given merely as a simple illustration. A more efficient square root algorithm is described in §7.6.

mid-point $x = (x_0 + x_1)/2$ and then decide whether x is greater than $\sqrt{a}$. Since we do not know $\sqrt{a}$, we compare x^2 and a. If $x^2 > a$ we replace the number x_1 by x; otherwise we replace x_0 by x. We stop when $x_1 - x_0 \leq 2\varepsilon$, since then we must have $x - \sqrt{a} \leq \varepsilon$, where ε is the accuracy we require. The loop in the flow diagram is called an *iteration*. The number of iterations required increases as ε is decreased. We must also make the arithmetic more accurate as ε is decreased, as otherwise the test whether $x^2 > a$ will give incorrect results and x_0 and x_1 will not remain distinct. If we are prepared to

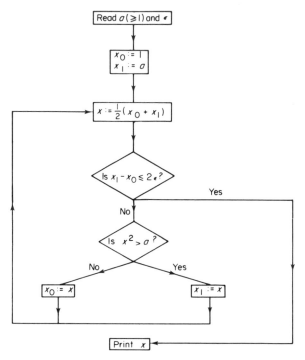

Fig. 1.1. Simple square root algorithm.†

do the arithmetic to arbitrary accuracy, we can make ε as small as we please, provided it is positive. We cannot take $\varepsilon = 0$ as then an infinite number of iterations are required (unless $a = 1$). □

In Example 1.2 we compute the first few members of a sequence of values (the mid-points x) which converge (see §2.3) to $\sqrt{a}$, the solution of the problem. As we can calculate only a finite number of elements of the sequence, we do not obtain the precise value of $\sqrt{a}$ (unless $a = 1$). Many algorithms consist of calculating a sequence and in this context we say that the algorithm is *convergent* if the sequence converges to the desired solution of the problem.

† The symbol := should be interpreted as "becomes equal to".

In the analysis of an algorithm we shall consider the error in the computed approximation to the solution of our problem. This error may be due to various factors, for example, rounding in arithmetic and the termination of an infinite process, as in Example 1.2. Errors, which are a deliberate but often unavoidable part of the solution process, should not be confused with *mistakes* which are due to failure on the part of the human user of an algorithm, particularly when he programs a digital computer. For each algorithm we look for an *error bound*, that is, a bound on the error in the computed solution. A bound which is not directly dependent on the solution and, therefore, can be computed in advance, is called an *a priori* error bound. When directly dependent on the solution, a bound cannot be computed until after the completion of the main part of the calculation and is called an *a posteriori* bound. If, as is often the case, an error bound is very much larger than the actual error, the bound cannot be used to provide a sensible estimate of the error. For this reason we also devise alternative methods of estimating the error for some problems. Error bounds and estimates are often very valuable in telling us about the asymptotic behaviour of the error as we improve approximations made in deriving an algorithm.

There are two ways in which we measure the magnitude of an error. If a is some real number and a^* is an approximation to a we define the *absolute error* to be $|a - a^*|$ and we define the *relative error* to be $|a - a^*|/|a|$. We can see that the absolute error may be quite meaningless unless we have some knowledge of the magnitude of a. We often give relative errors in terms of percentages.

One important point we must consider in looking at an algorithm is its efficiency. We usually measure this in terms of the number of basic arithmetical operations required for a given accuracy in the result. Sometimes we also consider other factors such as the amount of computer storage space required.

1.3 Properly posed and well-conditioned problems

There is one important question which we must investigate before applying a numerical method to a problem and that is "How sensitive is the solution to changes in data in the formulation of the problem?" For example, if we are solving a set of algebraic equations, the coefficients in the equations form data which must be provided. There will certainly be difficulties if small changes in these coefficients have a large effect on the solution. It is very unlikely that we can compute a solution without perturbing these coefficients if only because of errors due to rounding in the arithmetic.

Suppose that $S(d)$ represents † a solution of a problem for a given set of data d.

† $S(d)$ and d may be numbers, functions, matrices, vectors or combinations of these depending on the problem.

Now suppose that $d + \delta d$ is a perturbed set of data and $S(d + \delta d)$ is a corresponding solution. We write $\|\delta d\|$ to denote a non-negative number which is a measure of the magnitude of the perturbation in the data and we write $\|S(d + \delta d) - S(d)\|$ to denote a non-negative number which is a measure of the magnitude of the difference in the solutions. We say that the problem is *properly posed* for a given set of data d if the following two conditions are satisfied.

(i) A *unique* solution exists for the set of data d and for each set of data "near" d. That is, there is a real number $\varepsilon > 0$ such that $S(d + \delta d)$ exists and is unique for all δd with $\|\delta d\| < \varepsilon$.

(ii) The solution $S(d)$ is continuously dependent on the data at d, that is

$$\|S(d + \delta d) - S(d)\| \to 0 \quad \text{whenever} \quad \|\delta d\| \to 0.$$

If there were more than one solution for a given set of data, then we would not know which solution (if any) our numerical method obtains. Usually we can avoid existence of more than one solution by imposing extra conditions on the problem. For example, if we wish to solve a cubic equation we could stipulate that the largest real root is required.

If the condition (ii) above is not satisfied, it would be difficult to use a numerical algorithm as then there are data which are arbitrarily close to d and such that the corresponding solutions are quite different from the desired solution $S(d)$.

We say that a problem which is properly posed for a given set of data is *well-conditioned* if every small perturbation of that data results in a relatively small change in the solution. If the change in the solution is large we say that the problem is *ill-conditioned*. Of course these are relative terms.

Often the solution of a properly posed problem is *Lipschitz* dependent on the data (see §2.2). In this case there are constants $L > 0$ and $\varepsilon > 0$ such that

$$\|S(d + \delta d) - S(d)\| \leqslant L\|\delta d\| \tag{1.3}$$

for all δd with $\|\delta d\| < \varepsilon$. In this case we can see that the problem is well-conditioned if L is not too large. The restriction, $\|\delta d\| < \varepsilon$, only means that there are limits within which the data must lie if (1.3) is to hold. This restriction should not worry us unless ε is small. If L has to be made large before (1.3) is satisfied or if no such L exists, then a small change in the data can result in a large change in the solution and thus the problem is ill-conditioned.

As we remarked earlier, an ill-conditioned problem may be difficult to solve numerically as rounding errors may seriously affect the solution. In such a case we would at least need to use higher accuracy in the arithmetic than is usual. If the problem is not properly posed, rounding errors may make the numerical solution meaningless, regardless of the accuracy of the arithmetic. Sometimes the data contains *inherent* errors and is only known to a certain degree of accuracy. In such cases the amount of confidence we can place in the results depends on the conditioning of the problem. If it is ill-conditioned, the results may be subject to large errors and, if the problem is not properly posed, the results may be meaningless.

The next example illustrates that the properties properly posed and well-conditioned depend on both the problem and the data.

Example 1.3   We consider the computation of values of the quadratic

$$q(x) \equiv x^2 + x - 1150.$$

This is an ill-conditioned problem if x is near a root of the quadratic. For example, for $x = 100/3$, $q(100/3) = -50/9 \simeq -5.6$. However, if we take $x = 33$, which differs from $100/3$ by only 1%, we find that $q(33) = -28$ which is approximately five times the value of $q(100/3)$. To obtain a relation of the form (1.3) when $x = 100/3$ we need to take $L \simeq 70$. (See Problem 1.5.) □

Suppose that we wish to solve a problem which we will denote by P. We often find it necessary to replace P by another problem P_1 (say) which approximates to P and is more amenable to numerical computation. We now have two types of error. Firstly, there is the difference between P and P_1 and we write E_P to denote some measure of this error. Often this results from cutting off an infinite process such as forming the sum of an infinite series (see §2.3) after only a finite number of steps, in which case we call it the *truncation* error. Secondly, there is the error consisting of the difference between the solutions of P and P_1 and we denote some measure of this by E_S. This second type of error is sometimes called the *global* error. Of course there is a relation between E_P and E_S, but it does not follow that if E_P is small then so is E_S. We shall derive relations between E_P and E_S for different types of problem and, for a satisfactory method, will require E_S to become arbitrarily small as the approximation is improved, so that E_P tends to zero. If this property does hold we say that the resulting algorithm is *convergent*. Unfortunately there are problems and approximations such that E_S does not tend to zero as E_P is decreased.

Example 1.4   Suppose that f is some given real valued function of a real variable x and that we seek values of the derivative f' for given particular values of x in the interval $a \leqslant x \leqslant b$. We will assume that f' cannot be obtained explicitly as f is a very complicated function but that we can compute $f(x)$ for any given value of x.

Our algorithm consists of replacing f by another function g which can easily be differentiated. We then compute values of $g'(x)$ instead of $f'(x)$. In Chapter 5 we will describe one way in which g can be chosen and that provided f satisfies certain conditions we can make the error $E_P(x) = f(x) - g(x)$ arbitrarily small for all x with $a \leqslant x \leqslant b$. Unfortunately it is possible for $E_S(x) = f'(x) - g'(x)$ to be arbitrarily large even though $E_P(x)$ is arbitrarily small. For example, if $E_P(x) = (1/n) \sin(n^2 x)$, where $n = 1, 2, 3, \ldots$ corresponds to different choices of g, then

$$E_P(x) \to 0 \quad \text{as} \quad n \to \infty,$$

for all values of x. However,

$$E_S(x) = n \cos(n^2 x)$$

and thus $E_S(x) \to \infty$ as $n \to \infty$, for $x = 0$. $\square$

If we replace a problem P by another, more convenient problem P_1 which is an approximation to P, we shall normally expect both P and P_1 to be properly posed and reasonably well-conditioned. If P were not properly posed, in general we would not obtain convergence of the solution of P_1 to that of P. We are almost certain in effect to perturb the data of P in making the approximation P_1. We therefore require, as a necessary condition for the convergence of our algorithm, that the original problem P be properly posed. The convergence will be "slow" unless P is well-conditioned.

We also have to solve P_1 using numerical calculations and we are almost certain to perturb the data of P_1, which should therefore be properly posed and well-conditioned. If P is well-conditioned, we normally expect P_1 to be well-conditioned; if not, we should look for a better alternative than P_1.

In some numerical algorithms, especially those for differential equations, we compute elements of a sequence as in Example 1.2 but are interested in each element of the sequence and not just the limit. In both types of problem we use a *recurrence relation*, which relates each element of a sequence to earlier values. For example, we may compute elements of a sequence $\{y_n\}_{n=0}^{\infty}$ using a recurrence relation of the form

$$y_{n+1} = F(y_n, y_{n-1}, \ldots, y_{n-m}) \tag{1.4}$$

for $n = m, m + 1, m + 2, \ldots$, where F is a known function of $m + 1$ variables. We need starting values $y_0, y_1, \ldots, y_m$ and normally these must be given. In using (1.4) we are certain to make rounding errors. Now an error introduced in any one step (one value of n) will affect all subsequent values and, therefore, we need to investigate the propagated effect of rounding errors. These effects may be very important, especially as modern computers allow us to attempt to compute very many elements of a sequence and so introduce many rounding errors. In this context we use the term *stability* in referring to whether the process is properly posed and well-conditioned.

Example 1.5 If $\{y_n\}_{n=0}^{\infty}$ satisfies

$$y_{n+1} = 100.01 y_n - y_{n-1}, \qquad n \geqslant 1, \tag{1.5}$$

with $y_0 = 1$ and $y_1 = 0.01$ then $y_5 = 10^{-10}$, whereas if $y_0 = 1 + 10^{-6}$ and $y_1 = 0.01$ then $y_5 \simeq -1.0001$. The large difference in the value of y_5 shows that the process is very ill-conditioned. In fact we say that the recurrence relation is *numerically unstable*. Further details are given in §11.11, particularly Example 11.18. □

Problems

Section 1.1

1.1   Give both constructive and non-constructive proofs of the following statements.
   (i) Every quadratic equation has at most two distinct real roots.
   (ii) There are at least two real values of x such that

$$2 \cos(2 \cos^{-1} x) = 1.$$

   (iii) Given a straight line ℓ and a point P not on ℓ, there is a perpendicular to ℓ which passes through P.

Section 1.2

1.2   Show that an a priori error bound for the result of the bisection method for finding $\sqrt{a}$ with $0 < a < 1$ is

$$|x - \sqrt{a}| \leqslant (1 - a)/2^N,$$

where N is the number of iterations completed.

1.3   In the algorithm described by Fig. 1.1 for determining square roots, we have at each stage

$$|x - \sqrt{a}| \leqslant \frac{x_1 - x_0}{2}.$$

Show that if

$$(x_1 - x_0)^2 < 4a\varepsilon^2,$$

then the relative error in x as an approximation to $\sqrt{a}$ satisfies

$$\left| \frac{x - \sqrt{a}}{\sqrt{a}} \right| < \varepsilon.$$

Hence modify the algorithm so that the iterations stop when the relative error is less than ε.

Section 1.3

1.4   Show that the equations

$$x + y = 2$$
$$\tfrac{1}{6}x + \tfrac{1}{6}y = \tfrac{1}{3}$$
$$x + 2y = 3$$

have a unique solution, but that the problem of computing this solution is not properly posed, in the sense given in §1.3. Notice that, if we express coefficients as rounded decimal fractions, there is no solution.

1.5   If $q(x)$ is the quadratic of Example 1.3, show that

$$q(x + \delta x) - q(x) = (2x + \delta x + 1).\delta x$$

and thus that, if $33 \leqslant x \leqslant 34$ and $|\delta x| \leqslant 1$, then

$$|q(x + \delta x) - q(x)| \leqslant 70|\delta x|.$$

Chapter 2

Basic Analysis

2.1 Functions

Definition 2.1   A *set* is a collection of objects. The number of objects may be finite or infinite. □

We write

$$S = \{a, b, c, \ldots\},$$

where S denotes a set and $a, b, c, \ldots$ denote the objects belonging to the set. We call $a, b, c, \ldots$ the *elements* or *members* of the set S and we write $a \in S$ as a shorthand for "a belongs to S". If S and T are sets such that† $x \in T \Rightarrow x \in S$, we say that T is a subset of S and write $T \subset S$.

Example 2.1   The set of all numbers x such that $a \leqslant x \leqslant b$ is denoted by $[a, b]$ and is called a *closed* interval. The set of x such that $a < x < b$ is denoted by (a, b) and is called an *open* interval. We write $(a, b]$ to denote the set of x such that $a < x \leqslant b$ and $(-\infty, \infty)$ to denote the set of all real numbers. □

Definition 2.2   If, given two sets X and Y, there is a relation between them such that each member of X is related to exactly one member of Y, we call this relation a *function* or *mapping* from X to Y. □

Example 2.2   If, at some instant, X denotes the set of all London buses in use and Y the set of all human beings, the relation "y is the driver of x, with $x \in X$, $y \in Y$" is a function from X to Y. □

Example 2.3   If X is the set $[0, 1]$ and Y is the set of all real numbers y, then the relation

$$y = x + 2, \qquad 0 \leqslant x \leqslant 1,$$

defines a function from X to Y. □

† The symbol $\Rightarrow$ means "implies".

We sometimes write $y = f(x)$, where $x \in X$ and $y \in Y$ to denote a function by which x is mapped into y. We then refer to this as "the function f". Note that not every $y \in Y$ need be such that $y = f(x)$, for some $x \in X$. X is called the *domain* of f and the set of all $y \in Y$ such that $y = f(x)$, for some $x \in X$, is called the *range* of f or ran(f). Thus we have

$$\text{ran}(f) \subset Y.$$

In Example 2.3, the domain is the closed interval $[0, 1]$ and the range is $[2, 3]$.

Example 2.4 Let X denote the set of all vectors $\mathbf{x}$ with n real elements and $\mathbf{A}$ denote an $n \times n$ matrix with all elements real. Then (see Chapter 8) $\mathbf{A}\mathbf{x} \in X$ and the relation $\mathbf{y} = \mathbf{A}\mathbf{x}$ is a function from X to X. $\square$

We may depict a function schematically as in Fig. 2.1, which shows the sets X, Y and ran(f) and shows a "typical element" $x \in X$ being mapped into an element $y \in Y$ by f.

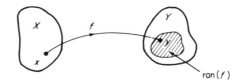

Fig. 2.1. Schematic representation of a function.

If X and Y are subsets of the real numbers and $y = f(x)$ denotes a function from X to Y, the reader will be used to representing the function by a graph, as in Fig. 2.2. Such a function is called a function of one (real) variable.

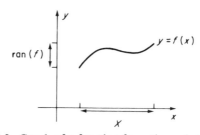

Fig. 2.2. Graph of a function from the reals to the reals.

Example 2.5 Let X denote the set of *ordered pairs* (x, y) where x and y are reals and let Y denote the set of all real numbers. A mapping from X to Y is called a function of two real variables. Similarly, we may define a function of n variables. The set of all ordered pairs (x, y) is called the *xy*-plane. Geometrically, the function $z = f(x, y)$ may be represented by a *surface* in

xyz-space. Thus, for instance, the function $z = 1 + x + y$ is represented by the plane which passes through the three points with coordinates $(-1, 0, 0)$, $(0, -1, 0)$, $(0, 0, 1)$. $\square$

Definition 2.3 A function f from X to $\operatorname{ran}(f) \subset Y$ is said to be *one-one* if to each $y \in \operatorname{ran}(f)$ there corresponds *exactly one* $x \in X$ such that $y = f(x)$. $\square$

Thus the function of Example 2.3 is one-one. The function defined by $y = x^2$, $-1 \leqslant x \leqslant 1$, is not one-one since each y such that $0 < y \leqslant 1$ corresponds to two values of x.

Definition 2.4 If f is a one-one function from X to $\operatorname{ran}(f) \subset Y$, we may define the *inverse function* of f by

$$x = f^{-1}(y). \quad \square$$

The inverse function f^{-1} maps $y \in \operatorname{ran}(f)$ into the unique $x \in X$ such that $y = f(x)$. Note that $\operatorname{ran}(f^{-1}) = X$.

Example 2.6 If $y = f(x)$, $x \in X$, is the function $y = x + 2$, $0 \leqslant x \leqslant 1$, of Example 2.3, then the inverse function $x = f^{-1}(y)$ exists and is given by $x = y - 2$, $2 \leqslant y \leqslant 3$. $\square$

For the remainder of this chapter we shall be concerned mainly with functions from the reals although most of the functions may also be defined for a complex variable. We start by briefly considering polynomials and circular functions. Later in the chapter we look at logarithmic and exponential functions.

Polynomials

The function

$$y = a_0 + a_1 x + \cdots + a_n x^n, \tag{2.1}$$

where x belongs to the reals and the *coefficients* $a_0, a_1, \ldots, a_n$ are fixed real numbers with $a_n \neq 0$, is called a *polynomial* in x of degree n.

The function in Example 2.3 is thus a polynomial of degree 1. Throughout the text, we use P_n to denote the set of all polynomials of degree n or less. If p is a polynomial of degree $n \geqslant 1$ and α is a *zero* of p, that is $p(\alpha) = 0$, then we have (see Problem 2.2)

$$p(x) = (x - \alpha)q(x), \tag{2.2}$$

where q is a polynomial of degree $n - 1$. If p is a polynomial of degree $n \geqslant r > 1$ and

$$p(x) = (x - \alpha)^r q(x),$$

Numerical Analysis

where q is a polynomial of degree $n - r$ such that $q(\alpha) \neq 0$, we say that α is a multiple zero of p of order r. If $r = 2$, we say that α is a double zero. For example, $p(x) \equiv (x - 1)^2(x - 2)$ has three real zeros, a single zero at $x = 2$ and a double zero at $x = 1$.

We can use (2.2) to prove that a polynomial of degree n cannot have more than n real zeros unless the polynomial is identically zero (that is, all its coefficients are zero). This result is clearly true for a polynomial of degree one (whose graph is a straight line) and using (2.2) and the principle of induction (see §2.3) we can extend this to the general case.

Circular functions

In Fig. 2.3, AB is an arc of a circle of radius 1 and centre O. In radian measure, the *angle* θ is defined as the length of the circular arc AB. Thus the

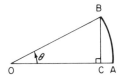

Fig. 2.3. OA = OB = 1, sin θ = BC, cos θ = OC.

angle of one complete revolution is 2π and a right angle is $\tfrac{1}{2}\pi$. Evidently BC and OC are functions of θ and these are denoted by

$$BC = \sin \theta, \qquad OC = \cos \theta, \qquad 0 < \theta < \tfrac{1}{2}\pi.$$

Since the sum of the angles of triangle OBC is π, we have

$$\cos \theta = \sin(\tfrac{1}{2}\pi - \theta), \qquad (2.3)$$

where $0 < \theta < \tfrac{1}{2}\pi$.

The domain of the sine function may be extended from $(0, \tfrac{1}{2}\pi)$ to all real θ as follows:

$$\sin 0 = 0, \ \sin \tfrac{1}{2}\pi = 1, \qquad (2.4)$$
$$\sin(\pi - \theta) = \sin \theta, \qquad 0 \leqslant \theta \leqslant \tfrac{1}{2}\pi,$$
$$\sin(-\theta) = -\sin \theta, \qquad 0 \leqslant \theta \leqslant \pi,$$
$$\sin(\theta + 2n\pi) = \sin \theta, \qquad n = 0, \pm 1, \pm 2, \ldots. \qquad (2.5)$$

The domain of cos θ is then extended to all real θ by allowing (2.3) to hold for all θ. Note that the values in (2.4) are chosen to make sin θ *continuous*. (See §2.2.) Equation (2.5) demonstrates the *periodic* nature of sin θ. Both sin θ and cos θ are said to be periodic, with period 2π. (See Fig. 2.4.) We note some identities satisfied by the sines and cosines:

$$\sin^2 \theta + \cos^2 \theta = 1, \quad \text{for all } \theta, \tag{2.6}$$

$$\sin \theta + \sin \phi = 2 \sin\left(\frac{\theta + \phi}{2}\right) \cos\left(\frac{\theta - \phi}{2}\right), \tag{2.7}$$

$$\cos \theta + \cos \phi = 2 \cos\left(\frac{\theta + \phi}{2}\right) \cos\left(\frac{\theta - \phi}{2}\right). \tag{2.8}$$

The identities (2.7) and (2.8) hold for all θ and ϕ. Other circular functions (for example, tan θ) may be defined in terms of the sine and cosine.

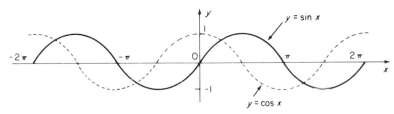

Fig. 2.4. The sine and cosine functions.

Further classes of functions

Definition 2.5 A function f from the reals to the reals is said to be *even* if $f(-x) = f(x)$ and to be *odd* if $f(-x) = -f(x)$, for all x. $\square$

Thus sin x is odd, cos x is even and $1 + x$ is neither.

Definition 2.6 A function f defined on $[a, b]$ is said to be *convex* if

$$f(\lambda x_1 + (1 - \lambda)x_2) < \lambda f(x_1) + (1 - \lambda)f(x_2)$$

for all $x_1, x_2 \in [a, b]$ and any $\lambda \in (0, 1)$. $\square$

Geometrically, this means that a chord joining any two points on the curve $y = f(x)$ lies above the curve, as for the function x^2 on $(-\infty, \infty)$. We can similarly define a *concave* function by reversing the inequality so that points on a chord must lie below the curve.

Definition 2.7 A function f is said to be *bounded above* on an interval I (which may be finite or infinite) if there exists a number M such that

$$f(x) \leqslant M, \quad \text{for all } x \in I.$$

Any such number M is said to be an *upper bound* of f on I. The function is said to be *bounded below* on I if there exists a number m such that

$$f(x) \geqslant m, \quad \text{for all } x \in I.$$

If a function is bounded above and below, it is said to be *bounded*. □

If f is bounded above on I, it can be shown that there exists a *least upper bound*, say M^*, such that $M^* \leqslant M$, for every upper bound M. The least upper bound, which is unique, is also called the *supremum*, denoted by

$$\sup_{x \in I} f(x).$$

The supremum is thus the smallest number which is not less than any $f(x)$, $x \in I$. Similarly, if f is bounded below on I, we have the *greatest lower bound* or *infimum*, denoted by

$$\inf_{x \in I} f(x),$$

which is the greatest number which is not greater than any $f(x)$, $x \in I$.

Example 2.7 For $\sin x$ on $[0, 2\pi]$, we have

$$\inf_{0 \leqslant x \leqslant 2\pi} \sin x = -1, \qquad \sup_{0 \leqslant x \leqslant 2\pi} \sin x = 1,$$

the infimum and supremum being attained at $x = 3\pi/2$ and $\pi/2$ respectively. □

Example 2.8 For $1/x^2$ on $(-\infty, \infty)$ there is no supremum, since $1/x^2$ is not bounded above. However, we have

$$\inf_{-\infty < x < \infty} 1/x^2 = 0$$

although the infimum is *not attained*, since there is no value of x for which $1/x^2 = 0$. □

If a supremum is attained, we write "max" (for maximum) in place of "sup". If an infimum is attained, we write "min" (for minimum) in place of "inf".

2.2 Limits and derivatives

We define

$$g(x) = \frac{\sin x}{x}, \qquad 0 < x \leqslant \tfrac{1}{2}\pi.$$

Note that we have not defined $g(0)$. What can we say about $g(x)$ *near* $x = 0$? Inspection of a table of values of $\sin x$ (see Table 2.1) suggests that $g(x)$

Table 2.1 Limit of $(\sin x)/x$ as $x \to 0$.

x	0.50	0.20	0.10	0.05	0.02
$(\sin x)/x$	0.959	0.993	0.998	1.000	1.000

approaches the value 1 as x approaches 0. To pursue this, we see from Fig. 2.5 that

area of $\triangle$ OAB < area of sector OAB < area of $\triangle$ OAC

or (see Problem 2.11)

$$\tfrac{1}{2}\sin x < \tfrac{1}{2}x < \tfrac{1}{2}\tan x, \qquad 0 < x < \tfrac{1}{2}\pi.$$

This yields

$$1 < \frac{x}{\sin x} < \frac{1}{\cos x}$$

and thus

$$\cos x < \frac{\sin x}{x} < 1. \qquad (2.9)$$

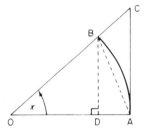

Fig. 2.5. OA = OB = 1; AB is a circular arc; OAC is a right angle.

Since $\cos x$ approaches the value 1 as x approaches 0, (2.9) shows that $(\sin x)/x$ also approaches 1 as x approaches 0. More precisely, we deduce from (2.9) (see Problem 2.12) that

$$0 < 1 - \frac{\sin x}{x} < \tfrac{1}{2}x^2. \qquad (2.10)$$

This shows that $(\sin x)/x$ may be made as near to 1 *as we please* for all values of x sufficiently near to zero. This is an example of a *limit*.

Definition 2.8 Suppose that $x_0 \in [a, b]$ and the function g is defined at every point of $[a, b]$ with the possible exception of x_0. We say that g has limit c at x_0 if to each $\varepsilon > 0$ there corresponds a number $\delta > 0$ such that

$$x \in [a, b] \quad \text{with} \quad |x - x_0| < \delta \quad \text{and} \quad x \neq x_0 \Rightarrow |g(x) - c| < \varepsilon.$$

As a short-hand for the classical ε-δ terminology, we write

$$\lim_{x \to x_0} g(x) = c. \quad \square$$

Example 2.9   The function $(\sin x)/x$ has the limit 1 at $x = 0$. Given $\varepsilon > 0$, we may take $\delta = (2\varepsilon)^{1/2}$ since, for all x such that $x \in (0, \pi/2)$ with $|x| < (2\varepsilon)^{1/2}$, we have $\frac{1}{2}x^2 < \varepsilon$ and thus from (2.10)

$$\left| \frac{\sin x}{x} - 1 \right| < \varepsilon,$$

as required by Definition 2.8. □

We say that

$$\lim_{x \to \infty} g(x) = c$$

if to any $\varepsilon > 0$ there corresponds a number M such that

$$|g(x) - c| < \varepsilon \quad \text{for all } x > M.$$

Properties of limits

If the functions g and h are both defined on some interval containing x_0, a point at which both have a limit, then

(a) $\lim_{x \to x_0} (\alpha g(x) + \beta h(x)) = \alpha \lim_{x \to x_0} g(x) + \beta \lim_{x \to x_0} h(x),$

where α and β are any constants;

(b) $\lim_{x \to x_0} g(x)h(x) = \left(\lim_{x \to x_0} g(x) \right) \left(\lim_{x \to x_0} h(x) \right);$

(c) $\lim_{x \to x_0} \frac{1}{g(x)} = 1 \Big/ \lim_{x \to x_0} g(x)$

provided $\lim_{x \to x_0} g(x) \neq 0$.

Continuity

Consider the function $g(x)$ defined by

$$g(x) = \begin{cases} -1, & -1 \leqslant x \leqslant 0 \\ 1, & 0 < x \leqslant 1. \end{cases}$$

We say that g has a *discontinuity* at $x = 0$, where the value of $g(x)$ jumps from -1 on the left to $+1$ on the right.

Definition 2.9   A function g defined on $[a, b]$ is said to be *continuous* at $x_0 \in [a, b]$ if g has a limit at x_0 and this limit is $g(x_0)$. □

Thus g is continuous at a point $x_0 \in [a, b]$ if, to each $\varepsilon > 0$, there corresponds a number $\delta > 0$ such that

$$x \in [a, b] \quad \text{with} \quad |x - x_0| < \delta \Rightarrow |g(x) - g(x_0)| < \varepsilon.$$

We say that g is continuous on $[a, b]$ if g is continuous at each $x_0 \in [a, b]$. The concept of continuity is applied also to functions defined on open and infinite intervals.

Definition 2.10 A function g, defined on $[a, b]$, is said to be *uniformly continuous* on $[a, b]$ if, to each $\varepsilon > 0$, there corresponds a number $\delta > 0$ such that

$$x \in [a, b] \quad \text{with} \quad |x - x_0| < \delta \Rightarrow |g(x) - g(x_0)| < \varepsilon \qquad (2.11)$$

for *every* $x_0 \in [a, b]$. $\square$

The point of this definition is that, given an $\varepsilon > 0$, the corresponding δ which appears in (2.11) is *independent* of x_0, so that the same δ will serve for each point of $[a, b]$. We state without proof that if a function g is continuous on the closed interval $[a, b]$ it is also uniformly continuous on $[a, b]$.

We quote the following theorems concerning continuous functions. The first of these is a simple consequence of the properties of limits.

Theorem 2.1 If f and g are continuous on an interval I, so also are fg and $\alpha f + \beta g$, where α and β are constants. If f is continuous on I and $f(x) \neq 0$ for all $x \in I$ then $1/f$ is continuous on I. $\square$

Theorem 2.2 If f is continuous on $[a, b]$ and η is a number lying between $f(a)$ and $f(b)$, then there is a number $\xi \in [a, b]$ such that $f(\xi) = \eta$. $\square$

Thus a function f which is continuous on $[a, b]$ assumes every value between $f(a)$ and $f(b)$. As a special case of this, we have:

Theorem 2.3 If f is continuous on $[a, b]$, there exists a number $\xi \in [a, b]$ such that

$$f(\xi) = \tfrac{1}{2}(f(a) + f(b)). \square$$

The notion of continuity is easily extended to functions of several variables. For example, for two variables we state:

Definition 2.11 A function $f(x, y)$ defined on some region R of the xy-plane is said to be continuous at a point (x_0, y_0) of R if, to each $\varepsilon > 0$, there corresponds a number $\delta > 0$ such that

$$(x, y) \in R \quad \text{with} \quad |x - x_0| < \delta \quad \text{and} \quad |y - y_0| < \delta \Rightarrow$$
$$|f(x, y) - f(x_0, y_0)| < \varepsilon. \square$$

Lipschitz condition

Definition 2.12   A function f, defined on $[a, b]$, is said to satisfy a Lipschitz condition on $[a, b]$ if there exists a constant $L > 0$ such that

$$|f(x_1) - f(x_2)| \leqslant L|x_1 - x_2|, \qquad\qquad (2.12)$$

for all $x_1, x_2 \in [a, b]$. L is called the Lipschitz constant. $\square$

We may deduce from (2.12) that if f satisfies a Lipschitz condition on $[a, b]$, then f is uniformly continuous on $[a, b]$.

Example 2.10   Consider $f(x) \equiv x^2$ on $[-1, 1]$. Then

$$f(x_1) - f(x_2) = x_1{}^2 - x_2{}^2 = (x_1 - x_2)(x_1 + x_2).$$

Thus on $[-1, 1]$

$$|f(x_1) - f(x_2)| \leqslant 2|x_1 - x_2|,$$

showing that $f(x) \equiv x^2$ satisfies a Lipschitz condition on $[-1, 1]$ with Lipschitz constant 2. $\square$

Example 2.11   Consider $f(x) \equiv x^{1/2}$, with $x \geqslant 0$. Then

$$f(x_1) - f(x_2) = x_1^{1/2} - x_2^{1/2} = (x_1 - x_2)/(x_1^{1/2} + x_2^{1/2}).$$

By taking x_1 and x_2 sufficiently close to 0 we can make $1/(x_1^{1/2} + x_2^{1/2})$ as large as we please. Thus the function $x^{1/2}$ does not satisfy a Lipschitz condition on any interval which includes $x = 0$. $\square$

This last example demonstrates that not every continuous function satisfies a Lipschitz condition.

Derivatives

We are often interested in how much $f(x)$ changes for a given change in x. As a measure of the *rate of change* as x varies from, say, x_0 to $x_0 + h$, we may calculate the quotient

$$\frac{f(x_0 + h) - f(x_0)}{(x_0 + h) - x_0} = \frac{\text{QR}}{\text{PR}},$$

as shown in Fig. 2.6. Geometrically, this ratio is the gradient of the chord PQ. The number h (shown as positive in Fig. 2.6) may be positive or negative. The rate of change of $f(x)$ with respect to x at x_0 is defined as

$$\lim_{h \to 0} \frac{f(x_0 + h) - f(x_0)}{h},$$

assuming that this limit exists. Geometrically, the limit gives the gradient of the tangent to the curve at **P**. Replacing x_0 by x, we write

$$f'(x) = \lim_{h \to 0} \frac{f(x + h) - f(x)}{h} \tag{2.13}$$

and call this limit the *derivative* of f with respect to x. Instead of $f'(x)$ we sometimes write $(d/dx)f(x)$, df/dx or dy/dx if $y = f(x)$. If the derivative of f exists for all x in some interval I, we say that f is differentiable on I and write f' for the resulting function on I.

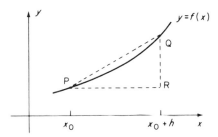

Fig. 2.6. Measurement of rate of change.

Properties of derivatives

(a) $\dfrac{d}{dx}(\alpha f(x) + \beta g(x)) = \alpha \dfrac{d}{dx} f(x) + \beta \dfrac{d}{dx} g(x).$ (2.14)

(b) $\dfrac{d}{dx}(f(x)\, g(x)) = f(x) \cdot \dfrac{d}{dx} g(x) + g(x) \cdot \dfrac{d}{dx} f(x).$ (2.15)

(c) If we write $u = g(x)$, then

$$\frac{d}{dx} f(g(x)) = \frac{df}{du} \cdot \frac{du}{dx} = f'(g(x)) \cdot g'(x). \tag{2.16}$$

It is assumed that all the derivatives occurring in the right sides of these equations exist. The verification of these properties uses the basic properties of limits.

Property (c) is called the "function of a function" rule or "chain rule". One application of this rule concerns inverse functions. If the function $y = f(x)$ has an inverse function $x = \phi(y)$, we may write

$$y = f(\phi(y))$$

and deduce from (2.16) that

$$1 = \frac{dy}{dy} = \frac{dy}{dx} \cdot \frac{dx}{dy},$$

so that

$$\frac{dx}{dy} = 1 \bigg/ \frac{dy}{dx}. \tag{2.17}$$

The derivative of an arbitrary polynomial is obtained by first showing directly from the definition of a derivative (see Problem 2.16) that

$$\frac{d}{dx} x^n = nx^{n-1}, \qquad n \text{ an integer.} \tag{2.18}$$

(This result holds also if n is any real number.) Applying (2.14) repeatedly and using (2.18) we find

$$\frac{d}{dx}(a_0 + a_1 x + a_2 x^2 + \cdots + a_n x^n) = a_1 + 2a_2 x + \cdots + na_n x^{n-1}.$$

For the derivatives of the circular functions, we find (Problem 2.18) that

$$\frac{d}{dx} \sin x = \cos x. \tag{2.19}$$

Making the substitution $t = \pi/2 - x$, using the chain rule and finally replacing t by x, we deduce from (2.19) that

$$\frac{d}{dx} \cos x = -\sin x.$$

We now state two important theorems involving derivatives.

Theorem 2.4 (Rolle's theorem.)   If f is continuous on $[a, b]$ and differentiable in (a, b) and $f(a) = f(b) = 0$, then there exists at least one point $\xi \in (a, b)$ such that $f'(\xi) = 0$. $\square$

Theorem 2.5 (Mean value theorem.)   If f is continuous on $[a, b]$ and differentiable in (a, b), then there exists at least one point $\xi \in (a, b)$ such that

$$f(b) - f(a) = (b - a)f'(\xi). \square \tag{2.20}$$

Rolle's theorem is a special case of the mean value theorem, which is in turn a particular case of Taylor's theorem 3.1. We shall need Rolle's theorem in the proof of Theorem 3.1.

From the mean value theorem, if $x_1, x_2 \in [a, b]$ then

$$f(x_1) - f(x_2) = (x_1 - x_2)f'(\xi),$$

where ξ depends on the choice of x_1 and x_2 and thus

$$|f(x_1) - f(x_2)| \leqslant L|x_1 - x_2|,$$

where

$$L = \sup_{a \leqslant x \leqslant b} |f'(x)|.$$

Hence

differentiability $\Rightarrow$ Lipschitz condition $\Rightarrow$ uniform continuity.

We cannot reverse these implication symbols, as may be seen from Example 2.11 and Problem 2.19.

Higher derivatives

If the derivative of f' exists at x, we call this the second derivative of f and write it as $f''(x)$. Similarly, we define third, fourth and higher order derivatives from

$$f^{(n)}(x) = \frac{d}{dx}(f^{(n-1)}(x)), \qquad n = 2, 3, \ldots.$$

The nth derivative is sometimes written as $d^n f/dx^n$ or $d^n y/dx^n$ if $y = f(x)$. To unify the notation, we sometimes write $f^{(0)}(x)$ and $f^{(1)}(x)$ to denote $f(x)$ and $f'(x)$ respectively.

We have already noted the geometrical interpretation of $f'(x)$ as the gradient of the tangent to $y = f(x)$ at the point $(x, f(x))$. Thus if $f'(x) > 0$, $f(x)$ is increasing and if $f'(x) < 0$, $f(x)$ is decreasing. The second derivative too is easily related to the graph of $y = f(x)$: if $f''(x) > 0$, f' is increasing and therefore f is convex. (See Problem 2.20.) If $f''(x) < 0$, $-f$ is convex and thus f is concave.

Maxima and minima

Definition 2.13 We say that $f(x_0)$ is a local maximum of a function f if there exists a number $\delta > 0$ such that $f(x_0) > f(x)$ for all x such that $x \neq x_0$ and $|x - x_0| < \delta$. $\square$

We often drop the adjective "local" and speak simply of a maximum of $f(x)$ at $x = x_0$, meaning that $f(x_0)$ is the greatest value of $f(x)$ in the vicinity of x_0. Similarly we say we have a minimum of $f(x)$ at $x = x_0$ if $f(x_0)$ is the smallest value of $f(x)$ in the vicinity of x_0. If $f(x_0)$ is the greatest (least) value of $f(x)$ for *all* $x \in [a, b]$, we call $f(x_0)$ the *global* maximum (minimum) on $[a, b]$.

Theorem 2.6 If f'' is continuous at x_0 and $f'(x_0) = 0$ then $f(x_0)$ is a local maximum (minimum) if $f''(x_0) < 0$ (> 0). $\square$

The truth of this result is intuitively obvious from our remark above that $f''(x_0) < 0$ (> 0) implies that the function $f(x)$ is concave (convex). The theorem is easily proved (Problem 3.5) with the aid of Taylor's theorem. Notice that Theorem 2.6 tells us nothing if $f'(x_0) = f''(x_0) = 0$.

Partial derivatives

For a function of two variables, $f(x, y)$, it is convenient to analyse the rate of change of f with respect to x and y *separately*. This involves the use of a *partial derivative*, defined by

$$\frac{\partial f}{\partial x} = \lim_{h \to 0} \frac{f(x + h, y) - f(x, y)}{h}, \tag{2.21}$$

if the limit exists. Note that $\partial f/\partial x$ is obtained by treating y as a constant and differentiating with respect to x in the usual way. Similarly we obtain $\partial f/\partial y$ on differentiating f with respect to y, treating x as a constant.

For given values $x = x_0$, $y = y_0$, $\partial f/\partial x$ is the gradient of the tangent *in the x-direction* to the surface $z = f(x, y)$ at the point (x_0, y_0). Sometimes we write f_x and f_y for $\partial f/\partial x$ and $\partial f/\partial y$ respectively. We also have higher partial derivatives, for example

$$f_{xx} = \frac{\partial^2 f}{\partial x^2} = \frac{\partial}{\partial x}\left(\frac{\partial f}{\partial x}\right),$$

$$f_{xy} = \frac{\partial^2 f}{\partial x\, \partial y} = \frac{\partial}{\partial x}\left(\frac{\partial f}{\partial y}\right).$$

Example 2.12 If $f(x, y) = x^2 \cos y$, we have

$$\begin{aligned}
f_x &= 2x \cos y, & f_y &= -x^2 \sin y, \\
f_{xx} &= 2 \cos y, & f_{xy} &= -2x \sin y, \\
f_{yx} &= -2x \sin y, & f_{yy} &= -x^2 \cos y. \ \square
\end{aligned}$$

If f is a function of two variables x and y and each of these variables is in turn a function of say t, we have the function of a function (or chain) rule

$$\frac{df}{dt} = \frac{\partial f}{\partial x}\frac{dx}{dt} + \frac{\partial f}{\partial y}\frac{dy}{dt}. \tag{2.22}$$

The definition of a local maximum or minimum of $f(x, y)$ at (x_0, y_0) is similar to that for a function of one variable. For a local maximum or minimum, we require $f_x = f_y = 0$ at (x_0, y_0) but these are not sufficient conditions.

2.3 Sequences and series

Sequences

Consider a function whose domain is the set of positive integers such that the integer r is mapped into an element u_r. The elements $u_1, u_2, u_3, \ldots$ are called collectively a *sequence*, which we will denote by $\{u_r\}$ or $\{u_r\}_{r=1}^{\infty}$. The sequences in which we will be most interested are those with real members (elements).

Definition 2.14   We say that the sequence of real numbers $\{u_r\}_{r=1}^{\infty}$ *converges* if there is a number, say u, such that

$$\lim_{r \to \infty} u_r = u, \tag{2.23}$$

and u is called the limit of the sequence. □

By the limit (2.23) we mean that to each $\varepsilon > 0$ there corresponds a number N such that

$$r > N \Rightarrow |u_r - u| < \varepsilon.$$

If a sequence does not converge, it is said to *diverge*.

Principle of induction

Let S_r denote a statement involving the positive integer r. If
(i) S_1 is true,
(ii) for each $k \geqslant 1$, the truth of S_k implies the truth of S_{k+1},
then *every* statement in the sequence $\{S_r\}_{r=1}^{\infty}$ is true. This is known as the principle of induction.

Example 2.13   If $0 \leqslant \alpha \leqslant 1$, show that

$$(1 - \alpha)^r \geqslant 1 - r\alpha, \qquad r = 1, 2, \ldots. \tag{2.24}$$

If S_r denotes the inequality (2.24), we see that S_1 is obviously true. Assuming S_k is true, we have

$$(1 - \alpha)^k \geqslant 1 - k\alpha$$

and, on multiplying each side of this inequality by $1 - \alpha \geqslant 0$, we have

$$(1 - \alpha)^{k+1} \geqslant (1 - \alpha)(1 - k\alpha)$$
$$= 1 - (k + 1)\alpha + k\alpha^2 \geqslant 1 - (k + 1)\alpha,$$

showing that S_{k+1} is true. By induction, the result is established. □

Convergence of sequences

Having defined convergence of a sequence (Definition 2.14) we now state a result which is often referred to as Cauchy's principle of convergence.

Theorem 2.7 (Cauchy)   The sequence $\{u_r\}_{r=1}^{\infty}$ converges if, and only if, to each $\varepsilon > 0$, there corresponds a number N such that

$$n > N \Rightarrow |u_{n+k} - u_n| < \varepsilon \quad \text{for } k = 1, 2, \ldots. \square$$

This result enables us to discuss convergence of a sequence without necessarily knowing the value of the limit to which the sequence converges.

Series

A sum

$$a_1 + a_2 + \cdots + a_n,$$

where $a_1, a_2, \ldots, a_n$ are real numbers, is called a *finite series*. Suppose that we have an infinite sequence of real numbers $\{a_n\}_{n=1}^{\infty}$ and that

$$u_n = a_1 + a_2 + \cdots + a_n. \tag{2.25}$$

If the sequence $\{u_n\}_{n=1}^{\infty}$ converges to a limit u, we write

$$u = a_1 + a_2 + \cdots \tag{2.26}$$

and say that the *infinite series* on the right side of (2.26) is convergent with sum u. We call u_n the nth *partial sum* of the infinite series.

Example 2.14 Consider the infinite series

$$1 + \frac{1}{2} + \frac{1}{3} + \cdots + \frac{1}{n} + \cdots.$$

From the inequalities

$$\frac{1}{3} + \frac{1}{4} > 2 \cdot \frac{1}{4} = \frac{1}{2},$$

$$\frac{1}{5} + \frac{1}{6} + \frac{1}{7} + \frac{1}{8} > 4 \cdot \frac{1}{8} = \frac{1}{2},$$

and so on, we have

$$1 + \frac{1}{2} + \frac{1}{3} + \cdots + \frac{1}{2^n} > 1 + \frac{1}{2} n.$$

Thus the partial sums can be made as large as we please by taking enough terms and the series is divergent. □

Uniform convergence

Consider now a sequence $\{u_r(x)\}_{r=0}^{\infty}$ of which each member $u_r(x)$ is a function of x, defined on the interval $[a, b]$.

Definition 2.15 The sequence $\{u_r(x)\}_{r=0}^{\infty}$ is said to converge *pointwise* on $[a, b]$ if, for each $x \in [a, b]$, the sequence converges. □

Definition 2.16 The sequence $\{u_r(x)\}_{r=0}^{\infty}$ is said to converge *uniformly* to a function $u(x)$ on $[a, b]$ if the sequence

$$\left\{ \sup_{a \leqslant x \leqslant b} |u_r(x) - u(x)| \right\}_{r=0}^{\infty} \tag{2.27}$$

converges to zero. □

It is clear that uniform convergence to $u(x)$ implies pointwise convergence to $u(x)$. We now see that the converse does not always hold.

Example 2.15 Consider the sequence $\{x^r\}$ on $[0, 1]$. For any $x \in [0, 1)$, $x^r \to 0$ as $r \to \infty$ and at $x = 1$, $x^r \to 1$ as $r \to \infty$. Thus $\{x^r\}$ converges pointwise on $[0, 1]$ to the limit function $u(x)$ defined by

$$u(x) = \begin{cases} 0, & 0 \leqslant x < 1 \\ 1, & x = 1. \end{cases}$$

However, for any $r \geqslant 0$,

$$\sup_{0 \leqslant x \leqslant 1} |x^r - u(x)| = \sup_{0 \leqslant x < 1} |x^r| = 1$$

and therefore the sequence $\{x^r\}$ does *not* converge uniformly. See Fig. 2.7. $\square$

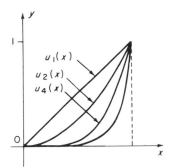

Fig. 2.7 Non-uniform convergence.

In Example 2.15, each member of the sequence is continuous on $[0, 1]$ but the limit function $u(x)$ is *not* continuous. This could not have occurred had the convergence been uniform, as the following theorem shows.

Theorem 2.8 If each element of the sequence $\{u_r(x)\}_{r=0}^{\infty}$ is continuous on $[a, b]$ and the sequence converges uniformly to $u(x)$ on $[a, b]$, then $u(x)$ is continuous on $[a, b]$. $\square$

If all the $u_r(x)$ and $u(x)$ are continuous on $[a, b]$ we can replace the "sup" in (2.27) by "max".

As an alternative definition of uniform convergence of $\{u_r(x)\}_{r=0}^{\infty}$, we may say that to each $\varepsilon > 0$ there corresponds a number N such that

$$n > N \Rightarrow |u_n(x) - u(x)| < \varepsilon \quad \text{for all } x \in [a, b].$$

Notice that N does not depend on x. This statement is represented in Fig. 2.8. Given ε, we draw a strip about $u(x)$ of width 2ε: for $n > N$, the graph of

$u_n(x)$ lies entirely inside this strip. We can make the strip as narrow as we please.

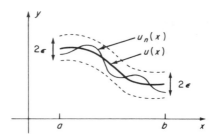

Fig. 2.8. Uniform convergence.

The following theorem (cf. Theorem 2.7) is also useful.

Theorem 2.9　The sequence $\{u_r(x)\}_{r=0}^{\infty}$ converges uniformly on $[a, b]$ if, to each $\varepsilon > 0$ there corresponds a number N such that

$$n > N \Rightarrow |u_{n+k}(x) - u_n(x)| < \varepsilon$$

for all $x \in [a, b]$ and $k = 1, 2, \ldots$. $\square$

2.4 Integration

We consider a function f, continuous on an interval $[a, b]$ which we divide into N equal sub-intervals with points of sub-division $x_r = x_0 + rh$, $r = 0, 1, \ldots, N$, with $x_0 = a$, so that $x_N = b$ and $h = (b - a)/N$. Then the sum

$$s_N = \sum_{r=1}^{N} hf(x_r) \tag{2.28}$$

is the sum of the areas of the rectangles depicted in Fig. 2.9. If $\{s_N\}_{N=1}^{\infty}$ converges, we write

$$\lim_{N \to \infty} s_N = \int_a^b f(x)\, dx. \tag{2.29}$$

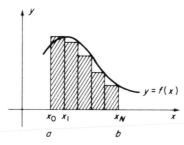

Fig. 2.9. Area approximation by a "sum" of rectangles.

We call this limit a *definite integral* and say that f is *integrable* on $[a, b]$. It can be shown that the limit always exists if f is continuous and the definite integral is thus interpreted as the area under the curve $y = f(x)$, above the x-axis (assuming f is positive) and between the limits $x = a$ and $x = b$. If we allow the rectangles, which contribute to the sum s_N, to vary in width and ask that the longest width tends to zero as $N \to \infty$, the sequence $\{s_N\}$ still converges to the same limit. Also, we remark that continuity of f is not a necessary condition for convergence of the sequence $\{s_N\}$.

In (2.29), the right side is a constant and x is therefore called a dummy variable. Thus we could equally express (2.29) as

$$\lim_{N \to \infty} s_N = \int_a^b f(t)\, dt.$$

Now we allow the upper limit b to vary, so that the integral becomes a function of b. Writing x for b, we define

$$F(x) = \int_a^x f(t)\, dt$$

and note that $F(a) = 0$. We have:

Theorem 2.10 If f is continuous on $[a, b]$, then

$$F'(x) = f(x), \qquad a \leqslant x \leqslant b. \quad \square$$

We write $\int f(x)\, dx$, called an *indefinite integral*, to denote any function ϕ such that $\phi'(x) = f(x)$. The function is sometimes called an *anti-derivative* of f. From Theorem 2.10 we deduce:

Theorem 2.11

$$\int_a^b f(t)\, dt = \phi(b) - \phi(a), \qquad\qquad (2.30)$$

where ϕ is any anti-derivative of f. $\square$

We usually abbreviate the right side of (2.30) to $[\phi(t)]_a^b$. This important theorem allows us often to evaluate an integral without explicitly relying on the limiting process (2.29). We shall also require the following results.

Theorem 2.12 (Mean value theorem for integrals.) If f is continuous and g is integrable on $[a, b]$ and $g(x) \geqslant 0$ for $a \leqslant x \leqslant b$, then there exists a number $\xi \in (a, b)$ such that

$$\int_a^b f(x)g(x)\, dx = f(\xi) \int_a^b g(x)\, dx. \quad \square$$

Theorem 2.13 (Integration by substitution.) If g' is continuous on $[a, b]$,

$$\int_a^b f(g(t))g'(t)\, dt = \int_{g(a)}^{g(b)} f(u)\, du.$$

(We obtain the integral on the right from that on the left by making the substitution $u = g(t)$.) $\square$

Theorem 2.14 (Integration by parts.) If f' and g' are continuous on $[a, b]$,

$$\int_a^b f(x)g'(x)\, dx = [f(x)g(x)]_a^b - \int_a^b f'(x)g(x)\, dx. \qquad (2.31)$$

This is proved by integrating (2.15). $\square$

2.5 Logarithmic and exponential functions

We define the function

$$\ln x = \int_1^x \frac{dt}{t} \qquad (2.32)$$

for $x > 0$ and call $\ln x$ the natural logarithm of x. The integral in (2.32) exists for all $x > 0$ as the integrand is continuous. From the definition and Theorem 2.10 we have

$$\frac{d}{dx} \ln x = 1/x. \qquad (2.33)$$

From (2.32) we also deduce (Problem 2.28) that

$$\ln x_1 x_2 = \ln x_1 + \ln x_2 \qquad (2.34)$$

for $x_1 > 0$, $x_2 > 0$. Logarithm tables owe their usefulness to this property, which allows multiplications to be replaced by additions. If $N > 2$ is a positive integer,

$$\ln N = \int_1^N \frac{dt}{t} = \sum_{r=2}^N \int_{r-1}^r \frac{dt}{t}$$

and thus

$$\ln N > \frac{1}{2} + \frac{1}{3} + \cdots + \frac{1}{N}. \qquad (2.35)$$

From this inequality and Example 2.14, it follows that $\ln x \to \infty$ as $x \to \infty$. Also, putting $x_1 = N$, $x_2 = 1/N$ in (2.34) and noting that $\ln 1 = 0$, we have

$$\ln(1/N) = -\ln N$$

and we see that $\ln x \to -\infty$ as $x \to 0$.

Since the logarithmic function is evidently one-one, we may define its inverse function. If $x = \ln y$, $0 < y < \infty$, we define the inverse function

$$y = \exp(x), \tag{2.36}$$

which is thus defined for $-\infty < x < \infty$. We see that $\exp(x) \to \infty$ as $x \to \infty$ and $\exp(x) \to 0$ as $x \to -\infty$. We deduce from (2.34) (see Problem 2.29) that

$$\exp(x_1) \exp(x_2) = \exp(x_1 + x_2), \tag{2.37}$$

where $-\infty < x_1, x_2 < \infty$.

To obtain the derivative of $y = \exp(x)$, we use (2.17) to give

$$\frac{dy}{dx} = 1 \bigg/ \frac{dx}{dy}$$

and

$$\frac{dx}{dy} = \frac{d}{dy}(\ln y) = 1/y = 1/\exp(x).$$

Thus

$$\frac{d}{dx}(\exp(x)) = \exp(x). \tag{2.38}$$

Hereafter, we shall write e^x in place of $\exp(x)$. The number $e = e^1$ evidently satisfies the equation

$$\int_1^e \frac{dt}{t} = 1$$

and, in fact, $e = 2.71828\ldots$ For any $a > 0$ we define

$$a^x = e^{x \ln a} \tag{2.39}$$

and this is also called an exponential function. It follows from (2.38) that

$$\frac{d}{dx} a^x = a^x \ln a,$$

and it is essentially the simplification which occurs for $a = e$, namely

$$\frac{d}{dx} e^x = e^x,$$

which makes e^x *the* exponential function.

Problems

Section 2.1

2.1   For the following functions f, state the range of f and say which functions have an inverse.

(i) $1 + x + x^2$, $x \in (-\infty, \infty)$,

(ii) $(1 + x)/(2 + x)$, $x \in [-1, 1]$,

(iii) $1 + x + x^2$, $x \in [0, \infty)$.

2.2　Show that, for any real numbers x and α,

$$x^n - \alpha^n = (x - \alpha)(x^{n-1} + \alpha x^{n-2} + \alpha^2 x^{n-3} + \cdots + \alpha^{n-1}).$$

Hence, if $p(x) = a_0 + a_1 x + \cdots + a_n x^n$, $a_n \neq 0$, show that

$$p(x) - p(\alpha) = (x - \alpha)q(x),$$

where q is a polynomial of degree $n - 1$, and thus obtain (2.2).

2.3　Suppose that $x_1 < x_2 < \cdots < x_k$ and that p is a polynomial which satisfies the conditions $(-1)p(x_1) > 0$, $(-1)^k p(x_k) > 0$ and $p(x_i) = 0$ for $2 \leqslant i \leqslant k - 1$. Show that p has at least $k - 1$ zeros in (x_1, x_k). (*Hint:* if k is even, $p(x_1).p(x_k) < 0$ and p has therefore an *odd* number of zeros in (x_1, x_k). Investigate also the case where k is odd.)

2.4　With the notation of the previous problem, show that if p satisfies the conditions $(-1)^i p(x_i) \geqslant 0$, $1 \leqslant i \leqslant k$, then p has at least $k - 1$ zeros in $[x_1, x_k]$.

2.5　If p and q are polynomials and

$$p(x) = (x - \alpha)^{r+1} q(x),$$

show that $p^{(k)}(\alpha) = 0$ for $k = 1, 2, \ldots, r$.

2.6　Assuming that (2.3) (for all values of θ) and (2.7) hold, deduce that (2.8) holds.

2.7　Assuming that (2.3) holds for all values of θ, deduce from the behaviour of the sine function that $\cos \theta$ satisfies the properties $\cos(\pi - \theta) = -\cos \theta$ and $\cos(-\theta) = \cos \theta$.

2.8　By writing $1 - \cos x = \cos 0 + \cos(\pi - x)$ and using (2.8), or otherwise, deduce that $1 - \cos x = 2 \sin^2 \frac{1}{2}x$.

2.9　Show directly from Definition 2.6 that $y = x^2$ is convex on $(-\infty, \infty)$.

2.10　Investigate which of the following suprema and infima exist and of those that exist, find which are attained.

(i)　$\sup\limits_{1 < x < \infty} (2 - x - x^2)$,　　(ii)　$\inf\limits_{-\infty < x < \infty} (2 + \sin x)e^x$,

(iii)　$\inf\limits_{0 < x < \infty} |\cos x \ln x|$,　　(iv)　$\sup\limits_{0 \leqslant x < \infty} (1 - x - x^2)$.

Section 2.2

2.11   In Fig. 2.5 show that

   (i) area of $\triangle$ OAB $= \frac{1}{2}$ OA.BD $- \frac{1}{2}$ sin x,
   (ii) area of $\triangle$ OAC $= \frac{1}{2}$ OA.AC $= \frac{1}{2}$ tan x,
   (iii) area of sector OAB $= \frac{1}{2}x$, assuming that the area of a circle with radius unity is π.

2.12   Deduce from (2.9) that

$$0 < 1 - \frac{\sin x}{x} < 1 - \cos x, \qquad 0 < x < \tfrac{1}{2}\pi.$$

Use the result of Problem 2.8 and the right inequality of (2.9) to show that

$$0 < 1 - \cos x < \tfrac{1}{2}x^2, \qquad 0 < x < \tfrac{1}{2}\pi,$$

thus verifying (2.10).

2.13   Which of the following functions are continuous on $[-1, 1]$?

   (i) x^n,   (ii) $|x|$,   (iii) $\sin\left(\dfrac{1}{1 + x}\right)$,   (iv) $\cot x$.

2.14   Verify Theorem 2.1.

2.15   Show that if f satisfies a Lipschitz condition on $[a, b]$, f is uniformly continuous on $[a, b]$.

2.16   Using the identity

$$a^n - b^n = (a - b)(a^{n-1} + a^{n-2}b + \cdots + ab^{n-2} + b^{n-1}),$$

n a positive integer, verify (2.18). (Treat separately the cases n positive and n negative in (2.18).)

2.17   From the *definition*, obtain the derivatives of

   (i) $2x^2 + 3x$,   (ii) $1/(x + 1)$,   (iii) $x^{1/2}$.

2.18   Writing $\sin(x + h) - \sin x = \sin(x + h) + \sin(-x)$ and using (2.7) and the result of Example 2.9, find the derivative of $\sin x$ directly from the definition of a derivative.

2.19   If $f(x) \equiv |x|$ on $[-1, 1]$, show that f satisfies a Lipschitz condition with Lipschitz constant $L = 1$ on $[-1, 1]$, but that $f'(x)$ does not exist at $x = 0$.

2.20   Suppose that f'' is continuous on $[x_1, x_2]$ and that $p_1(x) = 0$ denotes the equation of the chord joining the points $(x_r, f(x_r))$, $r = 1, 2$, so that $p_1(x_r) = f(x_r)$. Then, anticipating the result of (4.16), we have

$$f(x) - p_1(x) = \tfrac{1}{2}(x - x_1)(x - x_2)f''(\xi),$$

for some $\xi \in [x_1, x_2]$. Deduce that if $x \in [x_1, x_2]$ and $f''(x) > 0$ on $[x_1, x_2]$ then f is convex on $[x_1, x_2]$.

Section 2.3

2.21   Show that the sequence $\{u_n\}_{n=1}^{\infty}$ converges, where $u_n = (n + 1)/2n$.

2.22   Prove by induction that

$$1^2 + 2^2 + \cdots + n^2 = n(n + 1)(2n + 1)/6$$

and hence find the limit of the sequence $\{s_n\}$, where

$$s_n = \frac{1}{n}\left(\left(\frac{1}{n}\right)^2 + \left(\frac{2}{n}\right)^2 + \cdots + \left(\frac{n}{n}\right)^2\right).$$

2.23   If $u_n(x) = a_0 + a_1 x + \cdots + a_n x^n$ and $|a_r| \leqslant K$ for all r, use Theorem 2.9 to show that $\{u_n(x)\}_{n=0}^{\infty}$ converges uniformly in any interval $[-\rho, \rho]$, where $0 < \rho < 1$.

2.24   Show that the infinite series

$$1 + \tfrac{1}{2} + \tfrac{1}{4} + \tfrac{1}{8} + \tfrac{1}{16} + \cdots$$

has partial sums

$$u_r = 2 - \frac{1}{2^r}, \qquad r = 0, 1, 2, \ldots,$$

and deduce that the series is convergent with sum 2.

2.25   Show that the sequence $\{u_n(x)\}_{n=1}^{\infty}$ with

$$u_n(x) = \begin{cases} nx, & 0 \leqslant x \leqslant 1/n \\ 2 - nx, & 1/n < x < 2/n \\ 0, & 2/n \leqslant x \leqslant 2 \end{cases}$$

is pointwise convergent to $u(x) \equiv 0$ on $[0, 2]$ but that it is not uniformly convergent.

Section 2.4

2.26   Evaluate $\int_0^1 x^2 \, dx$ in two different ways:
  (a) by finding an anti-derivative (Theorem 2.11),
  (b) by treating the integral as the limit of a sum, as in (2.28), using the result of Problem 2.22.

2.27   Deduce from Theorem 2.12 that, if f is continuous on $[a, b]$,

$$m(b - a) \leqslant \int_a^b f(x) \, dx \leqslant M(b - a)$$

where m and M denote the minimum and maximum values of $f(x)$ on $[a, b]$.

Section 2.5

2.28   Write $\ln x_1 x_2$ as an integral, as in the definition (2.32). Split the range of integration $[1, x_1 x_2]$ into $[1, x_1]$ and $[x_1, x_1 x_2]$ and make the change of variable $t = x_1 u$ in the second integral. Thus establish (2.34).

2.29   If $y_i = \exp(x_i)$, $i = 1, 2$, use (2.34) to deduce (2.37).

Chapter 3

Taylor's Polynomial and Series

3.1 Function approximation

One very important aspect of numerical analysis is the study of approximations to functions. Suppose that f is a function defined on some interval of the reals. We now seek some other function g which "mimics" the behaviour of f on some interval. We say that g is an *approximation* to f on that interval. Usually we make such an approximation because we wish to carry out some numerical calculation or analytical operation involving f, but we find this difficult or impossible because of the nature of f. For example, we may wish to find the integral of f over some interval and there may be no explicit formula for such an integral. We replace f by a function g which may easily be integrated. Other problems include: evaluating f for a particular x, especially if f is defined to be the solution of some equation; differentiation of f; determining zeros of f (that is, finding roots of $f(x) = 0$); determining extrema of f. In all such cases we may need to replace f by some approximation g. We choose g so that it is more amenable than f to the type of operation involved. We can rarely then determine the exact solution of the original problem, but we do hope to obtain an approximation to the solution. There are, of course, many different ways in which an approximating function g may be chosen but we would normally restrict g to a certain class of functions. The most commonly used class is the polynomials. These are easily integrated and differentiated and are well-behaved, in that all derivatives exist and are continuous.

One very desirable property, which we expect of any class of approximating functions, is that, for any function f with certain basic properties on some interval, it is possible to approximate to f to within arbitrary accuracy on that interval. The class of all polynomials satisfies this property, with only the requirement that f be continuous. This remarkable result is embodied in Weierstrass' theorem† in Chapter 5. It must be stressed, however, that polynomials are not the only functions with this property but are certainly the most convenient for general manipulation.

Even if the approximating function g is restricted to be a polynomial of a

† Due to the German mathematician K. Weierstrass (1815–97).

given degree, there are still many ways of choosing g and the question of determining the most suitable g is very complicated. We will return to this point in Chapter 5.

3.2 Taylor's theorem

In this chapter we shall consider a polynomial approximation which mimics a function f near one given point. We will restrict our choice of polynomial to an element of P_n, the set of all polynomials of degree not exceeding n. We will seek coefficients $a_0, a_1, \ldots, a_n$ such that the polynomial

$$p_n(x) = a_0 + a_1 x + \cdots + a_n x^n$$

approximates to f near $x = x_0$. Assuming that f is n times differentiable at $x = x_0$, we try to choose the coefficients of p_n so that

$$p_n^{(j)}(x_0) = f^{(j)}(x_0), \qquad j = 0, 1, 2, \ldots, n, \tag{3.1}$$

where $p_n^{(0)}(x) \equiv p_n(x)$ and $f^{(0)}(x) \equiv f(x)$. Thus we seek p_n such that the values of p_n and f are equal at $x = x_0$ and the values of their first n derivatives are equal at $x = x_0$.

The equations (3.1) are

$$\left.\begin{aligned}
a_0 + a_1 x_0 + a_2 x_0^2 + \quad \cdots \quad + a_{n-1} x_0^{n-1} + a_n x_0^n &= f(x_0) \\
a_1 + 2a_2 x_0 + \cdots + (n-1)a_{n-1} x_0^{n-2} + n a_n x_0^{n-1} &= f'(x_0) \\
\vdots \qquad\qquad\qquad\qquad \vdots \\
(n-1)!\, a_{n-1} + n(n-1)\cdots 2 a_n x_0 &= f^{(n-1)}(x_0) \\
n!\, a_n &= f^{(n)}(x_0).
\end{aligned}\right\} \tag{3.2}$$

These are $n + 1$ equations† in the $n + 1$ unknown coefficients $a_0, a_1, \ldots, a_n$ and we see that these coefficients are determined uniquely by (3.2). The last equation in (3.2) gives a_n and then, from the penultimate equation, we may determine a_{n-1}. By working through the equations in reverse order we may determine $a_n, a_{n-1}, a_{n-2}, \ldots, a_0$ uniquely. Hence p_n is unique, that is, there is only one member of P_n satisfying (3.1).

We do not usually bother to obtain $a_0, a_1, \ldots, a_n$ explicitly (see Problem 3.1) but instead write p_n in the form

$$p_n(x) = f(x_0) + \frac{(x - x_0)}{1!} f'(x_0) + \frac{(x - x_0)^2}{2!} f''(x_0)$$

$$+ \cdots + \frac{(x - x_0)^n}{n!} f^{(n)}(x_0). \tag{3.3}$$

† These are triangular equations, whose solution is discussed in detail in Chapter 8.

It should be noted that (3.3) is a polynomial *in* x, as each term is a polynomial in x. (Remember that x_0 is a constant.) We call $p_n(x)$ of (3.3) the *Taylor polynomial*† of f constructed at the point $x = x_0$. The reader is probably familiar with (3.3) as the first $n + 1$ terms of the Taylor series of $f(x)$; however, he is urged to verify that the Taylor polynomial does indeed satisfy (3.1).

Example 3.1 For $f(x) = e^x$ (the exponential function), $f^{(r)}(x) = e^x$ and $f^{(r)}(0) = 1$ for $r = 0, 1, 2, \ldots$. Thus the Taylor polynomial of degree n constructed at $x = 0$ is

$$p_n(x) = 1 + x + \frac{x^2}{2!} + \frac{x^3}{3!} + \cdots + \frac{x^n}{n!}. \quad \square$$

Example 3.2 For $f(x) = \sin x$,

$$f^{(2r)}(x) = (-1)^r \sin x \quad \text{and} \quad f^{(2r+1)}(x) = (-1)^r \cos x, \qquad r = 0, 1, 2 \ldots.$$

Thus, for example, the Taylor polynomial of degree four constructed at $x = 0$ is

$$p_4(x) = x - \frac{x^3}{3!}.$$

In this case the coefficients of 1, x^2 and x^4 are zero.

At $x = \pi/4$ we obtain

$$p_4(x) = \frac{1}{\sqrt{2}} \left(1 + \frac{(x - \pi/4)}{1!} - \frac{(x - \pi/4)^2}{2!} - \frac{(x - \pi/4)^3}{3!} + \frac{(x - \pi/4)^4}{4!} \right). \cdot \square$$

The function $f(x) - p_n(x) = R_n(x)$, say, is the *error* introduced when we use $p_n(x)$ as an approximation to $f(x)$ and the following theorem, which is very important in classical mathematical analysis, gives an expression for $R_n(x)$.

Theorem 3.1 (Taylor's theorem.)   Let $f(x), f'(x), \ldots, f^{(n)}(x)$ exist and be continuous on some interval $[a, b]$, $f^{(n+1)}(x)$ exist for $a < x < b$ and $p_n(x)$ be the Taylor polynomial (3.3) for some $x_0 \in [a, b]$. Then, given any $x \in [a, b]$, there exists ξ_x (depending on x) which lies between x_0 and x and is such that

$$R_n(x) = f(x) - p_n(x) = \frac{(x - x_0)^{n+1}}{(n + 1)!} f^{(n+1)}(\xi_x). \qquad (3.4)$$

† After B. Taylor (1685–1731).

Alternatively, if we let $x = x_0 + h$, we may write (3.4) as

$$f(x_0 + h) = f(x_0) + \frac{h}{1!} f'(x_0) + \frac{h^2}{2!} f''(x_0) + \cdots$$

$$+ \frac{h^n}{n!} f^{(n)}(x_0) + \frac{h^{n+1}}{(n+1)!} f^{(n+1)}(x_0 + \theta_n h), \quad (3.5)$$

where $0 < \theta_h < 1$.

Proof Let $\alpha \in [a, b]$ be the point at which we wish to determine the error. We suppose (without loss of generality) that $\alpha > x_0$. Let

$$g(x) = f(\alpha) - f(x) - \frac{(\alpha - x)}{1!} f'(x) - \cdots - \frac{(\alpha - x)^n}{n!} f^{(n)}(x). \quad (3.6)$$

Then $g'(x)$ exists for $x \in (a, b)$ and

$$g'(x) = -\frac{(\alpha - x)^n}{n!} f^{(n+1)}(x).$$

Now consider the function

$$G(x) = g(x) - \left(\frac{\alpha - x}{\alpha - x_0}\right)^{n+1} g(x_0), \quad (3.7)$$

for which

$$G(x_0) = G(\alpha) = 0.$$

From the differentiability of g and $(\alpha - x)^{n+1}$ it follows that G is differentiable on any subinterval of (a, b) and we may apply Rolle's theorem 2.4 to G on the interval $[x_0, \alpha]$. Thus there exists $\xi_\alpha \in (x_0, \alpha)$ such that

$$G'(\xi_\alpha) = 0,$$

which yields

$$-\frac{(\alpha - \xi_\alpha)^n}{n!} f^{(n+1)}(\xi_\alpha) + (n + 1) \frac{(\alpha - \xi_\alpha)^n}{(\alpha - x_0)^{n+1}} g(x_0) = 0,$$

so that, since $\alpha - \xi_\alpha \neq 0$,

$$g(x_0) = \frac{(\alpha - x_0)^{n+1}}{(n + 1)!} f^{(n+1)}(\xi_\alpha).$$

Now

$$g(x_0) = f(\alpha) - p_n(\alpha) = R_n(\alpha)$$

and, therefore,

$$R_n(\alpha) = \frac{(\alpha - x_0)^{n+1}}{(n + 1)!} f^{(n+1)}(\xi_\alpha),$$

where $\xi_\alpha \in (x_0, \alpha)$. On replacing α by x, we obtain (3.4). □

There are various expressions for the error term R_n other than (3.4). Some of these may be found by making a slightly different choice of $G(x)$ than (3.7), as is seen in Problem 3.3. With $n = 0$, we obtain the mean value theorem and we sometimes call Theorem 3.1 the generalized mean value theorem.

3.3 Convergence of Taylor series

We have seen in the last section how we may construct a polynomial p_n which approximates to a function f near a given point. We now ask whether the error $R_n(x) = f(x) - p_n(x)$ converges to zero as n is increased. Since $p_n(x)$ is the nth partial sum of the *infinite series*

$$f(x_0) + \frac{(x - x_0)}{1!} f'(x_0) + \frac{(x - x_0)^2}{2!} f''(x_0) + \cdots + \frac{(x - x_0)^r}{r!} f^{(r)}(x_0) + \cdots, \tag{3.8}$$

we are equivalently concerned with the convergence of this series to $f(x)$. We call (3.8) the *Taylor series* of f constructed at $x = x_0$ and R_n, the error when the series is terminated after $n + 1$ terms, is called the *remainder*. Sometimes it is possible to investigate the convergence of (3.8) directly, but usually we consider the behaviour of $R_n(x)$ as $n \to \infty$. In Chapter 2, we gave two definitions of convergence of such a sequence of functions, each defined on an interval I. We say that the $R_n(x)$ converge (pointwise) to zero on I if

$$R_n(x) \to 0, \quad \text{as } n \to \infty, \quad \text{for all } x \in I \tag{3.9}$$

and the $R_n(x)$ converge uniformly to zero on I if

$$\sup_{x \in I} |R_n(x)| \to 0, \quad \text{as } n \to \infty. \tag{3.10}$$

Corresponding to (3.9) and (3.10), we say respectively that the Taylor series (3.8) is (pointwise) convergent or uniformly convergent to $f(x)$ on I. It is not always easy to determine the behaviour of $R_n(x)$ from (3.4), as ξ_x depends on x and the form of this dependence is not known explicitly.

Example 3.3 We investigate the convergence of the Taylor series of $f(x) = (1 + x)^{-1}$ at $x = 0$. Clearly

$$f^{(r)}(x) = \frac{(-1)^r r!}{(1 + x)^{r+1}}$$

and, therefore, the required Taylor polynomial of nth degree is

$$p_n(x) = 1 - x + x^2 - x^3 + \cdots + (-x)^n.$$

On any interval not containing $x = -1$, all the derivatives of f exist and are continuous and from (3.4) the remainder is

$$R_n(x) = \frac{x^{n+1}}{(n+1)!} \cdot \frac{(-1)^{n+1}(n+1)!}{(1+\xi_x)^{n+2}} = \frac{(-x)^{n+1}}{(1+\xi_x)^{n+2}}. \tag{3.11}$$

We show first that the sequence $\{R_n(x)\}_{n=0}^{\infty}$ is uniformly convergent to zero (and, therefore, that the Taylor series is uniformly convergent) on any interval $[0, b]$ with $0 < b < 1$. For $0 \leqslant x \leqslant b$ we have $0 \leqslant \xi_x \leqslant x$ with equality only if $x = 0$. Thus

$$\frac{1}{1+\xi_x} \leqslant 1 \tag{3.12}$$

and from (3.11)

$$|R_n(x)| \leqslant x^{n+1}. \tag{3.13}$$

Now $x \leqslant b$ and, therefore,

$$\sup_{x \in [0,b]} |R_n(x)| \leqslant b^{n+1}.$$

From $0 < b < 1$ it follows that $b^{n+1} \to 0$ as $n \to \infty$ and we have uniform convergence of $\{R_n(x)\}$. In this case we were able to remove ξ_x by using an inequality.

Secondly, we show that $\{R_n(x)\}$ is (pointwise) convergent to zero on $[0, 1)$. Again (3.12) and, therefore, (3.13) hold and for any *given* non-negative $x < 1$

$$|R_n(x)| \leqslant x^{n+1} \to 0 \quad \text{as } n \to \infty.$$

However, as we have seen before (Example 2.15),

$$\sup_{x \in [0,1)} x^{n+1} = 1 \quad \text{for all } n$$

and, therefore, we cannot prove that the suprema of the $|R_n(x)|$ converge to zero on $[0, 1)$.

In fact, the Taylor series is not uniformly convergent on $[0, 1)$ as can be seen by a more direct approach. For this is a geometric series and

$$p_n(x) = \frac{1 - (-x)^{n+1}}{1 + x}, \qquad x \neq -1, \tag{3.14}$$

so that

$$R_n(x) = f(x) - p_n(x) = \frac{(-x)^{n+1}}{1 + x}. \tag{3.15}$$

Thus

$$\sup_{x \in [0,1)} |R_n(x)| = \tfrac{1}{2} \quad \text{for all } n$$

and $\{|R_n(x)|\}$ does not converge uniformly to zero.

The investigation for all intervals $[a, 0]$ where $-1 < a \leqslant 0$ is not possible using the remainder form (3.4) and the more explicit result (3.15) is required. Alternatively, one of the remainder forms described in Problem 3.3 may be considered. Thus we find that the Taylor series

$$1 - x + x^2 - x^3 + \cdots$$

is uniformly convergent on any interval $[a, b]$ with $-1 < a \leqslant b < 1$. (See Problem 3.6.) $\square$

Example 3.4 For $f(x) = \sin x$ the Taylor series at $x = 0$ is (see also Example 3.2)

$$x - \frac{x^3}{3!} + \frac{x^5}{5!} - \frac{x^7}{7!} + \cdots$$

The remainder is

$$R_n(x) = \begin{cases} \dfrac{x^{n+1}(-1)^{n/2} \cos \xi_x}{(n+1)!}, & \text{for even } n \\[2ex] \dfrac{x^{n+1}(-1)^{(n+1)/2} \sin \xi_x}{(n+1)!}, & \text{for odd } n \end{cases}$$

where ξ_x lies between 0 and x. In both cases

$$|R_n(x)| \leqslant \frac{|x|^{n+1}}{(n+1)!}$$

and on any interval $[a, b]$, with $c = \max\{|a|, |b|\}$,

$$\sup_{x \in [a,b]} |R_n(x)| \leqslant \frac{c^{n+1}}{(n+1)!} \to 0 \quad \text{as } n \to \infty.$$

Thus $\{|R_n(x)|\}$ is uniformly convergent to zero (and therefore the Taylor series converges to $f(x) = \sin x$) on any closed interval. This is also true on any open interval as we can always embed such an interval in a larger closed interval. $\square$

If, on some interval $[a, b]$, all the derivatives of f exist and have a common bound M (say), so that

$$\max_{x \in [a,b]} |f^{(n)}(x)| \leqslant M \quad \text{for all } n, \tag{3.16}$$

then from (3.4), if $x_0 \in [a, b]$,

$$\sup_{x \in [a,b]} |R_n(x)| \leqslant \frac{(b-a)^{n+1}}{(n+1)!} M.$$

It follows that

$$\max_{x \in [a,b]} |f(x) - p_n(x)| = \sup_{x \in [a,b]} |R_n(x)| \to 0 \quad \text{as } n \to \infty.$$

Equivalently we can say that, given any $\varepsilon > 0$, there is an N such that

$$\max_{x \in [a,b]} |f(x) - p_n(x)| < \varepsilon \quad \text{for all } n \geqslant N.$$

Thus we can choose a polynomial $p_n(x)$ which approximates to f to within any arbitrary accuracy on any interval $[a, b]$, by choosing n sufficiently large, provided f satisfies (3.16). This is a proof of a weak form of Weierstrass' theorem which states that we can approximate to *any continuous function* to within arbitrary accuracy using a polynomial. The condition (3.16) is very severe though it is satisfied by many functions including that of Example 3.4.

3.4 Taylor series in two variables

We extend Theorem 3.1 to functions of two variables.

Theorem 3.2 Suppose that $f(x, y)$ is a function of two real variables such that all the $(n + 1)$th order partial derivatives of f exist and are continuous on some circular domain D with centre (x_0, y_0). If $(x_0 + h, y_0 + k) \in D$, then

$$f(x_0 + h, y_0 + k) = f(x_0, y_0) + \frac{1}{1!} \left(h \frac{\partial}{\partial x} + k \frac{\partial}{\partial y} \right) f(x_0, y_0) + \cdots$$

$$+ \frac{1}{n!} \left(h \frac{\partial}{\partial x} + k \frac{\partial}{\partial y} \right)^n f(x_0, y_0)$$

$$+ \frac{1}{(n+1)!} \left(h \frac{\partial}{\partial x} + k \frac{\partial}{\partial y} \right)^{n+1} f(x_0 + \theta h, y_0 + \theta k), \tag{3.17}$$

where $0 < \theta < 1$.

Proof Let $x = x_0 + ht$ and $y = y_0 + kt$, where $0 \leqslant t \leqslant 1$. Let

$$g(t) = f(x_0 + ht, y_0 + kt)$$

when, by the chain rule,

$$g' = \frac{dg}{dt} = \frac{dx}{dt} \cdot \frac{\partial f}{\partial x} + \frac{dy}{dt} \cdot \frac{\partial f}{\partial y} = \left(h \frac{\partial}{\partial x} + k \frac{\partial}{\partial y} \right) f.$$

Similarly

$$g'' = \frac{d^2g}{dt^2} = \left(h\frac{\partial}{\partial x} + k\frac{\partial}{\partial y}\right)\left(h\frac{\partial}{\partial x} + k\frac{\partial}{\partial y}\right)f = \left(h\frac{\partial}{\partial x} + k\frac{\partial}{\partial y}\right)^2 f$$

and

$$g^{(r)} = \frac{d^r g}{dt^r} = \left(h\frac{\partial}{\partial x} + k\frac{\partial}{\partial y}\right)^r f, \qquad r = 0, 1, \ldots, n + 1.$$

We now construct the Taylor series of $g(t)$ at $t = 0$ and apply Theorem 3.1 to obtain

$$g(1) = g(0) + \frac{1g'(0)}{1!} + \frac{1^2 g''(0)}{2!} + \cdots + \frac{1^n g^{(n)}(0)}{n!} + \frac{1^{n+1} g^{(n+1)}(\theta)}{(n+1)!}, \qquad (3.18)$$

where $0 < \theta < 1$. On substituting for g and its derivatives in (3.18), we obtain (3.17). □

3.5 Power series

In many applications of Taylor's theorem we take $x_0 = 0$. The Taylor series becomes

$$f(0) + \frac{x}{1!}f'(0) + \frac{x^2}{2!}f''(0) + \cdots \qquad (3.19)$$

and is sometimes known as a *Maclaurin* series in this form. The series (3.19) is a special case of a *power series*, that is, a series of the type

$$a_0 + a_1 x + a_2 x^2 + a_3 x^3 + \cdots,$$

where the coefficients $a_0, a_1, a_2, \ldots$ are real constants, independent of x. As we have seen earlier, such a series may be convergent for only certain values of x. We define the *radius of convergence* of such a series to be the largest positive number r (if it exists) such that the series is convergent for any x with $|x| < r$. Note that this does not imply anything about the convergence or otherwise when $|x| = r$.

Example 3.5   We reconsider the Maclaurin series of $(1 + x)^{-1}$ (see Example 3.3),

$$1 - x + x^2 - x^3 + \cdots.$$

The partial sum is (3.14),

$$p_n(x) = \frac{1 - (-x)^{n+1}}{1 + x},$$

and this is convergent for $|x| < 1$ and divergent for $|x| > 1$. The radius of convergence is $+1$. $\square$

Example 3.6 The Maclaurin series of the exponential function e^x is

$$1 + \frac{x}{1!} + \frac{x^2}{2!} + \frac{x^3}{3!} + \cdots$$

and this is convergent for any x. We say that the radius of convergence is infinite. $\square$

There are various tests for determining the convergence of a power series and details of these may be found in any good text on advanced calculus. It can also be shown that a power series is uniformly convergent on any interval $[-a, a]$, where $0 < a < r$. We have only considered real power series, but the definitions may easily be extended to such series with complex coefficients and complex argument.

If we use the partial sum of a convergent infinite series as an approximation to the sum of the infinite series, the error thus introduced is called the *truncation error*. In the case of a Taylor series expressed as a power series, this truncation error is the remainder, which takes the form

$$R_n(x_0 + h) = \frac{h^{n+1}}{(n+1)!} f^{(n+1)}(x_0 + \theta h)$$

if f satisfies the conditions of Theorem 3.1 on some interval $[a, b]$. If, in addition, for some fixed n we are given that $f^{(n+1)}$ is bounded on (a, b) (for example, this will be true if $f^{(n+1)}$ is continuous for $a \leqslant x \leqslant b$), then we write

$$R_n(x_0 + h) = O(h^{n+1})$$

and say that R_n is of order h^{n+1}. By this we mean that there exists M independent of h such that

$$|R_n(x_0 + h)| \leqslant Mh^{n+1} \tag{3.20}$$

provided h is sufficiently small. The inequality (3.20) tells us how rapidly $|R_n|$ decreases as $h \to 0$.

Problems

Section 3.2

3.1 Show that the solution of equations (3.2) is

$$a_r = \sum_{j=0}^{n-r} \frac{(-x_0)^j}{j!\, r!} f^{(r+j)}(x_0), \qquad r = 0, 1, \ldots, n.$$

(*Hint:* see (3.3).)

3.2 Construct the Taylor polynomials of degree 6 at $x = 0$ for the following functions:

$$\cos x, \quad (1 - x)^{-1}, \quad (1 + x)^{1/2}, \quad \ln(1 + x), \quad \ln(1 - x).$$

3.3 Let $f(x)$, $g(x)$, α and x_0 be as in Theorem 3.1. For some positive integer m replace (3.7) by

$$G_m(x) = g(x) - \left(\frac{\alpha - x}{\alpha - x_0}\right)^m g(x_0)$$

and thus show that for some ξ (depending on m and x) between x_0 and x,

$$R_n(x) = \frac{(x - x_0)^m (x - \xi)^{n-m+1}}{m . n!} f^{(n+1)}(\xi).$$

In particular, for $m = 1$, show that

$$R_n(x) = \frac{(x - x_0)(x - \xi)^n}{n!} f^{(n+1)}(\xi)$$

$$= \frac{h^{n+1}(1 - \theta)^n}{n!} f^{(n+1)}(x_0 + \theta h),$$

$$(3.21)$$

where $x = x_0 + h$ and $0 < \theta < 1$.

3.4 If $p_n(x)$ denotes the Taylor polynomial of degree n for e^x constructed at $x = 0$, find the smallest value of n for which

$$\max_{0 \leqslant x \leqslant 1} |e^x - p_n(x)| \leqslant 10^{-6}.$$

3.5 Show that, if f satisfies the conditions of Theorem 2.6 and h is sufficiently small,

$$f(x_0 + h) = f(x_0) + \frac{h^2}{2} f''(x_0 + \theta h)$$

and thus prove the theorem.

Section 3.3

3.6 By considering the remainder form (3.21), show that the Taylor series at $x = 0$ of $f(x) = (1 + x)^{-1}$ (see Example 3.3) is uniformly convergent on any interval $[a, 0]$ with $-1 < a < 0$. Combine this result with that of Example 3.3 to show that the series is uniformly convergent on any interval $[a, b]$ with $-1 < a \leqslant b < 1$.

3.7 Show that the Taylor series at $x = 0$ of $f(x) = (1 + x)^{1/2}$ is (pointwise) convergent for $-1 < x < 1$ and uniformly convergent on any interval [a, b] with $-1 < a \leqslant b < 1$. (*Hint:* show from (3.4) that

$$|R_n(x)| = \left| \frac{1.3.5\cdots(2n-1)}{2.4.6\cdots 2n} \frac{x^{n+1}}{2(n+1)(1+\xi)^{n+1/2}} \right|$$

$$< \frac{x^{n+1}}{(1+\xi)^{n+1/2}} \quad \text{for } x > 0$$

and thus that $R_n(x) \to 0$ as $n \to \infty$ for $0 \leqslant x < 1$. Similarly show from (3.21) that

$$|R_n(x)| = \left| \frac{1.3.5\cdots(2n-1)}{2.4.6\cdots 2n} \frac{x(x-\xi)^n}{2(1+\xi)^{n+1/2}} \right|$$

$$< \left| \frac{x(x-\xi)^n}{(1+\xi)^{n+1/2}} \right| \quad \text{for } x \neq 0$$

so that $R_n(x) \to 0$ as $n \to \infty$ for $-1 < x < 0$.)

3.8   Show that the Taylor series at $x = 0$ of $f(x) = \ln(1 + x)$ is (pointwise) convergent for $-1 < x \leqslant 1$. As in Problem 3.7 consider $-1 < x < 0$ and $0 \leqslant x \leqslant 1$ separately. What happens when $x = -1$? (See also Example 2.14.)

Section 3.5

3.9   Show that the Taylor series at $x = 0$ of $\ln(1 + x)$ has radius of convergence $+1$.

3.10   Show that the Taylor series at $x = 0$ of $\sin x$ has infinite radius of convergence.

3.11   Show, using Problem 3.7, that

$$(1 + x)^{1/2} = 1 + \tfrac{1}{2}x + O(x^2).$$

3.12   If A is a given constant and $h = A/n$, where n is a positive integer, show that

$$(1 + O(h^2))^n = 1 + O(h) \quad \text{as } n \to \infty$$

and

$$(1 + O(h^3))^n = 1 + O(h^2) \quad \text{as } n \to \infty.$$

3.13   Show that for $x \geqslant 0$ and any fixed integer $n \geqslant 1$:

$$e^x \geqslant 1 + x,$$
$$e^x = 1 + x + O(x^2),$$
$$e^{nx} \geqslant (1 + x)^n,$$
$$e^{nx} = (1 + x)^n + O(x^2).$$

Chapter 4

The Interpolating Polynomial

4.1 Linear interpolation

Having met the Taylor polynomials, we continue our study of polynomial approximation by examining a second class of polynomials which are used to approximate a given function f defined on some interval of the reals. These are called *interpolating polynomials*. We begin with a simple case.

Suppose we know the values of some function f at two distinct points $x = x_0$ and $x = x_1$. Geometrically (see Fig. 4.1) it is clear that there is a

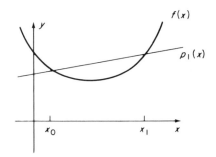

Fig. 4.1. Linear interpolation.

unique straight line which passes through the two points $(x_0, f(x_0))$ and $(x_1, f(x_1))$. To obtain the equation of this straight line, we may recall some coordinate geometry and write

$$p_1(x) = \left(\frac{x - x_1}{x_0 - x_1}\right) f(x_0) + \left(\frac{x - x_0}{x_1 - x_0}\right) f(x_1). \qquad (4.1)$$

In the next section, we will see that it is helpful to look at this in a slightly different way. We see from (4.1) that p_1 is linear in x (that is, of the form $a + bx$) and so has indeed a straight line graph. Also, by substituting $x = x_0$ and $x = x_1$ in turn into (4.1), we have that $p_1(x_0) = f(x_0)$ and $p_1(x_1) = f(x_1)$.

To dispel any doubts that geometrical intuition may have misled us, we can start afresh with the two function values at the distinct x values, $x = x_0$ and x_1. We now ask: does there exist a straight line $y = a + bx$ such that

48

$y = f(x_0)$ at $x = x_0$ and $y = f(x_1)$ at $x = x_1$? Secondly, if such a straight line exists, is it unique? The conditions at $x = x_0$ and x_1 give simultaneous equations

$$a + bx_0 = f(x_0), \qquad a + bx_1 = f(x_1) \tag{4.2}$$

which, on subtraction, yield

$$b = (f(x_1) - f(x_0))/(x_1 - x_0). \tag{4.3}$$

(Recall that $x_0 \neq x_1$.) Elimination of b from (4.2) gives

$$a = (x_1 f(x_0) - x_0 f(x_1))/(x_1 - x_0). \tag{4.4}$$

There is thus one (and only one) pair of values a, b which satisfies the equations (4.2), showing that there is a unique straight line which takes the same values as f at $x = x_0$ and $x = x_1$. A particular consequence of the uniqueness is that the function $a + bx$, with a and b given by (4.3) and (4.4), must be identical with that in (4.1). You may be wondering why we choose a straight line through the points $(x_0, f(x_0))$ and $(x_1, f(x_1))$ and not, for example, a parabola (that is, a polynomial of degree 2). We can always construct a parabola through two points; however, such a parabola is *not* unique. This is also true for polynomials of higher degree. (See Problem 4.3.)

It will help later to note that $p_1(x)$ may also be written as

$$p_1(x) = \frac{(x - x_0)f(x_1) - (x - x_1)f(x_0)}{x_1 - x_0} \tag{4.5}$$

or as

$$p_1(x) = f(x_0) + (x - x_0)\left(\frac{f(x_1) - f(x_0)}{x_1 - x_0}\right). \tag{4.6}$$

We will examine the form (4.6) a little more closely. From the mean value theorem 2.5, if f is continuous on $[x_0, x_1]$ and differentiable on (x_0, x_1), there exists a number $\xi \in (x_0, x_1)$ such that

$$\frac{f(x_1) - f(x_0)}{x_1 - x_0} = f'(\xi). \tag{4.7}$$

Hence (4.6) may be expressed in the form

$$p_1(x) = f(x_0) + (x - x_0)f'(\xi), \tag{4.8}$$

which shows that p_1 approximates to f at $x = x_0$ in a way similar to the approximation by the Taylor polynomial of degree one. If we keep x_0 fixed and let x_1 tend to x_0, the limit of (4.7) becomes $f'(x_0)$ (assuming continuity of f') and p_1 in (4.8) becomes precisely the Taylor polynomial.

We usually use p_1 to *interpolate*, that is, for some value of x we evaluate $p_1(x)$ in place of $f(x)$. This evaluation is called interpolation. If x is outside the interval $[x_0, x_1]$ (assuming $x_0 < x_1$), the term *extrapolation* is sometimes used, although we will not normally make this distinction. Since p_1 is a linear function of x, interpolation with p_1 is called *linear interpolation*. The question of how well p_1 approximates to f is one which we naturally ask and will be examined later in a manner similar to that used for the Taylor polynomial.

Example 4.1 From a set of four-figure tables of natural logarithms, we find that ln 3.1 is 1.1314 and ln 3.2 is 1.1632. We use linear interpolation on these two values to estimate ln 3.16 and see how the result compares with the entry ln 3.16 = 1.1506, given in the table itself. Here it is simplest to calculate $p_1(x)$ from (4.6), with $x_0 = 3.1$, $x_1 = 3.2$, $f(x_0) = 1.1314$, $f(x_1) = 1.1632$ and $x = 3.16$. Thus

$$p_1(3.16) = 1.1314 + 0.06 . \left(\frac{0.0318}{0.1}\right) = 1.15048.$$

Rounding this result to four figures gives the approximation 1.1505 for ln 3.16, which differs from the tabulated value by only one unit in the fourth decimal place. $\square$

4.2 Polynomial interpolation

We will now generalize most of the results of the last section to the case where we are given the values of the function f at $n + 1$ distinct points $x = x_0, x_1, \ldots, x_n$. These are not necessarily equally spaced or even arranged in increasing order; the only restriction is that they must be distinct. We want to construct a polynomial p_n which takes the same values as f at the $n + 1$ points $x_0, x_1, \ldots, x_n$. We might suspect that $n + 1$ coefficients could be determined by writing

$$p_n(x) = a_0 + a_1 x + \cdots + a_n x^n$$

and solving the $n + 1$ equations $p_n(x) = f(x)$, with $x = x_0, x_1, \ldots, x_n$ in turn. This suggests that p_n should be a polynomial of degree n. To generalize (4.1), we try to express p_n as

$$p_n(x) = L_0(x).f(x_0) + L_1(x).f(x_1) + \cdots + L_n(x).f(x_n), \qquad (4.9)$$

where the functions $L_0, L_1, \ldots, L_n$ are polynomials of degree at most n. The polynomial p_n will have the same values as f at $x = x_0, x_1, \ldots, x_n$ if

$$L_i(x_j) = \begin{cases} 0 & \text{if } j \neq i \\ 1 & \text{if } j = i \end{cases} \qquad (4.10)$$

for $0 \leqslant i \leqslant n$. So, for example, L_0 has the value zero at $x = x_1, x_2, \ldots, x_n$ and has the value 1 at $x = x_0$. Therefore, if we put

$$L_0(x) = C(x - x_1)(x - x_2)\cdots(x - x_n), \qquad (4.11)$$

where C is some constant, L_0 is seen to vanish at $x = x_1, x_2, \ldots, x_n$ and is a polynomial of degree n. Putting $x = x_0$ in (4.11), the condition $L_0(x) = 1$ gives

$$1 = C(x_0 - x_1)(x_0 - x_2) \cdots (x_0 - x_n), \tag{4.12}$$

which fixes the value of C. On substituting this value in (4.11) we obtain

$$L_0(x) = \frac{(x - x_1)(x - x_2) \cdots (x - x_n)}{(x_0 - x_1)(x_0 - x_2) \cdots (x_0 - x_n)},$$

which may be written more neatly as

$$L_0(x) = \prod_{j=1}^{n} \left(\frac{x - x_j}{x_0 - x_j} \right).$$

Similarly, for a general value of i, $0 \leqslant i \leqslant n$, we obtain

$$L_i(x) = \prod_{\substack{j=0 \\ j \neq i}}^{n} \left(\frac{x - x_j}{x_i - x_j} \right). \tag{4.13}$$

In (4.13), the product is taken over all values of j from 0 to n except for $j = i$. From (4.13) we see immediately that L_i is zero at all values $x = x_0$, $x_1, \ldots, x_n$, except for $x = x_i$ when L_i takes the value 1. This agrees with the requirement (4.10).

From (4.9), we now have

$$p_n(x) = \sum_{i=0}^{n} L_i(x) . f(x_i)$$

or, more fully, using (4.13) we write the interpolating polynomial for f at $x = x_0, x_1, \ldots, x_n$ as

$$p_n(x) = \sum_{i=0}^{n} \left\{ f(x_i) . \prod_{\substack{j=0 \\ j \neq i}}^{n} \left(\frac{x - x_j}{x_i - x_j} \right) \right\}. \tag{4.14}$$

This is known as the Lagrange form of the interpolating polynomial, after the French-Italian mathematician J. L. Lagrange (1736–1813).

Example 4.2   Given the values of $f(x)$ at distinct points $x = x_0$, x_1 and x_2, write out explicitly the interpolating polynomial p_2. We have

$$p_2(x) = \frac{(x - x_1)(x - x_2)}{(x_0 - x_1)(x_0 - x_2)} f(x_0) + \frac{(x - x_0)(x - x_2)}{(x_1 - x_0)(x_1 - x_2)} f(x_1)$$

$$+ \frac{(x - x_0)(x - x_1)}{(x_2 - x_0)(x_2 - x_1)} f(x_2). \quad \square \tag{4.15}$$

Example 4.3   Find the interpolating polynomial for a function f, given that $f(x) = 0, -3$ and 4 when $x = 1, -1$ and 2 respectively. (The reader should draw a graph.) Putting $x_0 = 1$, $x_1 = -1$ and $x_2 = 2$ in (4.15) (these three values could have been assigned in any order), we obtain

$$p_2(x) = \frac{(x + 1)(x - 2)}{(1 + 1)(1 - 2)} \times 0 + \frac{(x - 1)(x - 2)}{(-1 - 1)(-1 - 2)} \times -3$$
$$+ \frac{(x - 1)(x + 1)}{(2 - 1)(2 + 1)} \times 4,$$

which simplifies to give $p_2(x) = \frac{1}{6}(5x^2 + 9x - 14)$. As a check on our calculations, it may be verified from this last equation that $p_2(1) = 0$, $p_2(-1) = -3$ and $p_2(2) = 4$. $\square$

The above example invites the question: is the interpolating polynomial unique? With the question put in this way, the answer is obviously no. For example, the polynomial

$$q(x) = \frac{1}{6}(5x^2 + 9x - 14) + C(x - 1)(x + 1)(x - 2),$$

with any choice of the constant C, also interpolates the data of Example 4.3. However, the following is true.

Theorem 4.1   Given the values of f at the $n + 1$ distinct points $x_0, x_1, \ldots, x_n$, there is a unique polynomial of degree at most n which takes the same value as f at these points.

Proof   First, notice the crucial phrase "of degree at most n" in the statement of the theorem. We already know that there is at least one polynomial $p_n \in P_n$ which interpolates f at $x = x_0, x_1, \ldots, x_n$. This is displayed in (4.14). Now consider any $q_n \in P_n$ such that $q_n(x_i) = f(x_i)$, $0 \leqslant i \leqslant n$. It follows that $p_n - q_n \in P_n$ is zero at the $n + 1$ points $x_0, x_1, \ldots, x_n$. But a polynomial of degree at most n has no more than n zeros unless it is zero at every point, that is, it is identically zero. Thus $p_n(x) \equiv q_n(x)$, showing the uniqueness of the interpolating polynomial and completing the proof. $\square$

4.3 Accuracy of interpolation

In this section, we will examine the accuracy with which the interpolating polynomial approximates the function f. First, it is not going to be possible to estimate the size of the error $f - p_n$ from a knowledge of the values of f at $x = x_0, x_1, \ldots, x_n$ alone. Some further information about f is required. To see this, consider the $n + 1$ points $(x_0, f(x_0)), (x_1, f(x_1)), \ldots, (x_n, f(x_n))$, where $x_0, x_1, \ldots, x_n$ are distinct. These fix the polynomial p_n, but we are free to draw any curve which passes through these $n + 1$ points and let this define the function f. Hence we can arrange for $f(x) - p_n(x)$ to be arbitrarily large at any value of x, except $x = x_0, x_1, \ldots, x_n$, where $f(x) - p_n(x)$ is zero.

However, we can estimate the error $f - p_n$ in terms of the $(n + 1)$th derivative of f, if this exists.

Theorem 4.2 Let $[a, b]$ be any interval which contains all $n + 1$ points $x_0, x_1, \ldots, x_n$. Let $f, f', \ldots, f^{(n)}$ exist and be continuous on $[a, b]$ and let $f^{(n+1)}$ exist for $a < x < b$. Then, given any $x \in [a, b]$, there exists a number ξ_x (depending on x) in (a, b) such that

$$f(x) - p_n(x) = (x - x_0)\cdots(x - x_n)\frac{f^{(n+1)}(\xi_x)}{(n + 1)!}. \tag{4.16}$$

Proof The statement of this theorem is similar to that of Taylor's theorem 3.1, whose proof depends on Rolle's theorem 2.4. We use Rolle's theorem again in this proof. First, since $f - p_n$ and $(x - x_0)\cdots(x - x_n)$ have zeros at the $n + 1$ points $x_0, x_1, \ldots, x_n$, so also does the function

$$g(x) = f(x) - p_n(x) + \lambda(x - x_0)\cdots(x - x_n), \tag{4.17}$$

where λ is any constant. We now choose λ so as to ensure that g has at least $n + 2$ zeros. If we wish to estimate the error at the point $x = \alpha$ in $[a, b]$, we choose λ in (4.17) so that $g(\alpha) = 0$. This gives

$$0 = f(\alpha) - p_n(\alpha) + \lambda(\alpha - x_0)\cdots(\alpha - x_n).$$

With this choice of λ,

$$g(x) = f(x) - p_n(x) - \frac{(x - x_0)\cdots(x - x_n)}{(\alpha - x_0)\cdots(\alpha - x_n)}\cdot(f(\alpha) - p_n(\alpha)). \tag{4.18}$$

The function g is seen to be zero at not less than $n + 2$ points including $x = x_0, \ldots, x_n$ and $x = \alpha$. Let us imagine that these points are arranged in order. On applying Rolle's theorem to g on each of the $n + 1$ sub-intervals between pairs of adjoining zeros, we deduce that g' has at least $n + 1$ zeros. Next we apply Rolle's theorem to g' on each of the n sub-intervals between pairs of its zeros. It follows that g'' has at least n zeros. Continuing in this way, we deduce that $g^{(n+1)}$ has at least one zero in (a, b). Let one such zero be denoted by ξ_α. Differentiating (4.18) $n + 1$ times gives

$$g^{(n+1)}(x) = f^{(n+1)}(x) - \frac{(n + 1)!}{(\alpha - x_0)\cdots(\alpha - x_n)}\cdot(f(\alpha) - p_n(\alpha)), \tag{4.19}$$

since, on differentiating $n + 1$ times, p_n vanishes and the term x^{n+1} in the product $(x - x_0)\cdots(x - x_n)$ is reduced to $(n + 1)!$. Substituting the zero ξ_α in (4.19) we obtain

$$0 = f^{(n+1)}(\xi_\alpha) - \frac{(n + 1)!}{(\alpha - x_0)\cdots(\alpha - x_n)}(f(\alpha) - p_n(\alpha)).$$

Lastly, rearranging this last equation and replacing α by x gives

$$f(x) - p_n(x) = (x - x_0)\cdots(x - x_n)\frac{f^{(n+1)}(\xi_x)}{(n+1)!}. \quad \square$$

To use the error estimate (4.16), we need to estimate $f^{(n+1)}$. Usually we do not know the value of ξ_x in (4.16) and have to work with bounds for the $(n+1)$th derivative, as we show in the following example. In §4.8, we mention a method for estimating the $(n+1)$th derivative, when the points x_i are equally spaced.

Example 4.4 Use Theorem 4.2 to estimate the error in linear interpolation, with particular reference to Example 4.1, where we estimated $\ln 3.16$, given the values of $\ln 3.1$ and $\ln 3.2$. For linear interpolation of $f(x)$ between $x = x_0$ and x_1, (4.16) yields

$$f(x) - p_1(x) = (x - x_0)(x - x_1)\frac{f''(\xi_x)}{2!}. \quad (4.20)$$

Here $f(x) = \ln x, f'(x) = 1/x$ and $f''(x) = -1/x^2$. Putting $x = 3.16, x_0 = 3.1$ and $x_1 = 3.2$, we obtain from (4.20) an error $-(0.06) \times (-0.04)/(2\xi_x^2)$. Since ξ_x lies between 3.1 and 3.2, the error lies between 0.00011 and 0.00013. $\square$

4.4 Neville's algorithm

We have already encountered one method for evaluating the interpolating polynomial, which is by the Lagrange formula (4.14). We describe two further methods for evaluating p_n which are more practical than the Lagrange form and are also of theoretical interest. These are the Neville algorithm, which will be described in this section, and the method of divided differences, which will be examined in §4.6. A hint of these two methods has already been given by (4.5) and (4.6).

It will be convenient to use $p_{i,j}$, with i, j non-negative integers and $i < j$, to denote the interpolating polynomial for a function f constructed at $x = x_i, x_{i+1}, \ldots, x_j$. This notation will be used in this section only. Thus, for example, $p_{0,1}$ is constructed at $x = x_0$ and x_1, $p_{2,5}$ is constructed at $x = x_2$, x_3, x_4, x_5 and $p_{0,n}$ (which we usually call p_n) is constructed at $x = x_0, x_1, \ldots, x_n$. The algorithm consists of building up the polynomials $p_{i,j}$ from those of lower degree. First, we express $p_{0,2}$ in terms of $p_{0,1}$ and $p_{1,2}$. Since $p_{0,2} \in P_2$ and matches f at $x = x_0, x_1$ and x_2, we try putting

$$p_{0,2}(x) = a(x - x_2)p_{0,1}(x) + b(x - x_0)p_{1,2}(x). \quad (4.21)$$

At $x = x_0$, the second term is zero and $p_{0,1}(x_0) = f(x_0)$, so we require $a = 1/(x_0 - x_2)$ to give $p_{0,2}(x_0) = f(x_0)$. At $x = x_2$, the first term is zero and we must choose $b = 1/(x_2 - x_0)$ to make $p_{0,2}(x_2) = f(x_2)$. With this choice of a and b, we have

$$p_{0,2}(x) = \frac{(x - x_0)p_{1,2}(x) - (x - x_2)p_{0,1}(x)}{x_2 - x_0}. \tag{4.22}$$

On putting $x = x_1$ in (4.22) we obtain $p_{0,2}(x_1) = f(x_1)$ and thus, by uniqueness, $p_{0,2}$ must be the required polynomial. We note that the format of the right side of (4.22), which is called a *linear cross-mean*, is the same as for the linear interpolating polynomial in (4.5). Thus to evaluated $p_{0,2}(x)$ for any value of x, we require (4.22) together with the following formulae for $p_{0,1}$ and $p_{1,2}$:

$$p_{0,1}(x) = \frac{(x - x_0)f(x_1) - (x - x_1)f(x_0)}{x_1 - x_0}, \tag{4.23}$$

$$p_{1,2}(x) = \frac{(x - x_1)f(x_2) - (x - x_2)f(x_1)}{x_2 - x_1}. \tag{4.24}$$

The calculation of $p_{0,2}(x)$ is therefore reduced to calculating the three linear cross-means (4.23), (4.24) and (4.22). Table 4.1 shows the order of the calculations. The arrows in Table 4.1 link the sets of four numbers required to

Table 4.1   Evaluation of a 3-point interpolating polynomial.

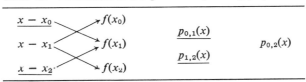

calculate $p_{0,1}(x)$ and $p_{1,2}(x)$; the four underlined numbers are used to calculate $p_{0,2}(x)$. It is also helpful to realize that the denominators in (4.22), (4.23) and (4.24) are obtainable by subtracting the multipliers in the numerators. For example, in (4.24)

$$x_2 - x_1 = (x - x_1) - (x - x_2).$$

This process may be extended to calculate interpolating polynomials which match f at still more points. We have

$$p_{0,3}(x) = \frac{(x - x_0)p_{1,3}(x) - (x - x_3)p_{0,2}(x)}{x_3 - x_0}, \tag{4.25}$$

since the right side of (4.25) is a polynomial of degree at most three and, as the reader may care to verify by substituting in turn $x = x_0$, x_1, x_2 and x_3, this polynomial matches f at these points. When making such a substitution in

(4.25), one must remember that $p_{1,3}$ matches f at $x = x_1, x_2, x_3$ and $p_{0,2}$ matches f at $x = x_0, x_1, x_2$. We see from (4.25) that $p_{0,2}(x)$ and $p_{1,3}(x)$ need to be evaluated before $p_{0,3}(x)$. We have already seen how to evaluate $p_{0,2}(x)$ by (4.22) and $p_{1,3}(x)$ is obtained by using (4.22) with every suffix increased by one. Thus

$$p_{1,3}(x) = \frac{(x - x_1)p_{2,3}(x) - (x - x_3)p_{1,2}(x)}{x_3 - x_1}.$$

The complete scheme for evaluating $p_{0,3}(x)$ is indicated by Table 4.2. It is important to realize that the polynomials in Table 4.2 are evaluated for a

Table 4.2   Evaluation of a 4-point interpolating polynomial.

$x - x_0$	$f(x_0)$			
		$p_{0,1}(x)$		
$x - x_1$	$f(x_1)$		$p_{0,2}(x)$	
		$p_{1,2}(x)$		$p_{0,3}(x)$
$x - x_2$	$f(x_2)$		$p_{1,3}(x)$	
		$p_{2,3}(x)$		
$x - x_3$	$f(x_3)$			

particular value of x, so that the final number obtained, $p_{0,3}(x)$, gives the approximation for $f(x)$ at that point, obtained by cubic interpolation on the chosen four points. This process may be extended to calculate any order of interpolating polynomial via interpolating polynomials of lower degree. (See Problem 4.13.) This method of constructing the interpolating polynomial by using linear cross-means is called Neville's algorithm. Another variant, which also uses linear cross-means but introduces the points in a different order, is called Aitken's algorithm.

Example 4.5   Given the following data at four points for the function e^{-x}, we use Neville's algorithm to evaluate the cubic interpolating polynomial at $x = 0.2$. We also test how well this approximates to e^{-x} at $x = 0.2$, by using the error formula (4.16).

x_i	0.10	0.15	0.25	0.30
e^{-x_i}	0.904837	0.860708	0.778801	0.740818

The calculation of the Neville table is set out in Table 4.3, which is simply Table 4.2 applied to the present example. The numbers in the last three columns have been rounded to seven decimal places, so that to six figures the result is 0.818730. To estimate the accuracy, we apply the error formula (4.16), where $n = 3$, $x = 0.2$, $x_0 = 0.1$, $x_1 = 0.15$, $x_2 = 0.25$, $x_3 = 0.3$ and the fourth derivative of f is the fourth derivative of e^{-x}, which is also e^{-x}.

Electric Car Project.   27/8/79

Budget £1000.

Find out:

(1) legality of converting means of propulsion.

(2). Appr. cost of one vehicle in MoT'able
condition ① Power transmission, materials
   practicability

(3) Vehicle weight — less engine & gearbox.

(4). Cost of electric-motors
    suggest Lucas & CAV.

Notes
(1) Power transmission without gears.

(2) labour costs within department
    are free-

-0.272177E+01

-0.322641E+01

-0.170712E+01
-0.322707B3E+01

-0.322708E+01   HT= 0.10000008E-01

-0.380662E+01

-0.437724E+01

-0.495069E+01

-0.322708E+01
-0.494928895E+01

-0.494929E+01   HT= 0.10000008E-01

-0.551688E+01

-0.606949E+01

-0.663110E+01

-0.662662638E+01

-0.662662E+01   HT= 0.10000008E-01

-0.717630

Table 4.3 Application of Neville's algorithm to estimate e^{-x} at $x = 0.2$.

$x - x_i$	e^{-x_i}			
0.10	0.904837			
		0.8165790		
0.05	0.860708		0.8186960	
		0.8197545		0.8187302
−0.05	0.778801		0.8187643	
		0.8167840		
−0.10	0.740818			

Therefore the error is

$$E = (0.1).(0.05).(-0.05).(-0.1)e^{-\xi}/24, \qquad (4.26)$$

where $0.1 < \xi < 0.3$ and so, from the data, $e^{-\xi}$ lies between 0.904837 and 0.740818. From this and (4.26) we see that the error E satisfies the inequalities

$$0.77 . 10^{-6} < E < 0.95 . 10^{-6}.$$

This shows that E is almost one unit in the sixth place. $\square$

4.5 Inverse interpolation

Let us write $y = f(x)$. So far we have been interpolating in the x direction. It is possible to consider interpolation in the y direction also. This is referred to as "inverse interpolation". In this case it is the numbers $f(x_i)$ which we require to be distinct. The roles of x and $f(x)$ are interchanged. Otherwise, the calculation of the interpolating polynomial, say by Neville's method, may be carried out as before. The resulting interpolating polynomial may be regarded as a polynomial in y.

Example 4.6 Use inverse interpolation at $x = 0, \frac{1}{2}$ and 1 to estimate the only real root of the equation

$$x^3 + x^2 + x - 1 = 0,$$

which lies between $x = 0$ and 1. Secondly, use inverse interpolation at only $x = 0.5$ and 0.6 to estimate this root and estimate the accuracy of the result.

We see that the first derivative of $y = x^3 + x^2 + x - 1$ is always positive, showing that there is exactly one real root of the given equation. At $x = 0, \frac{1}{2}$ and 1, y has the values -1, $-\frac{1}{8}$ and 2 respectively. Neville's method gives the scheme set out in Table 4.4. We have written $y - y_i$ in the first column, although $y = 0$, to emphasize the pattern of the calculation. From the table, the result for inverse interpolation at the three points is $199/357 = 0.557$ to three decimal places. Secondly, at $x = 0.5$ and 0.6, $y = -0.125$ and 0.176 respectively. Inverse interpolation this time gives, as the estimate of the root, $x = 163/301 = 0.5415$ to four decimal places. We now want to estimate the accuracy of the last result. Suppose that $x = \phi(y)$ and $y = f(x)$. Assuming that ϕ'' exists on $[-0.125, 0.176]$, the error formula (4.16) gives an error

$$(y - y_0)(y - y_1)\phi''(\eta)/2!. \tag{4.27}$$

Table 4.4   Inverse interpolation for Example 4.6.

$y - y_i$	x_i		
1	0		
		4/7	
$\frac{1}{8}$	$\frac{1}{2}$		199/357
		9/17	
-2	1		

In (4.27) η lies in an interval containing y, y_0 and y_1 which, in this case, are the values 0, -0.125 and 0.176. We have

$$\phi''(y) = \frac{d^2x}{dy^2} = \frac{d}{dy}\left(\frac{dx}{dy}\right) = \frac{d}{dy}\left(1 \bigg/ \frac{dy}{dx}\right) = \frac{d}{dy}(1/f'(x)),$$

provided $f'(x) \neq 0$. Thus

$$\phi''(y) = \frac{d}{dx}(1/f'(x)) \cdot \frac{dx}{dy} = -\frac{f''(x)}{[f'(x)]^2} \cdot \frac{1}{f'(x)},$$

so that

$$\phi''(y) = -\frac{f''(x)}{[f'(x)]^3}.$$

Since $f(x) = x^3 + x^2 + x - 1$, $f'(x) = 3x^2 + 2x + 1$, $f''(x) = 6x + 2$ and so

$$\phi''(y) = -\frac{(6x + 2)}{(3x^2 + 2x + 1)^3}.$$

For x between 0.5 and 0.6, $\phi''(y)$ lies between -0.25 and -0.15. From (4.27), it follows that the error lies between 0.001 and 0.003, so the result $x = 0.5415$ is certainly correct to two decimal places.

The problem of finding solutions of an algebraic equation is discussed more generally in Chapter 7. □

Uncritical use of inverse interpolation (as also for direct interpolation) can produce misleading results. For example, consider the function $y = x^4$ with $x \geqslant 0$ tabulated at the four points where $y = 0, 1, 16$ and 81. It is left to the reader to verify (in Problem 4.16) that cubic inverse interpolation based on these four points produces the approximation $x = -33.4$ at $y = 64$, where the correct result is $x = (64)^{1/4} = 2\sqrt{2}$. In this case, we are approximating to $x = y^{1/4}$ by a cubic polynomial in y. The error term includes the fourth derivative of $y^{1/4}$ with respect to y, which explains why such a large error can occur, as this derivative is infinite at $y = 0$.

4.6 Divided differences

Another approach to the interpolating polynomial is to use divided differences, which will now be discussed. This method is due to Isaac Newton. As well as being of theoretical interest, it presents the interpolating polynomial in a form which may be readily simplified, when we later choose the interpolating points to be equally spaced.

At the moment, however, we allow the numbers $x_0, x_1, \ldots, x_n$ to be any $n + 1$ distinct numbers and attempt to write p_n, the interpolating polynomial for f at these points, as

$$p_n(x) = a_0 + (x - x_0)a_1 + (x - x_0)(x - x_1)a_2 + \cdots$$
$$+ (x - x_0)(x - x_1)\cdots(x - x_{n-1})a_n. \tag{4.28}$$

Substituting $x = x_0, x_1, \ldots, x_n$ in turn into (4.28), we obtain the following simultaneous equations:

$$f(x_0) = a_0$$
$$f(x_1) = a_0 + (x_1 - x_0)a_1$$
$$f(x_2) = a_0 + (x_2 - x_0)a_1 + (x_2 - x_0)(x_2 - x_1)a_2 \tag{4.29}$$
$$\vdots$$
$$f(x_n) = a_0 + (x_n - x_0)a_1 + \cdots + (x_n - x_0)\cdots(x_n - x_{n-1})a_n.$$

We see that these equations† determine values for $a_0, \ldots, a_n$ uniquely. The first equation in (4.29) gives a_0 and the second gives a_1, as $(x_1 - x_0) \neq 0$. Knowing a_0 and a_1, the third equation gives a_2, as $(x_2 - x_0)(x_2 - x_1) \neq 0$. Finally, knowing $a_0, a_1, \ldots, a_{n-1}$, the last equation gives a_n, since $(x_n - x_0)\cdots$ $(x_n - x_{n-1}) \neq 0$. Hence $p_n(x)$ can be written, as in (4.28), in a unique way. Moreover, the coefficients which appear in (4.28) enjoy what is called *permanence*. If we add one further point x_{n+1}, distinct from $x_0, \ldots, x_n$, and write the interpolating polynomial constructed at all points $x_0, \ldots, x_{n+1}$ in the form

$$p_{n+1}(x) = b_0 + (x - x_0)b_1 + \cdots + (x - x_0)\cdots(x - x_{n-1})b_n$$
$$+ (x - x_0)\cdots(x - x_n)b_{n+1}, \tag{4.30}$$

we find that $b_0 = a_0, b_1 = a_1, \ldots, b_n = a_n$. For, on writing down the linear

† These are triangular equations, whose solution is discussed in detail in Chapter 8.

equations obtained for determining the b_i by substituting $x = x_0, x_1, \ldots, x_{n+1}$
into (4.30) in turn, we have

$$f(x_0) = b_0$$
$$f(x_1) = b_0 + (x_1 - x_0)b_1$$
$$\vdots \qquad\qquad\qquad\qquad\qquad\qquad\qquad\qquad\qquad (4.31)$$
$$f(x_n) = b_0 + (x_n - x_0)b_1 + \cdots + (x_n - x_0)\cdots(x_n - x_{n-1})b_n$$
$$f(x_{n+1}) = b_0 + (x_{n+1} - x_0)b_1 + \cdots + (x_{n+1} - x_0)\cdots(x_{n+1} - x_n)b_{n+1}.$$

An examination of the first $n + 1$ equations in (4.31) shows that they are the
same as the equations (4.29), which justifies the statement that $b_i = a_i$,
$0 \leqslant i \leqslant n$. In fact, we now see that a_0 involves only $f(x_0)$, a_1 involves only
$f(x_0)$ and $f(x_1)$, and so on. In general, we see that a_j involves $f(x_0), f(x_1), \ldots,$
$f(x_j)$ only. We now rewrite a_j as

$$a_j = f[x_0, x_1, \ldots, x_j]$$

to emphasize its dependence on these suffices. It is instructive to compare
(4.28) with the Lagrange form (4.14). If we equate coefficients of x^n, we obtain

$$f[x_0, \ldots, x_n] = \sum_{i=0}^{n} \frac{f(x_i)}{\prod\limits_{\substack{j=0 \\ j \neq i}}^{n} (x_i - x_j)}. \qquad (4.32)$$

Thus, for example,

$$f[x_0, x_1, x_2] = \frac{f(x_0)}{(x_0 - x_1)(x_0 - x_2)} + \frac{f(x_1)}{(x_1 - x_0)(x_1 - x_2)}$$

$$+ \frac{f(x_2)}{(x_2 - x_0)(x_2 - x_1)}, \qquad (4.33)$$

and in general the right side of (4.32) consists of a linear combination of the
function values $f(x_i)$. Can we express these coefficients more simply? For
example, we might try to find numbers α and β so that

$$f[x_0, x_1, x_2, x_3] = \alpha f[x_0, x_1, x_2] + \beta f[x_1, x_2, x_3], \qquad (4.34)$$

for certainly (from (4.32)) each side of this equation is a sum of multiples of
$f(x_0), f(x_1), f(x_2)$ and $f(x_3)$. Comparing the term in $f(x_0)$ in (4.32) and (4.34),
we need to choose $\alpha = 1/(x_0 - x_3)$, and a comparison of the term in $f(x_3)$
requires $\beta = 1/(x_3 - x_0)$. It may be verified that this choice of α and β agrees
also with the terms in $f(x_1)$ and $f(x_2)$, so that

$$f[x_0, x_1, x_2, x_3] = \frac{f[x_1, x_2, x_3] - f[x_0, x_1, x_2]}{x_3 - x_0}. \qquad (4.35)$$

This shows why these expressions are called *divided differences*, especially as a similar recurrence relation holds generally. We have

$$f[x_0, \ldots, x_n] = \frac{f[x_1, \ldots, x_n] - f[x_0, \ldots, x_{n-1}]}{x_n - x_0}. \tag{4.36}$$

The last relation may be verified by substituting for each divided difference its representation in the form (4.32). It is helpful to exhibit these divided differences, as in Table 4.5, to remind us of how they are calculated. To preserve

Table 4.5 Calculation of divided differences.

x_0	$f[x_0]$			
		$f[x_0, x_1]$		
x_1	$f[x_1]$		$f[x_0, x_1, x_2]$	
		$f[x_1, x_2]$		$f[x_0, x_1, x_2, x_3]$
x_2	$f[x_2]$		$f[x_1, x_2, x_3]$	
		$f[x_2, x_3]$		
x_3	$f[x_3]$			

the pattern, we have written $f[x_i]$ for $f(x_i)$. Thus, for example, we calculate $f[x_2, x_3]$ from

$$f[x_2, x_3] = \frac{f[x_3] - f[x_2]}{x_3 - x_2}$$

and $f[x_1, x_2, x_3]$ from

$$f[x_1, x_2, x_3] = \frac{f[x_2, x_3] - f[x_1, x_2]}{x_3 - x_1}.$$

Applying this notation to (4.28), we write the interpolating polynomial in the form

$$p_n(x) = f[x_0] + (x - x_0)f[x_0, x_1] + \cdots$$
$$+ (x - x_0) \cdots (x - x_{n-1})f[x_0, \ldots, x_n]. \tag{4.37}$$

Example 4.7 By using divided differences, obtain the interpolating polynomial which we sought in Example 4.3, where $f(x) = 0, -3$ and 4 at $x = 1$, -1 and 2 respectively. The calculation of the divided differences is shown in Table 4.6. We see from this table that $f[x_0, x_1] = \frac{3}{2}$ and $f[x_0, x_1, x_2] = \frac{5}{6}$ so that, from (4.37), the required interpolating polynomial is

$$p_2(x) = 0 + (x - 1)\tfrac{3}{2} + (x - 1)(x + 1)\tfrac{5}{6}.$$

This simplifies to give

$$p_2(x) = \tfrac{1}{6}(5x^2 + 9x - 14),$$

as we found in Example 4.3 by the Lagrange method. □

It may be noted that the method of divided differences enables us to exhibit the interpolating polynomial explicitly in the form (4.37). On the other hand, Neville's method is really only suitable for evaluating the interpolating polynomial at specified values of x, and is not convenient for obtaining the polynomial p_n explicitly. The reason for the superiority of the method of divided differences, in this respect, is that divided differences are independent of x, whereas the terms $p_{i,j}$, which appear in Neville's method, are functions of x. If one simply wants to evaluate p_n for some given value of x, the divided difference method can provide the result with fewer arithmetical operations,

Table 4.6   Calculation of divided differences for Example 4.7.

x	$f(x)$	Divided differences		
1	0			
		$\dfrac{-3-0}{-1-1}=\dfrac{3}{2}$		
-1	-3			$\dfrac{7/3-3/2}{2-1}=\dfrac{5}{6}$
		$\dfrac{4-(-3)}{2-(-1)}=\dfrac{7}{3}$		
2	4			

noting that p_n in (4.37) can be evaluated very compactly. For example, we calculate $p_2(x)$ from

$$p_2(x) = f[x_0] + (x - x_0)(f[x_0, x_1] + (x - x_1)f[x_0, x_1, x_2]).$$

Evaluating p_n similarly requires only n multiplications. This economical method for evaluating polynomials is called "nested multiplication". However, the Neville table has the advantage that every entry is an interpolating polynomial (evaluated at some point x) for some part of the data. This provides some empirical evidence about the accuracy of the interpolation, by noting the level of agreement between the numbers in the Neville table. (See, for example, Table 4.3.) If, as very commonly happens, the points x_i are to be equally spaced, simpler forms of the interpolating polynomial may be constructed.

4.7 Equally spaced points

Hitherto, we have considered interpolation for a function f whose value is known at $n + 1$ arbitrary distinct points. In this section, we assume these to be arranged in order and equally spaced, with spacing h, so that $x_j = x_0 + jh$, $0 \leqslant j \leqslant n$. It is interesting to see what effect this has on the table of divided differences. If we recalculate Table 4.5, we obtain the simplification shown in Table 4.7 where, for the sake of brevity, we have used f_i to denote $f(x_i)$. It is

Table 4.7   Divided differences with equally spaced points.

x_0	f_0			
		$\frac{1}{h}(f_1 - f_0)$		
x_1	f_1		$\frac{1}{2h^2}(f_2 - 2f_1 + f_0)$	
		$\frac{1}{h}(f_2 - f_1)$		$\frac{1}{6h^3}(f_3 - 3f_2 + 3f_1 - f_0)$
x_2	f_2		$\frac{1}{2h^2}(f_3 - 2f_2 + f_1)$	
		$\frac{1}{h}(f_3 - f_2)$		
x_3	f_3			

now convenient to introduce the *forward difference operator* Δ, which operates on f as follows:

$$\Delta f(x) = f(x + h) - f(x).$$

Thus

$$\Delta f_0 = f(x_0 + h) - f(x_0) = f_1 - f_0$$

and $\Delta f_1 = f_2 - f_1$. Second differences are written as $\Delta^2 f(x)$, defined by

$$\Delta^2 f(x) = \Delta(\Delta f(x)),$$

so that

$$\begin{aligned} \Delta^2 f(x) &= \Delta(f(x + h) - f(x)) \\ &= (f(x + 2h) - f(x + h)) - (f(x + h) - f(x)) \\ &= f(x + 2h) - 2f(x + h) + f(x). \end{aligned}$$

For example, $\Delta^2 f_1 = f_3 - 2f_2 + f_1$. Higher differences are defined similarly, so that $\Delta^3 f(x)$ means $\Delta(\Delta^2 f(x))$, giving

$$\Delta^3 f(x) = \Delta^2 f(x + h) - \Delta^2 f(x)$$

and, on expanding the second differences,

$$\Delta^3 f(x) = f(x + 3h) - 3f(x + 2h) + 3f(x + h) - f(x).$$

In general, $\Delta^{n+1}f(x)$ is defined as $\Delta(\Delta^n f(x))$, for $n = 1, 2, \ldots$. It is convenient to define $\Delta^0 f(x)$ to mean simply $f(x)$. In this notation, Table 4.7 becomes Table 4.8. On comparing Tables 4.8 and 4.5, we see that the effect of taking the x_i to be equally spaced is to replace divided differences by multiples of

Table 4.8 Divided differences with equally spaced points, expressed in terms of forward differences.

x_0	f_0			
		$\frac{1}{h}\Delta f_0$		
x_1	f_1		$\frac{1}{2h^2}\Delta^2 f_0$	
		$\frac{1}{h}\Delta f_1$		$\frac{1}{6h^3}\Delta^3 f_0$
x_2	f_2		$\frac{1}{2h^2}\Delta^2 f_1$	
		$\frac{1}{h}\Delta f_2$		
x_3	f_3			

differences, which are much simpler. The general connection between divided differences and finite differences is

$$f[x_0, x_1, \ldots, x_n] = \frac{\Delta^n f_0}{n!\, h^n}. \tag{4.38}$$

This relation, which is valid only for equally spaced points, may be verified by an inductive argument. First (4.38) holds for $n = 0$, since $0! = 1$ and $\Delta^0 f_0 = f_0$. (Anyone who prefers may begin with $n = 1$, for which (4.38) is also easily verified.) Second let us assume that a relation of the form (4.38) holds for k interpolating points. We know from (4.36) that

$$f[x_0, \ldots, x_k] = \frac{f[x_1, \ldots, x_k] - f[x_0, \ldots, x_{k-1}]}{x_k - x_0}. \tag{4.39}$$

Replacing $x_k - x_0$ in (4.39) by kh and using the hypothesis that (4.38) holds for k interpolating points, we have

$$f[x_0, \ldots, x_k] = \frac{1}{kh}\left(\frac{\Delta^{k-1} f_1}{(k-1)!\, h^{k-1}} - \frac{\Delta^{k-1} f_0}{(k-1)!\, h^{k-1}}\right)$$

$$= \frac{1}{k!\, h^k}(\Delta^{k-1} f_1 - \Delta^{k-1} f_0) = \frac{\Delta^k f_0}{k!\, h^k},$$

which completes the proof.

We can now simplify (4.37), which expresses the interpolating polynomial in terms of divided differences. It is helpful first to make a change of variable, putting $x = x_0 + sh$, so that we measure the interpolating point x in units of h from the point x_0. With this change of variable, a factor $x - x_i$ becomes $(s - i)h$ and, using (4.38), a typical term in (4.37) becomes

$$(x - x_0)\cdots(x - x_{j-1})f[x_0, \ldots, x_j]$$
$$= h^j s(s - 1)\cdots(s - j + 1)\frac{\Delta^j f_0}{j!\, h^j} = \binom{s}{j} \Delta^j f_0.$$

We have written $\binom{s}{j}$ to denote the binomial coefficient

$$\frac{s(s - 1)\cdots(s - j + 1)}{j!}.$$

Then (4.37) becomes

$$p_n(x) = p_n(x_0 + sh) = f_0 + \binom{s}{1}\Delta f_0 + \binom{s}{2}\Delta^2 f_0 + \cdots + \binom{s}{n}\Delta^n f_0, \qquad (4.40)$$

which is referred to as the *forward difference formula*. This result was first discovered† by James Gregory (1638–1675), a Scots contemporary of Isaac Newton. In this notation, we may rewrite (4.16) to give interpolating polynomial and error term as

$$f(x_0 + sh) = \sum_{j=0}^{n} \binom{s}{j}\Delta^j f_0 + h^{n+1}\binom{s}{n+1}f^{(n+1)}(\xi_x). \qquad (4.41)$$

The last term is the error and the summation is the interpolating polynomial.

Example 4.8 Given the following table, use the forward difference formula to estimate $\sin x$ at $x = 0.63$ and determine the accuracy of the result.

x	0.6	0.7	0.8	0.9	1.0
$\sin x$	0.564642	0.644218	0.717356	0.783327	0.841471

The differences for this data are set out in Table 4.9. For convenience, the decimal point has been omitted from the differences. This is normal practice. In this case, $x_0 = 0.6$ and $h = 0.1$, so in using (4.40) to interpolate at $x = 0.63$, we choose $s = 0.3$ and $n = 4$. We see from Table 4.9 that $f_0, \Delta f_0, \Delta^2 f_0, \Delta^3 f_0$ and $\Delta^4 f_0$ have the values 0.564642, 0.079576, -0.006438, -0.000729 and 0.000069 respectively, which are underlined in the table. With these figures, the estimate for $\sin(0.63)$ provided by (4.40) is 0.589145, to six decimal places. Using (4.41) to estimate the error, the fifth derivative of $\sin x$ is $\cos x$, which never exceeds 1 in modulus. Hence the error is not greater than the modulus of

$$10^{-5}(0.3)(-0.7)(-1.7)(-2.7)(-3.7)/5!,$$

which is smaller than 0.3×10^{-6}. □

† Hildebrand (p. 91) makes an interesting comment on the naming of interpolation formulae.

Numerical Analysis

In Chapter 11, we require a companion formula to (4.40) which starts at the right-hand end (i.e. at $x = x_n$) of a set of equally spaced data and uses differences constructed from function values to the left of $x = x_n$. To obtain this, we return to the divided difference representation of $p_n(x)$, in (4.37). In this formula, the numbers $x_0, \ldots, x_n$ are distinct, but otherwise arbitrary.

Table 4.9   A table of differences for sin x.

x	sin x	Differences			
0.6	0.564642				
		79576			
0.7	0.644218		-6438		
		73138		-729	
0.8	0.717356		-7167		69
		65971		-660	
0.9	0.783327		-7827		
		58144			
1.0	0.841471				

In particular, we could rename $x_0, x_1, \ldots, x_n$ as $x_n, x_{n-1}, \ldots, x_0$ which would give

$$p_n(x) = f[x_n] + (x - x_n)f[x_n, x_{n-1}] + \cdots$$
$$+ (x - x_n)(x - x_{n-1})\cdots(x - x_1)f[x_n, \ldots, x_0]. \quad (4.42)$$

Now let us see what happens when the x_j are spaced equally. In this case, it is natural to express the interpolating polynomial in terms of *backward differences*. We define the backward difference operator ∇ from

$$\nabla f(x) = f(x) - f(x - h).$$

Higher differences are defined as for the operator Δ. Thus

$$\nabla^2 f(x) = f(x) - 2f(x - h) + f(x - 2h).$$

Returning to (4.42), we have

$$f[x_n, x_{n-1}] = \frac{1}{h} \nabla f_n.$$

To simplify the third term in (4.42), we write

$$f[x_n, x_{n-1}, x_{n-2}] = \frac{f[x_n, x_{n-1}] - f[x_{n-1}, x_{n-2}]}{x_n - x_{n-2}}$$

$$= \frac{1}{2h^2} (\nabla f_n - \nabla f_{n-1}) = \frac{1}{2h^2} \nabla^2 f_n.$$

By induction, we can show that

$$f[x_n, \ldots, x_0] = \frac{1}{n! \, h^n} \nabla^n f_n,$$

which allows us to simplify the last term in (4.42). Finally, since in (4.42) we are "measuring" from the point $x = x_n$, we make the change of variable $x = x_n + sh$. A factor $x - x_{n-i}$ becomes $(s + i)h$, so that a typical term in (4.42) becomes

$$(x - x_n)(x - x_{n-1}) \cdots (x - x_{n-j+1}) f[x_n, x_{n-1}, \ldots, x_{n-j}]$$

$$= h^j s(s + 1) \cdots (s + j - 1) \frac{\nabla^j f_n}{j! \, h^j}$$

$$= (-1)^j (-s)(-s - 1) \cdots (-s - j + 1) \frac{\nabla^j f_n}{j!}$$

$$= (-1)^j \binom{-s}{j} \nabla^j f_n.$$

From this, we see that (4.42) may be written as

$$p_n(x_n + sh) = \sum_{j=0}^{n} (-1)^j \binom{-s}{j} \nabla^j f_n. \tag{4.43}$$

This is the *backward difference formula*.

It should be emphasized that the forward and backward difference formulae are merely different ways of representing the *same* polynomial. It may also be noted that it is not strictly necessary to have both operators Δ and ∇. For example, ∇f_n, $\nabla^2 f_n$ and $\nabla^3 f_n$ denote the same numbers as Δf_{n-1}, $\Delta^2 f_{n-2}$ and $\Delta^3 f_{n-3}$ respectively. As another illustration, consider Tables 4.10(a) and

Table 4.10   Forward and backward differences.

(a)						(b)					
x_0	f_0					x_0	f_0				
		Δf_0						∇f_1			
x_1	f_1		$\Delta^2 f_0$			x_1	f_1		$\nabla^2 f_2$		
		Δf_1		$\Delta^3 f_0$				∇f_2		$\nabla^3 f_3$	
x_2	f_2		$\Delta^2 f_1$			x_2	f_2		$\nabla^2 f_3$		
		Δf_2						∇f_3			
x_3	f_3					x_3	f_3				

4.10(b) whose corresponding entries are numerically the same; only the notation is different. Nevertheless, it is convenient to have these alternative ways of expressing differences. For expressing differences symmetrically, there is a third operator δ, called the central difference operator, defined by

$$\delta f(x) = f(x + \tfrac{1}{2}h) - f(x - \tfrac{1}{2}h).$$

Also

$$\delta^2 f(x) = \delta(f(x + \tfrac{1}{2}h) - f(x - \tfrac{1}{2}h))$$

$$= f(x + h) - 2f(x) + f(x - h).$$

Hildebrand discusses several interpolating formulae which use central differences. However, we will not pursue this here, since such formulae are primarily of use in highly accurate interpolation of tables, which is rarely required nowadays, due to the availability of computers.

4.8 Derivatives and differences

One often observes that differences of a function (tabulated at equal intervals) tend to decrease as the order of the differences increases. This appears to be the case in Table 4.9. To explain this, we make use once more of divided differences. Let us write

$$f[x, x_0] = \frac{f(x_0) - f(x)}{x_0 - x},$$

which may be rearranged as

$$f(x) = f(x_0) + (x - x_0)f[x, x_0]. \tag{4.44}$$

Also,

$$f[x, x_0, x_1] = \frac{f[x_0, x_1] - f[x, x_0]}{x_1 - x},$$

which gives the equation

$$f[x, x_0] = f[x_0, x_1] + (x - x_1)f[x, x_0, x_1]. \tag{4.45}$$

Substituting for $f[x, x_0]$ from (4.45) into (4.44) gives

$$f(x) = f[x_0] + (x - x_0)f[x_0, x_1] + (x - x_0)(x - x_1)f[x, x_0, x_1],$$

where we have written $f[x_0]$ instead of $f(x_0)$ for the sake of uniformity of notation. This may be extended, by considering $f[x, x_0, x_1, x_2]$, and so on, until we obtain by induction

$$f(x) = f[x_0] + (x - x_0)f[x_0, x_1] + \cdots$$

$$+ (x - x_0)\cdots(x - x_{n-1})f[x_0, \ldots, x_n]$$

$$+ (x - x_0)\cdots(x - x_n)f[x, x_0, \ldots, x_n]. \tag{4.46}$$

Comparing this with (4.37), we see that (4.46) is of the form

$$f(x) = p_n(x) + (x - x_0)\cdots(x - x_n)f[x, x_0, \ldots, x_n] \tag{4.47}$$

and, on comparison with the error formula (4.16), we find that

$$f[x, x_0, \ldots, x_n] = \frac{f^{(n+1)}(\xi_x)}{(n+1)!}, \tag{4.48}$$

where ξ_x is some point in an interval containing $x, x_0, \ldots, x_n$. This last formula holds for any n and any distinct numbers $x, x_0, \ldots, x_n$, provided that the $(n + 1)$th derivative of f exists. In particular, omitting the point x_0 in (4.48), we have

$$f[x, x_1, \ldots, x_n] = \frac{f^{(n)}(\eta_x)}{n!}, \tag{4.49}$$

where η_x lies in an interval containing $x, x_1, \ldots, x_n$. Now put $x = x_0$ in (4.49) and let $x_0, \ldots, x_n$ be equally spaced. On using (4.38), we obtain

$$\Delta^n f_0 = h^n f^{(n)}(\eta_0), \tag{4.50}$$

where η_0 lies in the interval $(x_0, x_0 + nh)$. In the same way, we have

$$\Delta^n f_1 = h^n f^{(n)}(\eta_1),$$

where η_1 lies in the interval $(x_1, x_1 + nh)$, and so on. This shows that nth differences behave as h^n times the nth derivative of f. It is now clear why in Table 4.9, where $h = 0.1$, differences are decreasing by approximately a factor of 10, from one column to the next.

As an aside, if in (4.49) we put $x = x_0$ and retain the divided difference notation we have

$$f[x_0, \ldots, x_n] = \frac{f^{(n)}(\eta_0)}{n!}, \tag{4.51}$$

which shows the close connection between the divided difference representation for f, (4.46), and the Taylor polynomial representation. Indeed, allowing $x_0, \ldots, x_n$ to be equally spaced in (4.51) and letting $h \to 0$, we obtain

$$\lim_{h \to 0} f[x_0, \ldots, x_n] = \frac{f^{(n)}(x_0)}{n!} \tag{4.52}$$

and (4.46) becomes the Taylor polynomial for f about x_0, with remainder. What we have given is a proof of Taylor's theorem.

In practice, it is usually difficult to evaluate $f^{(n+1)}(x)$. It is then desirable to use an $(n + 1)$th difference in place of $f^{(n+1)}(\xi_x)$ to estimate the error term (4.16) for polynomial interpolation. If in Example 4.4 we are given $\ln 3.3 = 1.1939$ we find that, with $h = 0.1$ and $f(x) = \ln x$, $\Delta^2 f(3.1) = -0.0011$. (For x between 3.1 and 3.3, $h^2 f''(x)$ actually lies between -0.001041 and -0.000918. The discrepancy is due to rounding errors in the four-figure values for $\ln x$.)

The relation (4.50), which connects differences with derivatives, reveals one other property of polynomials. If $p \in P_n$, its nth derivative is constant and higher derivatives are zero. Substituting $p(x)$ into (4.50), we see that the same statement holds for differences so that, for $p \in P_n$, differences higher than the nth must be zero. We can make use of this property to tabulate a polynomial.

Example 4.9　In Table 4.11 the polynomial $p(x) = x^3 - 3x + 5$ is evaluated at $x = 0$, 1, 2, 3 and differenced, to produce all the numbers which appear

Table 4.11　Evaluation of a polynomial by "building-up" from its differences. Here $p(x) = x^3 - 3x + 5$, so third differences are constant.

x	$p(x)$	1st	2nd	3rd
		\multicolumn{3}{c}{Differences}		
0	5			
		-2		
1	3		6	
		4		6
2	7		12	
		16		6
3	23		18	
		34		6
4	57		24	
		58		
5	115			

above the line shown in the table. Further entries are constructed by using the property that third differences are constant. So, below the line, we insert the entries 6, 18, 34, 57 in that order. Next, returning to the constant third differences, we insert the entries 6, 24, 58, 115, and so on, to extend the table. □

4.9　Difference tables

Another feature of difference tables which the reader may have noticed is that the differences start to grow again after the stage at which they have diminished to nearly zero. This is seen in Table 4.12, which shows differences of e^x. From (4.50), since every derivative of e^x is also e^x and $h = 0.1$ here, the first, second and third differences should not be greater than (to four figures) 0.2718, 0.02718 and 0.002718 respectively. On rounding these numbers to three decimal places, the entries under 1st, 2nd and 3rd differences in Table 4.12 should be bounded by 272, 27 and 3 respectively, as is indeed the case. However, to three decimal places, all subsequent differences should be

Table 4.12   Differences of e^x tabulated to three decimal places.

x	e^x	1st	2nd	3rd	4th	5th	6th	7th
0.0	1.000							
		105						
0.1	1.105		11					
		116		2				
0.2	1.221		13		−2			
		129		0		4		
0.3	1.350		13		2		−7	
		142		2		−3		13
0.4	1.492		15		−1		6	
		157		1		3		−13
0.5	1.649		16		2		−7	
		173		3		−4		14
0.6	1.822		19		−2		7	
		192		1		3		−11
0.7	2.014		20		1		−4	
		212		2		−1		
0.8	2.226		22		0			
		234		2				
0.9	2.460		24					
		258						
1.0	2.718							

zero; it is because of rounding errors that they are not. There are errors of at most $\frac{1}{2}.10^{-3}$ in the values given for e^x in the table. To determine how badly these can affect the accuracy of the difference table, let us study Table 4.13. If a tabulated function has errors $\pm\varepsilon$, with the signs alternating, the errors will spread through the difference table as shown in Table 4.13. It can be seen that, in any table whose entries are in error by at most ε in modulus, the jth differences as in Table 4.13 cannot be in error by more than $2^j.\varepsilon$.

Table 4.13   The worst rate of growth of rounding errors in a difference table.

Errors	1st	2nd	3rd	4th
ε				
	-2ε			
$-\varepsilon$		4ε		
	2ε		-8ε	
ε		-4ε		16ε
	-2ε		8ε	
$-\varepsilon$		4ε		
	2ε			
ε				

For the data in Table 4.12, $\varepsilon = \frac{1}{2}.10^{-3}$, so we should expect errors in the seventh differences which are not greater than $2^7.\frac{1}{2}.10^{-3}$ or 64 units in the column of seventh differences. This is in accord with the largest discrepancy present of 14 units.

Table 4.13 also suggests that we should consider how the accuracy of function values affect the accuracy of interpolation. For example, suppose we wish to use linear interpolation and that function values are to be obtained from a set of four-figure tables. Now, a four-figure table for the function f may be regarded as a table of *exact values* of some other function, say $f*$. Since the entries in the table have been rounded to give the best four-digit approximation to $f(x)$ at each tabulated value, we have the inequality

$$|f(x) - f*(x)| \leqslant \frac{1}{2}.10^{-4} \qquad (4.53)$$

at each tabulated x. When we attempt to calculate the interpolating straight line p_1, as in (4.1), what we actually calculate is some other straight line, say p_1^*, given by

$$p_1^*(x) = \left(\frac{x - x_1}{x_0 - x_1}\right) f*(x_0) + \left(\frac{x - x_0}{x_1 - x_0}\right) f*(x_1). \qquad (4.54)$$

Subtracting (4.54) from (4.1) gives

$$p_1(x) - p_1^*(x) = \left(\frac{x - x_1}{x_0 - x_1}\right)(f(x_0) - f*(x_0))$$

$$+ \left(\frac{x - x_0}{x_1 - x_0}\right)(f(x_1) - f*(x_1)). \qquad (4.55)$$

We must be careful of the signs here. Let us assume that $x_0 < x_1$ and that x lies within the interval $[x_0, x_1]$. Then from (4.55), making use of (4.53), we have

$$|p_1(x) - p_1^*(x)| \leqslant \left(\frac{x_1 - x}{x_1 - x_0}\right).\frac{1}{2}.10^{-4} + \left(\frac{x - x_0}{x_1 - x_0}\right).\frac{1}{2}.10^{-4}$$

$$= \frac{(x_1 - x) + (x - x_0)}{(x_1 - x_0)}.\frac{1}{2}.10^{-4}.$$

Thus

$$|p_1(x) - p_1^*(x)| \leqslant \frac{1}{2}.10^{-4}, \qquad (4.56)$$

so that the discrepancy between the two interpolating straight lines cannot be greater than the maximum possible error in the table. This result can be deduced geometrically from Fig. 4.1, by drawing above and below p_1, at a distance $\frac{1}{2}.10^{-4}$, two straight lines parallel to p_1. We see that p_1^* must lie between these two parallel lines.

4.10 Choice of interpolating points

We now return to the formula (4.16), obtained for the error incurred by using the interpolating polynomial p_n in place of f:

$$f(x) - p_n(x) = (x - x_0)\cdots(x - x_n)\frac{f^{(n+1)}(\xi_x)}{(n+1)!}. \qquad (4.57)$$

It is natural to consider what are the best points $x_0, x_1, \ldots, x_n$ at which to interpolate, if we have the choice. One possibility is to seek numbers $x_0, \ldots, x_n$ so that the maximum modulus of the right side of (4.57), over some interval $[a, b]$, is made as small as possible. In the following chapter, we will return to this problem. At present, one obvious difficulty is that the term $f^{(n+1)}(\xi_x)$ contains the unknown ξ_x. We will accept for the moment that we can do nothing about this term and suppose that it is not too important. We will minimize the maximum modulus of $(x - x_0)\cdots(x - x_n)$. To be more specific, we take the interval $[a, b]$ as $[-1, 1]$ and find numbers $x_0, \ldots, x_n$ which minimize

$$\max_{-1 \leqslant x \leqslant 1} |(x - x_0)\cdots(x - x_n)|. \qquad (4.58)$$

This is called a *minimax* problem; the solution is not obvious. We first examine the simplest cases.

Case (*i*) $n = 0$. Here we choose x_0 to minimize $\max_{-1 \leqslant x \leqslant 1} |x - x_0|$. The solution is $x_0 = 0$, which gives $\max_{-1 \leqslant x \leqslant 1} |x| = 1$. Any other choice of x_0 gives a greater maximum modulus. This is seen by studying the graph of $x - x_0$ over the interval $[-1, 1]$.

Case (*ii*) $n = 1$. This time we choose x_0 and x_1 to minimize

$$\max_{-1 \leqslant x \leqslant 1} |(x - x_0)(x - x_1)|. \qquad (4.59)$$

Taking $x_0 \leqslant x_1$, we see from the graph of the quadratic $(x - x_0)(x - x_1)$ that the maximum modulus on $[-1, 1]$ can occur only at $x = -1$, $x = 1$ or at the turning value between the zeros $x = x_0$ and x_1. (See Fig. 4.2.) With $f(x) = (x - x_0)(x - x_1)$, $f'(x) = 2x - (x_0 + x_1)$, so that $f'(x) = 0$ at $x = (x_0 + x_1)/2$, which is the turning value. At this point, $f(x) = -(x_1 - x_0)^2/4$. We now claim that the best choice of x_0 and x_1 is one for which the maximum modulus of $(x - x_0)(x - x_1)$ on $[-1, 1]$ occurs at $x = -1$, $x = 1$ and also at the turning value, $(x_0 + x_1)/2$. Let x_0^* and x_1^* be such values. The value of $(x - x_0^*)(x - x_1^*)$ will be positive and equal at $x = \pm 1$ and be negative, but with the same modulus, at $(x_0^* + x_1^*)/2$. (See Fig. 4.3.) We observe that the difference between $(x - x_0^*)(x - x_1^*)$ and any

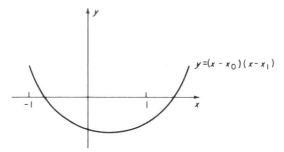

Fig. 4.2   Graph of the quadratic $(x - x_0)(x - x_1)$.

"rival" quadratic $(x - x_0)(x - x_1)$ is a linear function of x. Hence the only way one could improve on the choice $(x - x_0^*)(x - x_1^*)$ would be to add some linear expression $a + bx$. Geometrically, it may be seen that there is no straight line $y = a + bx$ which can be "added" to $(x - x_0^*)(x - x_1^*)$ so as to reduce simultaneously its modulus at $x = -1$, 1 and $(x_0^* + x_1^*)/2$. This justifies the claim made above. To find the values of x_0^* and x_1^*, let the maximum modulus of $(x - x_0^*)(x - x_1^*)$ on $[-1, 1]$ be δ. We have the equations

$$(-1 - x_0^*)(-1 - x_1^*) = \delta \qquad (4.60)$$

$$-(x_1^* - x_0^*)^2/4 = -\delta \qquad (4.61)$$

$$(1 - x_0^*)(1 - x_1^*) = \delta. \qquad (4.62)$$

Subtracting (4.60) from (4.62) we find that $x_0^* + x_1^* = 0$. Then the first two equations may be added to show that $x_1^* = 1/\sqrt{2}$ and hence $x_0^* = -1/\sqrt{2}$ and $\delta = \frac{1}{2}$.

We shall see presently (Theorem 4.3) that the general solution to the problem of minimizing the maximum modulus in (4.58) is obtained by choosing $x_0, \ldots, x_n$ to be the zeros of a polynomial denoted by T_{n+1}, where

$$T_n(x) = \cos n(\cos^{-1} x). \qquad (4.63)$$

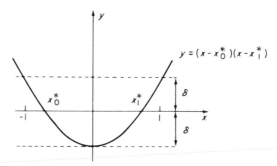

Fig. 4.3   Minimax approximation.

Using the trigonometrical identity (2.8), we have

$$\cos(n + 1)\theta + \cos(n - 1)\theta = 2 \cos n\theta \cos \theta$$

which gives, on putting $\theta = \cos^{-1} x$ and rearranging,

$$T_{n+1}(x) = 2xT_n(x) - T_{n-1}(x). \tag{4.64}$$

Since, from (4.63), $T_0(x) = 1$ and $T_1(x) = x$, it follows by induction on (4.64) that T_n is a polynomial of degree n with leading term $2^{n-1}x^n$ for $n > 0$. The T_n are called the *Chebyshev* polynomials, after the Russian mathematician P. L. Chebyshev (1821–1894), who was the first to solve the problem with which we are concerned in this section.

From the recurrence relation (4.64) we find that

$$T_2(x) = 2x^2 - 1, \qquad T_3(x) = 4x^3 - 3x,$$
$$T_4(x) = 8x^4 - 8x^2 + 1, \qquad T_5(x) = 16x^5 - 20x^3 + 5x.$$

All n zeros of T_n are real and lie in $[-1, 1]$. For if $\theta = \cos^{-1} x$ then

$$T_n(x) = 0 \Rightarrow \cos n\theta = 0,$$

whence $n\theta = (2k + 1)\pi/2$, with k an integer, and

$$x = \cos \theta = \cos[(2k + 1)\pi/(2n)].$$

We obtain exactly n distinct values of x on $[-1, 1]$ by taking $k = 0, 1, \ldots, n - 1$.

To determine the turning values of T_n, we find the derivative,

$$\frac{d}{dx} T_n(x) = \frac{d}{d\theta} \cos n\theta \frac{d\theta}{dx} = \frac{n \sin n\theta}{\sin \theta}, \tag{4.65}$$

since

$$\frac{d\theta}{dx} = 1 \Big/ \frac{dx}{d\theta} = -1/\sin \theta.$$

At the turning values, $\sin n\theta = 0$ so that $n\theta = k\pi$ and thus $\theta = k\pi/n$, where k is an integer. We see that the $n - 1$ zeros of $T_n' \in P_{n-1}$ are $x = \cos(k\pi/n)$, $1 \leqslant k \leqslant n - 1$, and that each lies between two adjacent zeros of T_n. At these $n - 1$ turning values, T_n has the values $\cos n(\cos^{-1} \cos (k\pi/n)) = \cos k\pi = (-1)^k$, $1 \leqslant k \leqslant n - 1$. Also, at the end points $x = -1$ and $x = 1$, $\theta = \pi$ and $\theta = 0$ respectively and T_n has the values $(-1)^n$ and 1 there. From the definition (4.63) it is also clear that the modulus of $T_n(x)$ on $[-1, 1]$ does not exceed 1. We now see that T_n has all its n zeros in the interval $[-1, 1]$ and attains its maximum modulus 1 on that interval at the $n + 1$ points $x = \cos (k\pi/n)$, $0 \leqslant k \leqslant n$, (which include both end points). At these $n + 1$ so-called *extreme points*, the polynomial T_n assumes the values ± 1 alternately.

Theorem 4.3 The expression $\max\limits_{-1 \leqslant x \leqslant 1} |(x - x_0)\cdots(x - x_n)|$ attains its minimum value when the numbers $x_0, \ldots, x_n$ are chosen to take the values of the zeros of the Chebyshev polynomial T_{n+1}.

Proof Since $T_{n+1}(x)$ has leading term $2^n x^{n+1}$, the statement of Theorem 4.3 is concerned with the polynomial

$$(x - x_0)\cdots(x - x_n) = \frac{1}{2^n} T_{n+1}(x).$$

It will be sufficient to prove that, of all polynomials of degree $n + 1$ with leading coefficient 1, the polynomial $T_{n+1}/2^n$ has the smallest maximum modulus on $[-1, 1]$, which is $1/2^n$. Let us suppose that there exists a polynomial $q_{n+1} \in P_{n+1}$ with leading term x^{n+1} such that

$$\max\limits_{-1 \leqslant x \leqslant 1} |q_{n+1}(x)| < \max\limits_{-1 \leqslant x \leqslant 1} \frac{1}{2^n} |T_{n+1}(x)| = \frac{1}{2^n}. \qquad (4.66)$$

Consider the polynomial

$$r_n(x) = \frac{1}{2^n} T_{n+1}(x) - q_{n+1}(x).$$

The polynomial $r_n \in P_n$ must (from (4.66)) take the same sign as T_{n+1} at each of the $n + 2$ extreme points where T_{n+1} assumes its maximum modulus. Since T_{n+1} alternates in sign at these points, so also must r_n. This implies that $r_n \in P_n$ has at least $n + 1$ zeros, one between each pair of adjoining extreme points of T_{n+1}. Hence r_n must be identically zero, which completes the proof by providing a contradiction. $\square$

We now return to interpolation. Suppose that f is a function such that

$$|f^{(n+1)}(x)| \leqslant M_{n+1},$$

for $-1 \leqslant x \leqslant 1$. If p_n^* denotes the interpolating polynomial for f constructed at the zeros of T_{n+1}, it follows from the error formula (4.16) that

$$\max\limits_{-1 \leqslant x \leqslant 1} |f(x) - p_n^*(x)| \leqslant \frac{M_{n+1}}{2^n(n + 1)!}. \qquad (4.67)$$

Thus we see that the class of functions for which M_{n+1} increases less rapidly than $2^n(n + 1)!$, as n tends to infinity, may be approximated as accurately as we wish by means of polynomials. This is a rather weak form of Weierstrass' theorem, with which we will be concerned in the following chapter.

Problems

Section 4.1

4.1   If $f(x) = 1$ and 5 at $x = 0$ and 1 respectively, construct the interpolating polynomial p_1 which matches f at these points.

4.2   Verify that the function $a + bx$, with a and b given by (4.4) and (4.3), is identical to (4.1).

4.3   Show that for any real number $\lambda \neq 0$ and any positive integers r and s the polynomial

$$q(x) = \lambda(x - x_0)^r(x - x_1)^s + \left(\frac{x - x_1}{x_0 - x_1}\right)f(x_0) + \left(\frac{x - x_0}{x_1 - x_0}\right)f(x_1),$$

which is of degree $r + s$, passes through the points $(x_0, f(x_0))$ and $(x_1, f(x_1))$.

4.4   Use linear interpolation to estimate $\cos 50°$, given that $\cos 45° = 1/\sqrt{2}$ (0.7071 to four decimal places) and $\cos 60° = \frac{1}{2}$.

4.5   By using interpolation at $x = 1$ and 4, find approximations to the function $y = x^{1/2}$ at $x = 2$ and $x = 3$.

Section 4.2

4.6   Find the interpolating polynomial for the function f which agrees with the following data: $f(x) = 1$, -1 and 1 at $x = -1$, 0 and 1.

4.7   Obtain the interpolating polynomial for the function $x^2 + 1$ at $x = 0$, 1, 2 and 3.

4.8   Construct a cubic polynomial p such that, at $x = 0$ and 1, p takes the same values as f (0 and 1 respectively) and p' has the same values as f' $(-3$ and 9 respectively). (*Hint:* let $p(x) \equiv ax^3 + bx^2 + cx + d$, derive four equations in a, b, c, d and solve them. See also §5.10 on Hermite approximations.)

Section 4.3

4.9   Estimate the accuracy of the results obtained in Problem 4.4, noting first that each cosine must be expressed in radian measure: for example, $\cos 50°$ as $\cos(50\pi/180)$.

4.10   Suppose that a function f is tabulated at equal intervals of length h and that $|f''(x)| \leqslant M$ throughout the table. Show that the modulus of the error due to the use of linear interpolation between any two adjoining entries in the table cannot be greater than $\frac{1}{8}Mh^2$.

4.11   At what step size h ought the function $\sin x$ to be tabulated so that linear interpolation will produce an error of not more than $\frac{1}{2}.10^{-6}$? (Use the result quoted in Problem 4.10.)

Section 4.4

4.12   Construct a Neville table for interpolation at $x = 2$, based on the function $x^{1/2}$ evaluated at $x = 0, 1$ and 4. Use the error formula (4.16) to help explain why it is possible for the result of interpolation based on all three points to be less accurate than that of interpolation based on $x = 1$ and 4 only.

4.13   In the Neville algorithm, we calculate

$$p_{0,n}(x) = \frac{(x - x_0)p_{1,n}(x) - (x - x_n)p_{0,n-1}(x)}{x_n - x_0},$$

where $p_{0,n}$, $p_{1,n}$ and $p_{0,n-1}$ are the interpolating polynomials defined in §4.4. Show from the above formula that $p_{0,n} \in P_n$ and that $p_{0,n}(x_i) = f(x_i)$, $0 \leqslant i \leqslant n$. (*Hint:* treat separately the three cases $i = 0$, $i = n$ and $0 < i < n$. Use an induction argument to verify that the above formula is valid for any number of interpolating points.)

Section 4.5

4.14   In the following table, y denotes the function $x \sin x - 1$ (to two decimal places). Use inverse interpolation to estimate the smallest positive root of the equation $x \sin x = 1$.

x	1.0	1.1	1.2
y	-0.16	-0.02	0.12

Section 4.6

4.15   It is alleged that a certain cubic polynomial in x matches the following data. Construct the polynomial by using divided differences.

x	-3	-1	0	2	3
y	-9	5	3	11	33

4.16   By using divided differences, construct the cubic polynomial in y which matches the function $x = y^{1/4}$ at $y = 0, 1, 16$ and 81. Evaluate the polynomial at $y = 64$. (See p. 58.)

4.17 Write down the divided difference form of the interpolating polynomial (4.37) constructed at the three points x_0, x_1 and x_2. In this, put $x_1 = x_0 + h$, $x_2 = x_0 - h$ to obtain

$$p_2(x_0 + sh) = f_0 + s(f_1 - f_0) + \frac{s(s-1)}{2}(f_1 - 2f_0 + f_{-1}),$$

where f_{-1} denotes $f(x_0 - h)$. Repeat this procedure with the three points taken in the order x_0, x_2, x_1 to obtain

$$p_2(x_0 + sh) = f_0 + s(f_0 - f_{-1}) + \frac{s(s+1)}{2}(f_1 - 2f_0 + f_{-1}).$$

Take the average of these to obtain the symmetric form

$$p_2(x_0 + sh) = f_0 + \tfrac{1}{2}s(f_1 - f_{-1}) + \tfrac{1}{2}s^2(f_1 - 2f_0 + f_{-1}).$$

Section 4.7

4.18 For the following data, calculate the differences, write down the forward and backward difference formulae and show that they give the same polynomial in x.

x	0.0	0.1	0.2	0.3	0.4	0.5
$f(x)$	1.00	1.32	1.68	2.08	2.52	3.00

4.19 In the following table, $S(r)$ refers to the sum of the squares of the first r positive integers. From a table of differences of $S(r)$, form a conjecture that $S(r)$ can be represented exactly by a polynomial of some degree in r. Use the forward difference formula to construct this polynomial in r.

r	1	2	3	4	5	6	7	8
$S(r)$	1	5	14	30	55	91	140	204

4.20 Show that $\Delta(f(x).g(x)) = f(x).\Delta g(x) + \Delta f(x).g(x + h)$, and note the similarity this result bears to an analogous result for derivatives.

4.21 Evaluate the polynomial $p(x) = x^2 + x + 7$ for $x = 0, 1, \ldots, 10$ by first evaluating $p(x)$ at $x = -1, 0$ and 1 and then "building up" from the constant second differences.

4.22 Construct the difference table for the following data. What is the lowest degree of polynomial which matches the data exactly?

x	0.0	0.1	0.2	0.3	0.4	0.5	0.6	0.7	0.8	0.9	1.0
y	0.000	0.541	1.168	1.887	2.704	3.625	4.656	5.803	7.072	8.469	10.000

Section 4.9

4.23   Suppose that, in the data of Problem 4.22, the value of y at $x = 0.5$ is incorrectly given as 3.265. Recalculate the differences, up to the sixth differences, noting how the error spreads.

Section 4.10

4.24   Use the recurrence relation (4.64) to find the polynomials $T_n(x)$ for $0 \leqslant n \leqslant 6$, given that $T_0(x) = 1$ and $T_1(x) = x$. As a check on the results, use the properties that $T_n(1) = 1$ and $T_n(-1) = (-1)^n$ for all n.

4.25   Verify that $T_{2n}(-x) = T_{2n}(x)$ and that $T_{2n+1}(-x) = -T_{2n+1}(x)$, for all x and $n = 0, 1, 2, \ldots$ (*Hint:* use the recurrence relation.)

4.26   By considering the function $x(1 - x)$, show that, if n is a positive integer, $r(n - r)/n^2 = (r/n)(1 - r/n)$ and

$$0 \leqslant \frac{r(n - r)}{n^2} \leqslant \frac{1}{4}$$

for $0 \leqslant r \leqslant n$. Next show that

$$\max_{0 \leqslant x \leqslant 1} \left| \left(x - \frac{r}{n} \right) \left(x - \frac{n - r}{n} \right) \right| \leqslant \frac{1}{4}$$

and hence that

$$\max_{0 \leqslant x \leqslant 1} \left| x \left(x - \frac{1}{n} \right) \cdots \left(x - \frac{n - 1}{n} \right) (x - 1) \right| \leqslant \frac{1}{2^{n+1}}.$$

4.27   Use the final result of Problem 4.26 to show that, if p_n denotes the interpolating polynomial constructed for e^x at the $n + 1$ equally spaced points $x = r/n$, $0 \leqslant r \leqslant n$, then

$$\max_{0 \leqslant x \leqslant 1} |e^x - p_n(x)| \leqslant \frac{e}{2^{n+1} \cdot (n + 1)!}.$$

What value of n will ensure that the error of interpolation is less than 10^{-6} in modulus? Compare your result with that of Problem 3.4.

4.28   Show that the change of variable

$$x = \left(\frac{b - a}{2} \right) t + \left(\frac{b + a}{2} \right)$$

maps the interval $-1 \leqslant t \leqslant 1$ onto the interval $a \leqslant x \leqslant b$ and that $x - x_r$

$= \frac{1}{2}(b - a)(t - t_r)$, where x_r and t_r are connected by the above change of variable. Hence show that

$$\min \max_{a \leqslant x \leqslant b} |(x - x_0) \cdots (x \quad x_n)| = 2\left(\frac{b - a}{4}\right)^{n+1},$$

the minimum being attained when $x_r = \frac{1}{2}(b - a)t_r + \frac{1}{2}(b + a)$ and the t_r are the zeros of T_{n+1}.

4.29   Apply the result of Problem 4.28 to map $[-1, 1]$ onto $[0, 1]$. From this, show that if we construct the interpolating polynomial q_n for e^x at the $n + 1$ points $\frac{1}{2}(1 + t_r)$, $0 \leqslant r \leqslant n$, where the t_r are the zeros of T_{n+1},

$$\max_{0 \leqslant x \leqslant 1} |e^x - q_n(x)| \leqslant \frac{e}{2^{2n+1}(n + 1)!}.$$

Compare this result with that in Problems 3.4 and 4.27.

Chapter 5

"Best" Approximation

5.1 Introduction

We now discuss further types of approximation for a given function, concentrating mainly on polynomial approximation. Initially, we consider approximation by first degree polynomials and ask: what choices of a and b will make $y = a + bx$ a good approximation to a given continuous function f on the interval $[0, 1]$? The work of Chapters 3 and 4 gives immediately two possible answers to this question.

(i) The Taylor polynomial at $x = 0$ (assuming $f'(0)$ exists),

$$y = f(0) + xf'(0).$$

(ii) The interpolating polynomial constructed at $x = 0$ and 1,

$$y = f(0) + x(f(1) - f(0)).$$

Perhaps a Taylor or interpolating polynomial constructed at some other points would be more suitable. However, these approximations are designed to imitate the behaviour of f at only one or two points. We are more concerned here with finding functions $y = a + bx$ which will be good approximations to f throughout the entire interval $[0, 1]$. The following is one possible interpretation of "good approximation".

(iii) Find values of a and b so that

$$\max_{0 \leqslant x \leqslant 1} |f(x) - (a + bx)|$$

is minimized, over all choices of a and b. This is called a *minimax* (or Chebyshev) approximation.

Instead of minimizing the maximum error between the (continuous) function f and the approximating straight line, we could minimize the "sum" of the moduli of the errors, as follows.

82

(iv) Find a and b so that

$$\int_0^1 |f(x) - (a + bx)| \, dx$$

is minimized. This is called a best L_1 approximation.

The L_1 approximation gives equal weight to all the errors, while the minimax approximation concentrates on the error of largest modulus. We define other approximations to f which, in a sense, lie between the extremes of L_1 and minimax approximations.

(v) For a fixed value of $p \geqslant 1$, find a and b so that

$$\int_0^1 |f(x) - (a + bx)|^p \, dx$$

is minimized. This is known as the best L_p approximation.

Due to the presence of the pth power, the error of largest modulus tends to dominate as p increases. With f continuous, it can be shown that, as $p \to \infty$, the best L_p approximation tends to the minimax approximation, which is therefore sometimes called the best L_∞ approximation. Thus the L_p approximations consist of a spectrum ranging from the L_1 to the minimax approximations. For $1 < p < \infty$, L_2 approximation is the only one commonly used and is better known as the *least squares* approximation.

Example 5.1 The five approximating straight lines listed above (taking $p = 2$ in (v)) for $f(x) = x^{1/2}$ on $[0, 1]$ are as follows.

(i) The Taylor polynomial does not exist, since $f'(0)$ does not exist.
(ii) $y = x$. (Interpolating polynomial based on $x = 0$ and 1.)
(iii) $y = \frac{1}{8} + x$. (Minimax approximation.)
(iv) $y = \frac{1}{4}(3 - \sqrt{3}) + (\sqrt{3} - 1)x$. (Best L_1 approximation.)
(v) $y = \frac{4}{15} + \frac{4}{5}x$. (Least squares approximation.)

The reader may find it helpful to sketch a graph of these approximating straight lines, together with a graph of $x^{1/2}$ on $[0, 1]$. The straight lines in (iii) and (v) may be derived using methods to be developed later in this chapter. For (iv), see Problem 5.19. Our point in displaying these approximations is to emphasize that, in general, different criteria for best approximations lead to different solutions. $\square$

Two classes of approximation problems can be distinguished. The first is approximation to a function on a finite interval, for example, the problem of choosing a polynomial which approximates to within some given accuracy to e^x on $[0, 1]$. Such a polynomial could be used in a computer for evaluating e^x. For this purpose, a minimax approximation seems the most appropriate to use, amongst all the types of approximations which we have mentioned so far. In practice, as we will see later, it is preferable to use a polynomial which is near to the minimax but is easier to determine.

The second class of problems deals with approximation to a function whose values are given at only a finite number of points. An example is a set of experimental results where the value of some function f has been determined at points $x = x_0, x_1, \ldots, x_N$. If the points are distinct, we could construct the interpolating polynomial which matches f at each point. However, experimental data generally contain errors. It is therefore usually more appropriate to construct a polynomial of lower degree $n < N$ which does not necessarily pass through any of the points, but has the effect of *smoothing* out the errors in the data. We now consider what type of approximation will do this satisfactorily. Suppose we try a minimax approximation and minimize

$$\max_{0 \leqslant i \leqslant N} |f(x_i) - p(x_i)|$$

over all polynomials $p \in P_n$, the set of polynomials of degree at most n. In fact, this can be most unsatisfactory as we now demonstrate by taking an especially bad case. We suppose that in Fig. 5.1 the extreme right-hand point

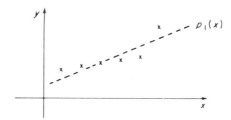

Fig. 5.1. Minimax straight line approximation.

would be in a straight line with the others, but for experimental error. The minimax straight line approximation for this data is that labelled $p_1(x)$ in Fig. 5.1. Notice how the presence of only one inaccurate point has substantially shifted the minimax approximation. Having seen an obvious disadvantage of minimax approximations for a function defined on a finite set of points, we naturally turn to one of the L_p approximations. We will consider in detail L_2 approximations, which are the simplest to compute.

5.2 Least squares approximation

In this section, it is expedient to generalize the set of approximating functions. A polynomial of degree n may be thought of as a linear combination of the functions $1, x, x^2, \ldots, x^n$. A key property of these monomials x^r is that no linear combination of them (that is, no polynomial of degree at most n) has more than n zeros, except for the polynomial which is identically zero. It is useful to have a special term to describe functions which have this property.

Definition 5.1 A set of functions $\{\psi_0, \psi_1, \ldots, \psi_n\}$, each continuous on an interval I, is said to be a *Chebyshev set* (or, equivalently, is said to satisfy a Haar condition) on I if, for any choice of $a_0, a_1, \ldots, a_n$ *not all zero*, the function

$$\Phi_a(x) = \sum_{r=0}^{n} a_r \psi_r(x) \tag{5.1}$$

has not more than n zeros on I. $\square$

Thus the monomials $\{1, x, \ldots, x^n\}$, for any n, form a Chebyshev set on any interval. Another example of a Chebyshev set is $\{1, \cos x, \sin x, \ldots, \cos kx, \sin kx\}$ on $0 \leqslant x < 2\pi$, where k is any positive integer.

Suppose we wish to approximate to a function f whose values are known at $x = x_0, \ldots, x_N$ by a function Φ_a defined by (5.1), with $n < N$. Let I be some interval which contains all the x_i. Then the condition that the functions ψ_r form a Chebyshev set on I certainly ensures that the set $\{\psi_0, \ldots, \psi_n\}$ contains no redundant members. For we cannot express any ψ_s as

$$\psi_s(x) = \sum_{\substack{r=0 \\ r \neq s}}^{n} b_r \psi_r(x), \quad \text{for all } x \in [a, b], \tag{5.2}$$

since this implies that a linear combination of the ψ_r has more than n (in fact, an infinite number) of zeros. A set of functions which satisfy a relationship of the form (5.2) is called *linearly dependent*. It is useful to give a formal definition.

Definition 5.2 A set of functions $\{\psi_0, \ldots, \psi_n\}$ is said to be *linearly independent* on some interval I if

$$\sum_{r=0}^{n} a_r \psi_r(x) = 0 \quad \text{for all } x \in I$$

only if $a_0 = a_1 = \cdots = a_n = 0$. Otherwise, the functions are said to be linearly dependent. $\square$

The concept of linear independence will be met again in Chapter 8. It follows from Definitions 5.1 and 5.2 that if a set of functions is a Chebyshev set on I, it is also linearly independent on I. The converse does not always hold, as we now show.

Example 5.2 The set $\{1, x, x^3\}$ is *not* a Chebyshev set on $[-1, 1]$, since the function $0.1 + (-1).x + 1.x^3$ has *three* zeros belonging† to $[-1, 1]$. However, these three functions are linearly independent on $[-1, 1]$. $\square$

---

† A simpler example is the set $\{x^3\}$, since x^3 has a triple zero at $x = 0$.

We now wish to find the least squares approximation to a function f on a finite point set $\{x_0, x_1, \ldots, x_N\}$ by a function of the form Φ_a defined by (5.1). We assume that $\{\psi_0, \ldots, \psi_n\}$, with $n < N$, is a Chebyshev set on some interval $[a, b]$ which contains all the x_i. We require the minimum of

$$E(a_0, \ldots, a_n) = \sum_{i=0}^{N} (f(x_i) - \Phi_a(x_i))^2$$

over all values of $a_0, \ldots, a_n$. A necessary condition for E to have a minimum is $\partial E / \partial a_r = 0$ for $r = 0, 1, \ldots, n$ at the minimum. This gives

$$-2 \sum_{i=0}^{N} [f(x_i) - (a_0 \psi_0(x_i) + \cdots + a_n \psi_n(x_i))]\psi_r(x_i) = 0$$

which yields the equations

$$a_0 \sum_{i=0}^{N} \psi_0(x_i)\psi_r(x_i) + \cdots + a_n \sum_{i=0}^{N} \psi_n(x_i)\psi_r(x_i) = \sum_{i=0}^{N} f(x_i)\psi_r(x_i), \quad (5.3)$$

for $0 \leqslant r \leqslant n$. These $n + 1$ linear equations in the $n + 1$ unknowns $a_0, \ldots, a_n$ are called the *normal equations*. We state, without proof here,† that these linear equations have a unique solution if there is no set of numbers $b_0, \ldots, b_n$ (except $b_0 = b_1 = \cdots = b_n = 0$) for which

$$\sum_{r=0}^{n} b_r \psi_r(x_i) = 0, \qquad 0 \leqslant i \leqslant N.$$

Since $N > n$ and the ψ_r form a Chebyshev set, this condition is satisfied. It is for this reason that we use functions ψ_r which form a Chebyshev set.

We have seen so far that a *necessary* condition for a minimum of $E(a_0, \ldots, a_n)$ is that the a_r satisfy the normal equations (5.3) and we know that these equations have a unique solution. It still remains to show that the solution of these equations does, in fact, provide the minimum. To see this, let $a_0^*, \ldots, a_n^*$ denote the solution of (5.3). First, we consider the case where there are only two functions ψ_0 and ψ_1. Consider the difference

$$E(a_0^* + \delta_0, a_1^* + \delta_1) - E(a_0^*, a_1^*)$$
$$= \sum_{i=0}^{N} [f(x_i) - (a_0^* + \delta_0)\psi_0(x_i) - (a_1^* + \delta_1)\psi_1(x_i)]^2$$
$$- \sum_{i=0}^{N} [f(x_i) - a_0^*\psi_0(x_i) - a_1^*\psi_1(x_i)]^2$$
$$= \sum_{i=0}^{N} [\delta_0 \psi_0(x_i) + \delta_1 \psi_1(x_i)]^2 - 2\delta_0 \sum_{i=0}^{N} \psi_0(x_i)[f(x_i) - a_0^*\psi_0(x_i) - a_1^*\psi_1(x_i)]$$
$$- 2\delta_1 \sum_{i=0}^{N} \psi_1(x_i)[f(x_i) - a_0^*\psi_0(x_i) - a_1^*\psi_1(x_i)].$$

† See Problem 8.37 of Chapter 8.

The last two summations are both zero, since a_0^* and a_1^* satisfy the normal equations (5.3) for the case $n = 1$. Therefore

$$E(a_0^* + \delta_0, a_1^* + \delta_1) - E(a_0^*, a_1^*) = \sum_{i=0}^{N} [\delta_0 \psi_0(x_i) + \delta_1 \psi_1(x_i)]^2. \quad (5.4)$$

The right side of (5.4) can only be zero if

$$\delta_0 \psi_0(x_i) + \delta_1 \psi_1(x_i) = 0$$

for $0 \leqslant i \leqslant N$, which cannot be so, unless $\delta_0 = \delta_1 = 0$, since ψ_0 and ψ_1 form a Chebyshev set. Thus the right side of (5.4) is always strictly positive unless $\delta_0 = \delta_1 = 0$, showing that $E(a_0, a_1)$ has indeed a minimum when $a_0 = a_0^*$, $a_1 = a_1^*$. For a general value of $n < N$, we may similarly show that

$$E(a_0^* + \delta_0, \ldots, a_n^* + \delta_n) - E(a_0^*, \ldots, a_n^*) = \sum_{i=0}^{N} [\delta_0 \psi_0(x_i) + \cdots + \delta_n \psi_n(x_i)]^2,$$

showing that the minimum value of $E(a_0, \ldots, a_n)$ occurs when $a_0 = a_0^*, \ldots,$ $a_n = a_n^*$.

Example 5.3 Find the least squares straight line approximation for the following data.

x	0.0	0.2	0.4	0.6	0.8
$f(x)$	0.9	1.9	2.8	3.3	4.2

If we take $n = 1$, $\psi_0(x) \equiv 1$ and $\psi_1(x) \equiv x$, the functions Φ_a defined by (5.1) will all be straight lines. To find the least squares straight line we must solve the normal equations (5.3), which in this case are

$$a_0(N + 1) + a_1 \sum_{i=0}^{N} x_i = \sum_{i=0}^{N} f(x_i)$$

$$a_0 \sum_{i=0}^{N} x_i + a_1 \sum_{i=0}^{N} x_i^2 = \sum_{i=0}^{N} f(x_i) . x_i. \quad (5.5)$$

For the above data, these equations are

$$5a_0 + 2a_1 = 13.1$$
$$2a_0 + 1.2a_1 = 6.84$$

with solution $a_0 = 1.02$, $a_1 = 4$. The required straight line is therefore $1.02 + 4x$. $\square$

The foregoing method is easily extended to cater for approximation to a function of several variables. To approximate to a function $f(x, y)$ at the points $(x_0, y_0), \ldots, (x_N, y_N)$ by an approximating function of the form $\sum_{r=0}^{n} a_r \psi_r(x, y)$, again assuming $N > n$, we minimize

$$\sum_{i=0}^{N} [f(x_i, y_i) - (a_0 \psi_0(x_i, y_i) + \cdots + a_n \psi_n(x_i, y_i))]^2.$$

By taking partial derivatives with respect to the a_r, we again obtain a set of linear equations, which are like (5.3) with (x_i, y_i) written in place of (x_i) throughout. For example, if $n = 2$, $\psi_0(x, y) = 1$, $\psi_1(x, y) = x$, $\psi_2(x, y) = y$, we have the system of three linear equations

$$a_0 \sum_{i=0}^{N} 1 + a_1 \sum_{i=0}^{N} x_i + a_2 \sum_{i=0}^{N} y_i = \sum_{i=0}^{N} f(x_i, y_i)$$

$$a_0 \sum_{i=0}^{N} x_i + a_1 \sum_{i=0}^{N} x_i^2 + a_2 \sum_{i=0}^{N} y_i x_i = \sum_{i=0}^{N} f(x_i, y_i) . x_i$$

$$a_0 \sum_{i=0}^{N} y_i + a_1 \sum_{i=0}^{N} x_i y_i + a_2 \sum_{i=0}^{N} y_i^2 = \sum_{i=0}^{N} f(x_i, y_i) . y_i$$

to determine a_0, a_1 and a_2.

Although we have discussed the existence and uniqueness of least squares approximations constructed from a Chebyshev set of functions, in practice we tend to use mainly polynomial approximations. It is sometimes quite difficult to decide what degree of approximating polynomial is most appropriate. If we are dealing with experimental results, we can often see that a polynomial of a certain degree appears suitable, judging from the number of large-scale fluctuations in the data and disregarding small-scale fluctuations which may be attributed to experimental error. Thus two large-scale fluctuations suggests a cubic approximation, since a cubic can have two turning values. (Ralston describes a more rigorous statistical approach to the question of choosing the degree of approximating polynomial.)

Increasing the degree of the approximating polynomial does not always increase the accuracy of the approximation. If the degree is large, the polynomial may have a large number of maxima and minima. It is then possible for the polynomial to fluctuate considerably more than the data, particularly if this contains irregularities due to errors. For this reason it is often more appropriate to use different low degree polynomials to approximate to different sections of the data. One way is to approximate to consecutive batches of the data, say points 1 to 5, then points 6 to 10, and so on. This is called *piecewise approximation*. Another way (especially for equally spaced data) is to construct polynomials for points 1 to 5, say, then 2 to 6, and so on, and to use each polynomial to approximate to the point on which it is centred. This leads to smoothing formulae, which we discuss in the next section.

5.3 Smoothing formulae

In applying the results of a calculation such as that in Example 5.3, we will usually want to know the values of the approximating function at the data points. These are called *smoothed values*. Those of $\Phi_a(x)$ relevant to Example 5.3 are (to one decimal place) as follows.

x	0.0	0.2	0.4	0.6	0.8
$\Phi_a(x)$	1.0	1.8	2.6	3.4	4.2

The smoothed values are usually of more interest to us than the coefficients (a_0 and a_1 in Example 5.3) of the approximating function.

Notice that these calculations are simplified if we replace the x_i by, say, $-2, -1, 0, 1, 2$. For, as the reader may convince himself, any linear transformation applied to the independent variable x (that is, a shift of origin and "stretching" of the x-axis) does not affect the smoothing process. With this choice of abscissae, $\sum x_i = 0$ and equations (5.5) become

$$5a_0 = \sum_{i=-2}^{+2} f_i$$

$$10a_1 = \sum_{i=-2}^{+2} if_i$$

where f_i, $-2 \leqslant i \leqslant 2$, denote the values of $f(x)$ in the table. Thus, for example, the smoothed value corresponding to the centre point is

$$y_0 = a_0 = (f_{-2} + f_{-1} + f_0 + f_1 + f_2)/5$$

and that corresponding to the extreme right-hand point is

$$y_2 = a_0 + 2a_1 = (-f_{-2} + f_0 + 2f_1 + 3f_2)/5.$$

Similarly, we may write down expressions for y_{-2}, y_{-1} and y_1. (The reader should verify that this yields the same results as given above.) These expressions, known collectively as a *smoothing formula*, give explicitly the smoothed values obtained from a least squares straight line approximation to *any* set of five points whose abscissae are equally spaced. Because straight line approximation is employed, we also refer to the expressions as the 5-point *linear* smoothing formula.

The simplest smoothing formula is the 3-point linear formula. To derive this, we return to (5.5) and put $N + 1 = 3$. Choosing the abscissae as $-1, 0$ and 1, we obtain

$$3a_0 = f_{-1} + f_0 + f_1$$
$$2a_1 = -f_{-1} + f_1.$$

The least squares straight line is

$$y = (f_{-1} + f_0 + f_1)/3 + x(f_1 - f_{-1})/2.$$

At the centre point, $x = 0$, the smoothed value is

$$y_0 = (f_{-1} + f_0 + f_1)/3. \tag{5.6}$$

At the right-hand point, $x = 1$, the smoothed value is

$$y_1 = (f_{-1} + f_0 + f_1)/3 + (f_1 - f_{-1})/2,$$

which may be written more simply as

$$y_1 = f_1 - \tfrac{1}{6}\delta^2 f_0, \tag{5.7}$$

where $\delta^2 f_0$ denotes the second difference $f_1 - 2f_0 + f_{-1}$. Similarly, at the left-hand point we obtain

$$y_{-1} = f_{-1} - \tfrac{1}{6}\delta^2 f_0. \tag{5.8}$$

Example 5.4   Smooth the following data.

x	0.0	0.1	0.2	0.3	0.4	0.5	0.6	0.7	0.8	0.9	1.0
$f(x)$	2	3	5	7	11	13	17	19	23	29	31

There is not a unique solution to this problem. Although the function f is clearly not linear in x, we could argue that each set of three consecutive points may reasonably be approximated by a straight line. For each point, except the first and last, we can therefore obtain a smoothed value by using (5.6). To smooth the values corresponding to $x = 0.0$ and $x = 1.0$, we use (5.8) and (5.7) respectively. Denoting the smoothed value at x by $y(x)$, we obtain the following (to 1 decimal place).

x	0.0	0.1	0.2	0.3	0.4	0.5	0.6	0.7	0.8	0.9	1.0
$y(x)$	1.8	3.3	5.0	7.7	10.3	13.7	16.3	19.7	23.7	27.7	31.7

□

Other smoothing formulae may be derived corresponding to different choices of the number of points used at a time (three in the example just given) and different degrees of least squares approximating polynomial. For example, we could construct a 5-point cubic formula. For this formula and others, see Ralston.

5.4 Orthogonal functions

We now consider the problem of finding least squares approximations to a function f on an interval $[a, b]$, rather than on a finite point set. Again, we seek approximations of the form $\sum_{r=0}^{n} a_r\psi_r(x)$ constructed from certain functions ψ_r, $0 \leqslant r \leqslant n$. We now assume that the ψ_r are linearly independent on $[a, b]$ and that the ψ_r and f are continuous on $[a, b]$. We wish to minimize

$$\int_a^b \left[f(x) - \sum_{r=0}^{n} a_r\psi_r(x) \right]^2 dx$$

with respect to $a_0, \ldots, a_n$. As before, we set to zero the partial derivatives with respect to each a_r and obtain the (linear) equations

$$a_0 \int_a^b \psi_0(x)\psi_r(x)\,dx + \cdots + a_n \int_a^b \psi_n(x)\psi_r(x)\,dx = \int_a^b f(x)\psi_r(x)\,dx, \quad (5.9)$$

for $0 \leqslant r \leqslant n$. Notice that these equations may be obtained from their counterparts (5.3) in the point set case, on replacing summations by integrals. It can be shown that the equations (5.9) have a unique solution if the ψ_r are linearly independent. (See Problem 8.20.)

In the particular case where $\psi_r(x) = x^r$ and $[a, b]$ is $[0, 1]$, the equations (5.9) become

$$a_0 + \frac{1}{2} a_1 + \cdots + \frac{1}{n+1} a_n = \int_0^1 f(x)\,dx$$

$$\frac{1}{2} a_0 + \frac{1}{3} a_1 + \cdots + \frac{1}{n+2} a_n = \int_0^1 f(x).x\,dx \qquad (5.10)$$

$$\vdots \qquad\qquad\qquad\qquad \vdots$$

$$\frac{1}{n+1} a_0 + \frac{1}{n+2} a_1 + \cdots + \frac{1}{2n+1} a_n = \int_0^1 f(x).x^n\,dx.$$

Unless n is quite small, these equations are of a type for which it is difficult to calculate an accurate solution. Such equations are called *ill-conditioned*, a term which will be made more precise in Chapter 9. As n increases, the accuracy of computed solutions of these linear equations tends to deteriorate rapidly, due to rounding errors. (See Example 9.3.) Similar behaviour is exhibited by the linear equations which arise in computing least squares polynomial approximations to f on a finite point set. The remedy is to use only low degree approximating polynomials, perhaps $n \leqslant 6$ or so, or to use orthogonal polynomials, which is described in §5.5.

It is natural to ask whether we can make a more appropriate choice of the functions ψ_r than the monomials x^r. We observe that, if the functions ψ_r satisfy

$$\int_a^b \psi_r(x)\psi_s(x)\,dx = 0, \qquad r \neq s, \qquad (5.11)$$

for all r and s, $0 \leqslant r, s \leqslant n$, then the linear equations (5.9) "uncouple" to give

$$a_r \int_a^b (\psi_r(x))^2 \, dx = \int_a^b f(x)\psi_r(x) \, dx, \tag{5.12}$$

for $0 \leqslant r \leqslant n$. Note that both integrals in (5.12) exist, since f and the ψ_r are continuous on $[a, b]$. Further, since the ψ_r are assumed to be linearly independent, the integral on the left of (5.12) is non-zero and therefore a_r exists. This provides the solution immediately.

Definition 5.3   A set of functions $\{\psi_0, \ldots, \psi_n\}$, which does not include the zero function, is said to be orthogonal on $[a, b]$ if the ψ_r satisfy (5.11). $\square$

It can be shown that a set of orthogonal functions is necessarily linearly independent. (See Problem 5.11.) Before studying particular sets of orthogonal functions, we note the *permanence* property of least squares approximations using orthogonal functions. Suppose we have found the coefficients $a_0, \ldots, a_n$ from (5.12) and wish to *add* a further continuous function ψ_{n+1}, orthogonal to the others, to find the best approximation of the form $\sum_{r=0}^{n+1} a_r \psi_r(x)$. We see that we need only calculate a_{n+1} (from (5.12) with $r = n + 1$); the values $a_0, \ldots, a_n$ will be as we have already obtained.

Define

$$E_n^* = \int_a^b \left[f(x) - \sum_{r=0}^n a_r \psi_r(x) \right]^2 dx, \tag{5.13}$$

$$E_n = \max_{a \leqslant x \leqslant b} \left| f(x) - \sum_{r=0}^n a_r \psi_r(x) \right|, \tag{5.14}$$

where the continuous, linearly independent functions ψ_r are orthogonal and the a_r are given by (5.12). As we have seen, the series $\sum_{r=0}^n a_r \psi_r(x)$ is the least squares approximation for the continuous function f. Therefore the right side of (5.13) will decrease (more precisely, will not increase) as n increases. Thus the sequence $\{E_n^*\}$ satisfies the property $E_{n+1}^* \leqslant E_n^*$, $n = 0, 1, \ldots$. It is natural to ask whether $E_n^* \to 0$ as $n \to \infty$. The infinite series $\sum_{r=0}^\infty a_r \psi_r(x)$ is called an *orthogonal series* (or expansion) or *Fourier series* for f and we shall refer to the a_r as the orthogonal or Fourier coefficients. To denote the relationship between f and an orthogonal series for f, we will write

$$f(x) \sim \sum_{r=0}^\infty a_r \psi_r(x).$$

If $E_n^* \to 0$ as $n \to \infty$, we say that the orthogonal series converges *in the least squares sense* to f. Of even greater interest, looking at (5.14), is the behaviour of the sequence $\{E_n\}$. If $E_n \to 0$ as $n \to \infty$, we say that the orthogonal series converges uniformly to f on $[a, b]$. Although uniform convergence implies convergence in the least squares sense, the converse is not true, as the following example illustrates.

Example 5.5   Let $f(x) \equiv 0$ on $[0, 1]$ and consider the functions

$$f_n(x) = x^n$$

as approximations to f on $[0, 1]$. We have

$$\int_0^1 [f(x) - f_n(x)]^2 \, dx = \frac{1}{2n + 1}$$

and

$$\max_{0 \leqslant x \leqslant 1} |f(x) - f_n(x)| = 1.$$

Thus $\{f_n\}$ converges in the least squares sense to f, but does not converge uniformly to f on $[0, 1]$. $\square$

Example 5.6   The functions $1, \cos x, \sin x, \cos 2x, \sin 2x, \ldots, \cos kx, \sin kx$ are orthogonal on $[-\pi, \pi]$. We need to verify that the appropriate integrals are zero. (See Problem 5.12.) $\square$

With the set of orthogonal functions of Example 5.6, the best approximation to f on $[-\pi, \pi]$ is

$$\tfrac{1}{2}a_0 + \sum_{r=1}^{k} (a_r \cos rx + b_r \sin rx), \tag{5.15}$$

where from (5.12)

$$(\tfrac{1}{2}a_0) \int_{-\pi}^{\pi} dx = \int_{-\pi}^{\pi} f(x) \, dx,$$

$$a_r \int_{-\pi}^{\pi} \cos^2 rx \, dx = \int_{-\pi}^{\pi} f(x) \cos rx \, dx, \qquad 1 \leqslant r \leqslant k,$$

$$b_r \int_{-\pi}^{\pi} \sin^2 rx \, dx - \int_{-\pi}^{\pi} f(x) \sin rx \, dx, \qquad 1 \leqslant r \leqslant k.$$

We find that the integrals on the left sides of the last two equations have the value π (see Problem 5.12), so that

$$a_r = \frac{1}{\pi} \int_{-\pi}^{\pi} f(x) \cos rx \, dx, \qquad b_r = \frac{1}{\pi} \int_{-\pi}^{\pi} f(x) \sin rx \, dx. \tag{5.16}$$

In (5.15), we write $\tfrac{1}{2}a_0$ rather than a_0 so that (5.16) also holds for $r = 0$.

Letting $k \to \infty$ in (5.15), we obtain an orthogonal series for f which we will refer to as the classical Fourier† series. Some authors use the term "Fourier

† Named after the French mathematician J. B. J. Fourier (1768–1830).

series" solely for this particular orthogonal series. If f is continuous on $[-\pi, \pi]$, it can be shown that the classical Fourier series converges in the least squares sense to f. Much investigation has been done on the *uniform* convergence of the classical Fourier series. This investigation is concerned with finding conditions on f which ensure that $E_n \to 0$ as $n \to \infty$. (See (5.14).) The assumption that f is continuous on $[-\pi, \pi]$ is not sufficient to guarantee uniform convergence. If, however, f is periodic, with period 2π, and the classical Fourier series converges uniformly to f on $[-\pi, \pi]$ then, due to the periodicity of all the functions involved, the series will converge uniformly to f everywhere. For more information on these convergence questions, see Davis.

Example 5.7   Find the classical Fourier series for the function $f(x) = x$. We have

$$a_r = \frac{1}{\pi} \int_{-\pi}^{\pi} x \cos rx \, dx = 0$$

for all r, since $x \cos rx$ is an *odd* function, and

$$b_r = \frac{1}{\pi} \int_{-\pi}^{\pi} x \sin rx \, dx = \frac{1}{\pi} \int_{-\pi}^{\pi} x \frac{d}{dx} \left(-\frac{1}{r} \cos rx \right) dx$$

$$= -\frac{1}{\pi r} [x \cos rx]_{-\pi}^{\pi} + \frac{1}{\pi r} \int_{-\pi}^{\pi} \cos rx \, dx$$

$$= (-1)^{r-1} 2/r,$$

since $\cos r\pi = \cos(-r\pi) = (-1)^r$ and the last integral is zero. Hence

$$x \sim 2 \sum_{r=1}^{\infty} \frac{(-1)^{r-1}}{r} \sin rx. \quad \square$$

Example 5.8   Find the classical Fourier series for the "square wave" function defined by

$$f(x) = \begin{cases} -\frac{1}{2}, & -\pi < x < 0 \\ 0, & x = 0 \\ \frac{1}{2}, & 0 < x \leqslant \pi \end{cases}$$

and f periodic elsewhere, with period 2π. (See Fig. 5.2.) Since f is an odd function, $a_r = 0$ for all r and

$$b_r = \frac{1}{\pi} \int_{-\pi}^{\pi} f(x) \sin rx \, dx = \frac{2}{\pi} \int_{0}^{\pi} f(x) \sin rx \, dx$$

$$= \frac{1}{\pi} \int_{0}^{\pi} \sin rx \, dx.$$

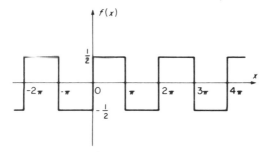

Fig. 5.2. The "square wave" function.

We find that

$$b_r = \begin{cases} 0, & r \text{ even} \\ 2/(\pi r), & r \text{ odd} \end{cases}$$

and therefore

$$f(x) \sim \frac{2}{\pi} \sum_{r=0}^{\infty} \frac{\sin(2r+1)x}{2r+1}. \quad \square$$

5.5 Orthogonal polynomials

To give a unified treatment of approximation, it is convenient to generalize least squares approximations by introducing a *weight function*. Let ω denote some function which is integrable, non-negative and not identically zero on a given interval $[a, b]$. Given functions $f, \psi_0, \ldots, \psi_n$ as before, we now wish to minimize

$$\int_a^b \omega(x) \left[f(x) - \sum_{r=0}^{n} a_r \psi_r(x) \right]^2 dx$$

over all choices of $a_0, \ldots, a_n$. Proceeding exactly as before, we find that the introduction of ω does not cause any dramatic changes. The a_r still satisfy linear equations and if we now choose the functions ψ_r so that

$$\int_a^b \omega(x) \psi_r(x) \psi_s(x) \, dx = 0, \qquad r \neq s, \tag{5.17}$$

the normal equations uncouple, as before, to give

$$a_r \int_a^b \omega(x) (\psi_r(x))^2 \, dx = \int_a^b \omega(x) f(x) \psi_r(x) \, dx. \tag{5.18}$$

Note that, when we take $\omega(x) \equiv 1$, (5.18) reduces to (5.12) as we should expect.

Functions which satisfy (5.17) are said to be orthogonal on $[a, b]$ with respect to the weight function ω. The resulting infinite series $\sum_{r=0}^{\infty} a_r \psi_r(x)$ is referred to as an orthogonal series, as before. We now show that we can construct *polynomials* which satisfy (5.17). Since we are now to deal with polynomials, we will write p_r in place of ψ_r. We will construct polynomials p_r, $r = 0, 1, \ldots$, such that $p_r(x)$ is a polynomial of degree r with leading term x^r (with coefficient 1) and

$$\int_a^b \omega(x) p_r(x) p_s(x)\, dx = 0, \qquad r \neq s. \tag{5.19}$$

This is done recursively. Suppose that, for some integer $k \geqslant 0$, we have constructed polynomials $p_0, p_1, \ldots, p_k$, each with leading coefficient 1, which are mutually orthogonal on $[a, b]$ with respect to ω. (This is easily achieved for $k = 0$, where we select $p_0(x) \equiv 1$.) We write

$$p_{k+1}(x) = x p_k(x) + \sum_{r=0}^{k} c_r p_r(x), \tag{5.20}$$

which obviously satisfies the requirement that p_{k+1} should have leading coefficient 1 and seek values of the c_r so that p_{k+1} satisfies the orthogonality property. Multiplying each term in (5.20) by $\omega(x) p_s(x)$ and integrating over $[a, b]$, we obtain

$$\int_a^b \omega(x) p_{k+1}(x) p_s(x)\, dx = \int_a^b \omega(x) x p_k(x) p_s(x)\, dx$$
$$+ \sum_{r=0}^{k} c_r \int_a^b \omega(x) p_r(x) p_s(x)\, dx. \tag{5.21}$$

We leave this equation for a moment to prove:

Lemma 5.1 If $0 \leqslant s \leqslant k - 2$,

$$\int_a^b \omega(x) x p_k(x) p_s(x)\, dx = 0.$$

Proof We may write

$$x p_s(x) = p_{s+1}(x) + \sum_{j=0}^{s} d_j p_j(x), \tag{5.22}$$

for some constants d_j, since $x p_s(x)$ and $p_{s+1}(x)$ are both polynomials of degree $s + 1$ with leading coefficient 1. (See Problem 5.15.) On multiplying each term of (5.22) by $\omega(x) p_k(x)$ and integrating over $[a, b]$, the proof of the lemma follows from the orthogonality property (5.19). $\square$

We see from the lemma that, for $0 \leqslant s \leqslant k - 2$, the first term on the right side of (5.21) is zero. If we choose $c_0 = \cdots = c_{k-2} = 0$, the second term on the right side of (5.21) will be zero for $0 \leqslant s \leqslant k - 2$ and thus p_{k+1} will be orthogonal to $p_0, \ldots, p_{k-2}$. It remains to arrange that p_{k+1} is orthogonal to p_{k-1} and p_k. Putting $s = k - 1$ in (5.21) and using the orthogonality property, we require that

$$0 = \int_a^b \omega(x)xp_k(x)p_{k-1}(x)\, dx + c_{k-1}\int_a^b \omega(x)(p_{k-1}(x))^2\, dx. \quad (5.23)$$

Rewriting $xp_{k-1}(x)$ as we did with $xp_s(x)$ in (5.22), we see that in (5.23) $xp_{k-1}(x)$ may be replaced by $p_k(x)$ and we obtain

$$-c_{k-1} = \int_a^b \omega(x)(p_k(x))^2\, dx \Big/ \int_a^b \omega(x)(p_{k-1}(x))^2\, dx = \gamma_k, \quad (5.24)$$

say. Secondly, putting $s = k$ in (5.21), we require for orthogonality that

$$-c_k = \int_a^b \omega(x)x(p_k(x))^2\, dx \Big/ \int_a^b \omega(x)(p_k(x))^2\, dx = \beta_k, \quad (5.25)$$

say. Thus, with c_{k-1} and c_k given by (5.24) and (5.25) and $c_0 = \cdots = c_{k-2} = 0$, the polynomial p_{k+1} defined by (5.20) is orthogonal to $p_0, \ldots, p_k$. We have

$$p_{k+1}(x) = (x - \beta_k)p_k(x) - \gamma_k p_{k-1}(x), \quad (5.26)$$

where $\beta_k = -c_k$, $\gamma_k = -c_{k-1}$, given by (5.25) and (5.24). This entails that the orthogonal polynomials p_r may be constructed from the recurrence relation (5.26), which holds for $k \geqslant 0$ if, for $k = 0$, we define $\gamma_0 = 0$. Obviously any multiples of orthogonal polynomials are also orthogonal. Our construction has obtained polynomials p_r with leading term x^r.

Example 5.9 As an example of polynomials orthogonal on an interval, take $a = -1$, $b = 1$ and $\omega(x) = (1 - x^2)^{-1/2}$. We find that the Chebyshev polynomials $T_r(x)$, introduced in Chapter 4, are orthogonal on $[-1, 1]$ with respect to $(1 - x^2)^{-1/2}$. This is seen on making the substitution $x = \cos\theta$ and using the orthogonality of the cosines. (See Problem 5.12.) To obtain polynomials with leading coefficient 1, we take $T_r(x)/2^{r-1}$ $(r > 0)$. We now look at the recurrence relation (5.26) for this case. Since (see Problem 4.25), $T_r(-x) = (-1)^r T_r(x)$, it follows that the numerator in (5.25) is zero for this case and thus $\beta_k = 0$. To find γ_k, first write

$$\int_a^b \omega(x)(p_k(x))^2\, dx = \int_{-1}^1 (1 - x^2)^{-1/2}(T_k(x)/2^{k-1})^2\, dx,$$

put $x = \cos \theta$ and integrate to give

$$\int_a^b \omega(x)(p_k(x))^2 \, dx = \pi/2^{2k-1},$$

for $k \geqslant 1$. Thus from (5.24) we see that $\gamma_k = \frac{1}{4}$ for $k \geqslant 2$ and therefore from (5.26)

$$T_{k+1}(x)/2^k = x(T_k(x)/2^{k-1}) - \frac{1}{4}(T_{k-1}(x)/2^{k-2}),$$

for $k \geqslant 2$. This gives

$$T_{k+1}(x) = 2xT_k(x) - T_{k-1}(x)$$

as we found independently in Chapter 4. (We find that $\gamma_1 = \frac{1}{2}$ and that the recurrence relation holds also for $k = 1$.) $\square$

Example 5.10   Derive the orthogonal polynomials p_r, $0 \leqslant r \leqslant 4$, if $a = -1$, $b = 1$ and $\omega(x) \equiv 1$. We have $p_0(x) = 1$ and, since $\beta_0 = 0$, $p_1(x) = x$ from (5.26). Hence $\beta_1 = 0$, $\gamma_1 = \frac{1}{3}$ and $p_2(x) = x^2 - \frac{1}{3}$. Then $\beta_2 = 0$, $\gamma_2 = \frac{4}{15}$ and $p_3(x) = x^3 - \frac{3}{5}x$. Lastly, $\beta_3 = 0$, $\gamma_3 = \frac{9}{35}$ and $p_4(x) = x^4 - \frac{6}{7}x^2 + \frac{3}{35}$. These are called the *Legendre polynomials*. Strictly, this name is reserved for the multiples of these polynomials for which $p_r(1) = 1$ for all r. Thus the first few Legendre polynomials proper are 1, x, $(3x^2 - 1)/2$, $(5x^3 - 3x)/2$, $(35x^4 - 30x^2 + 3)/8$. $\square$

A third set of orthogonal polynomials which is frequently encountered (although not as commonly as the two sets we have just mentioned) are the Chebyshev polynomials *of the second kind*, defined by

$$U_k(x) = \sin(k + 1)\theta \, / \sin \theta, \tag{5.27}$$

where $x = \cos \theta$. It is left as an exercise for the reader (in Problem 5.16) to show that these polynomials are orthogonal on $[-1, 1]$ with respect to weight function $(1 - x^2)^{1/2}$. Note that, as defined by (5.27), these polynomials do not have leading coefficient 1. The polynomials U_r play an important role in best L_1 approximations. We state without proof the following.

Theorem 5.1   If f is continuous on $[-1, 1]$ and $f - p$ has at most $n + 1$ zeros in $[-1, 1]$ for every $p \in P_n$, then the best L_1 approximation from P_n for f on $[-1, 1]$ is simply the interpolating polynomial for f constructed at the zeros of U_{n+1}. $\square$

Thus L_1 approximations are surprisingly easy to compute, for functions f which satisfy the conditions of this theorem. This result enables us to justify that the best L_1 approximation for $x^{1/2}$ on $[0, 1]$ is that given in Example 5.1(iv). (See Problem 5.19.) We will not pursue L_1 approximations any further.

Of all the orthogonal polynomials, those which concern us most here are the polynomials T_r. The orthogonal series based on these, called the *Chebyshev series*, is of considerable theoretical and practical interest. It is usual to write the series as

$$\sum_{r=0}^{\infty}{}' a_r T_r(x), \tag{5.28}$$

where $\sum'$ denotes a summation whose first term is halved and, from (5.18) with $a = -1$, $b = 1$, $\omega(x) = (1 - x^2)^{-1/2}$, $\psi_r = T_r$, we find that

$$a_r = \frac{2}{\pi} \int_{-1}^{1} (1 - x^2)^{-1/2} f(x) T_r(x) \, dx. \tag{5.29}$$

Note that, in this case, we have already effectively evaluated the denominator on the right side of (5.25) in Example 5.9. This denominator has the value $\pi/2$ for $r \geqslant 1$ and the value π for $r = 0$. The use of the $\sum'$ notation allows the single formula (5.29) to stand for all values of r. On making the substitution $x = \cos\theta$ in (5.29), we obtain

$$a_r = \frac{2}{\pi} \int_{0}^{\pi} f(\cos\theta) \cos r\theta \, d\theta, \tag{5.30}$$

which reminds us of the classical Fourier coefficients. Denoting the classical Fourier coefficients of $f(\cos\theta)$ by a_r^* and b_r^*, we have from (5.16) that

$$a_r^* = \frac{1}{\pi} \int_{-\pi}^{\pi} f(\cos\theta) \cos r\theta \, d\theta, \qquad b_r^* = \frac{1}{\pi} \int_{-\pi}^{\pi} f(\cos\theta) \sin r\theta \, d\theta.$$

Putting $u = -\theta$, we have

$$\int_{-\pi}^{0} f(\cos\theta) \sin r\theta \, d\theta = \int_{\pi}^{0} f(\cos(-u)) \sin(-ru) \cdot -du$$

$$= -\int_{0}^{\pi} f(\cos u) \sin ru \, du = -\int_{0}^{\pi} f(\cos\theta) \sin r\theta \, d\theta$$

and so $b_r^* = 0$. We similarly find that

$$\int_{-\pi}^{0} f(\cos\theta) \cos r\theta \, d\theta = \int_{0}^{\pi} f(\cos\theta) \cos r\theta \, d\theta,$$

so that

$$a_r^* = \frac{2}{\pi} \int_{0}^{\pi} f(\cos\theta) \cos r\theta \, d\theta = a_r.$$

Thus the Chebyshev series for $f(x)$ is just the classical Fourier series for $f(\cos\theta)$ and therefore we may apply results concerning convergence of the latter series. For example (see Davis), if f' is piecewise continuous on $[-1, 1]$, the Chebyshev series converges uniformly to f on $[-1, 1]$.

Example 5.11   Find the Chebyshev series for $f(x) = ((1 + x)/2)^{1/2}$. Since, with $x = \cos \theta$, $((1 + x)/2)^{1/2} = \cos \frac{1}{2}\theta$, we obtain

$$a_r = \frac{2}{\pi} \int_0^\pi \cos \tfrac{1}{2}\theta \cos r\theta \, d\theta.$$

From (2.8) we have

$$2 \cos \tfrac{1}{2}\theta \cos r\theta = \cos(r + \tfrac{1}{2})\theta + \cos(r - \tfrac{1}{2})\theta$$

and so obtain

$$a_r = (-1)^{r-1} \frac{4}{\pi(4r^2 - 1)}$$

for all r. This gives the Chebyshev series

$$\left(\frac{1 + x}{2}\right)^{1/2} \sim \frac{2}{\pi} + \frac{4}{\pi} \sum_{r=1}^\infty (-1)^{r-1} T_r(x)/(4r^2 - 1). \quad \square$$

We may use orthogonal polynomials analogously to construct least squares approximations for f defined on a point set $\{x_0, \ldots, x_N\}$ with respect to weights $\{\omega_0, \ldots, \omega_N\}$, with each $\omega_i > 0$. If we believe that the data at some x_i are more reliable than at other points, we can choose values for the weights ω_i accordingly. Otherwise, we can set each $\omega_i = 1$ for equal weighting. As for approximation on an interval, we find that the orthogonal polynomials satisfy the same form of recurrence relation (see (5.26)),

$$p_{k+1}(x) = (x - \beta_k)p_k(x) - \gamma_k p_{k-1}(x). \tag{5.31}$$

In this case, we compute β_k and γ_k from

$$\beta_k = \sum_{i=0}^N \omega_i x_i (p_k(x_i))^2 \bigg/ \sum_{i=0}^N \omega_i (p_k(x_i))^2, \tag{5.32}$$

$$\gamma_k = \sum_{i=0}^N \omega_i (p_k(x_i))^2 \bigg/ \sum_{i=0}^N \omega_i (p_{k-1}(x_i))^2 \tag{5.33}$$

which are analogous to (5.25) and (5.24) respectively. If the denominator in (5.32) is zero, β_k is not defined. This can only happen if $p_k(x_i) = 0$, $i = 0$, $\ldots, N$, and therefore only if $k \geqslant N + 1$, since p_k is a polynomial of degree k. The least squares polynomial for f of degree $\leqslant n$ is

$$\sum_{r=0}^n a_r p_r(x), \tag{5.34}$$

where

$$a_r = \sum_{i=0}^N \omega_i f(x_i) p_r(x_i) \bigg/ \sum_{i=0}^N \omega_i (p_r(x_i))^2, \tag{5.35}$$

which is analogous to (5.18). If in (5.34) we choose $n = N$, the least squares polynomial must coincide with the interpolating polynomial for f constructed at $x = x_0, \ldots, x_N$ and, therefore, we terminate the orthogonal series for f after $N + 1$ terms. Thus the orthogonal series for a function on a finite point set is a *finite* series.

In practice, we let $p_{-1}(x) \equiv 0$ (this is included so that we may use the recurrence relation (5.31) with $k = 0$), and choose $p_0(x) \equiv 1$ and $\gamma_0 = 0$. We find β_0 from (5.32) and a_0 from (5.35) with $r = 0$. Suppose that, at some stage, we have computed β_r, γ_r, a_r and $p_r(x_i)$, $0 \leqslant i \leqslant N$, for $r = 0, 1, \ldots$, $k \leqslant n$. If $k = n$, we terminate the process. Otherwise, we continue by computing $p_{k+1}(x_i)$, $i = 0, 1, \ldots, N$, from (5.31) and thus compute β_{k+1}, γ_{k+1} and a_{k+1}. Thus we obtain (5.34) as the best approximation and to evaluate this for a particular value of x we use the recurrence relation (5.31).

This process for computing least squares approximations avoids the ill-conditioning, referred to in §5.4, which is encountered in the direct method of solving the normal equations.

Example 5.12 Construct least squares approximations of degrees 1, 2 and 3 for the following data.

x_i	-2	-1	0	1	2
f_i	-1	-1	0	1	1

We take every $\omega_i = 1$, $p_{-1}(x) \equiv 0$, $p_0(x) \equiv 1$, $\gamma_0 = 0$ and find that $\beta_0 = 0$. Thus $p_1(x) = x$, $\beta_1 = 0$, $\gamma_1 = 2$; $p_2(x) = x^2 - 2$, $\beta_2 = 0$, $\gamma_2 = \frac{7}{5}$; $p_3(x) = x^3 - \frac{17}{5}x$. Hence $a_0 = 0$, $a_1 = \frac{3}{5}$, $a_2 = 0$, $a_3 = -\frac{1}{6}$, so that the best approximation of degrees 1 and 2 is $\frac{3}{5}x$; that of degree 3 is $(7x - x^3)/6$, which interpolates f at all five points. $\sqcup$

Example 5.13 Let x_j denote $\cos(\pi j/N)$. The point set $\{x_0, \ldots, x_N\}$ consists of the extreme points (that is, points of maximum modulus) of T_N. The Chebyshev polynomials T_r, for $0 \leqslant r \leqslant N$, are orthogonal on this point set with respect to weights $\{\frac{1}{2}, 1, 1, \ldots, 1, 1, \frac{1}{2}\}$. More precisely, we can show (see Problem 5.24) that

$$\sum_{j=0}^{N}{''} T_r(x_j)T_s(x_j) = \begin{cases} 0, & r \neq s, \quad 0 \leqslant r, s \leqslant N \\ N/2, & r = s, \quad 0 < r < N \\ N, & r = s = 0 \text{ or } N \end{cases} \tag{5.36}$$

where $\sum''$ denotes a summation whose first and last terms are halved. We now use our results on orthogonal series to obtain least squares approximations for a function f for the given sets of points and weights. To obtain polynomials with leading coefficient 1, we write $p_r = T_r/2^{r-1}$ (for $r \geqslant 1$) in (5.34) and define $\alpha_r = a_r/2^{r-1}$, where a_r is given by the appropriate form of (5.35). On checking the coefficients a_r for the three cases $r = 0$, N and $0 < r < N$

and using (5.36), we find that the least squares polynomial of degree $n < N$ on the given point set is

$$\sum_{r=0}^{n}{}' \alpha_r T_r(x), \qquad (5.37)$$

where

$$\alpha_r = \frac{2}{N} \sum_{j=0}^{N}{}'' f(x_j) T_r(x_j). \qquad (5.38)$$

If we take $n = N$, the least squares polynomial, which coincides with the interpolating polynomial for f at the x_j, is given by

$$\sum_{r=0}^{N}{}'' \alpha_r T_r(x). \quad \square \qquad (5.39)$$

5.6 Minimax approximation

As we have just seen, the Chebyshev series for a given function f is the solution of a certain least squares problem. We shall see later that, surprisingly, this series is closely related to the best polynomials in the minimax sense.

If f is continuous on $[a, b]$, we write

$$E_n(f) = \inf_{p \in P_n} \max_{a \leq x \leq b} |f(x) - p(x)|. \qquad (5.40)$$

Recall that "inf", an abbreviation for "infimum", means the greatest lower bound. It is important to realize that a greatest lower bound is not always attained. (See Example 2.8.) In this case, we want to know whether the infimum is attained by any polynomial $p \in P_n$. If there *is* such a polynomial p, we will say that the best approximation exists. If anyone still thinks it obvious that the best approximation exists, let him consider another approximation problem: find $\inf |\alpha - \sqrt{2}|$, where the infimum is over all rational numbers α. Obviously the infimum has the value zero, but there is no rational number closest to the irrational number $\sqrt{2}$. This is an example of an infimum not being attained. Besides the question of existence of a best approximating polynomial, we naturally want to know if it is unique, whether it has any characteristic properties and how it can be computed. We should also ask what happens as the degree of the approximating polynomial is increased: can we obtain a sequence of approximating polynomials which converges uniformly to f on $[a, b]$? We have already discussed these questions for least squares approximations; in that case we found conditions for existence and uniqueness and found, as a characteristic property, that the coefficients of the best approximating polynomial satisfy the normal equations. We also referred to the convergence of orthogonal series.

In fact, if f is continuous on $[a, b]$, the infimum in (5.40) *is* attained and so there is a best approximating polynomial. The proof of this is beyond the scope of this text. The proof shows first that, for a polynomial to be a "good" approximation to f, its coefficients must not be "too large" in modulus. This confines the search for the best polynomial to a finite region of the $(n + 1)$ dimensional space of points $(a_0, \ldots, a_n)$ and it is shown that the infimum is attained by some polynomial whose coefficients $(a_0, \ldots, a_n)$ belong to this finite region. The interested reader will find the complete proof in Davis. Now, to give a hint of what to expect as answers to some of the other questions on minimax approximation, we consider a simple example.

Example 5.14 Find the minimax straight line approximation $a + bx$ to the function $x^{1/2}$ on $[\frac{1}{4}, 1]$. It is convenient to draw a graph of $x^{1/2}$ on $[\frac{1}{4}, 1]$. On the graph, we can visualize a straight line segment $y = a + bx$ which we move around in different positions, seeking

$$E = \min_{a,b} \ \max_{1/4 \leqslant x \leqslant 1} |x^{1/2} - (a + bx)|,$$

the minimum being over all a and b. It is obvious geometrically that, as shown in Fig. 5.3, the best straight line is that for which the maximum error

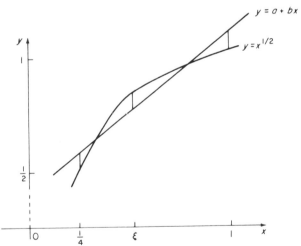

Fig. 5.3. Minimax straight line approximation for $x^{1/2}$ on $[\frac{1}{4}, 1]$.

occurs at $x = \frac{1}{4}$, $x = 1$ and at some interior point, say $x = \xi$. We therefore have

$$a + b \cdot \tfrac{1}{4} - (\tfrac{1}{4})^{1/2} = E \tag{5.41}$$

$$a + b\xi - \xi^{1/2} = -E \tag{5.42}$$

$$a + b \cdot 1 - 1 = E. \tag{5.43}$$

So far, to find four unknowns a, b, E and ξ, we have only three equations. From Fig. 5.3 it is clear that the error $y = x^{1/2} - (a + bx)$ has a turning value at $x = \xi$ and so has zero derivative there. This gives, as a fourth equation,

$$\tfrac{1}{2}\xi^{-1/2} - b = 0. \tag{5.44}$$

On subtracting (5.41) from (5.43) we obtain $b = \tfrac{2}{3}$ and so, from (5.44), $\xi = \tfrac{9}{16}$. Adding (5.42) and (5.43) yields $a = \tfrac{17}{48}$ and finally we obtain, say from (5.43), $E = \tfrac{1}{48}$. Thus the best straight line is $y = \tfrac{17}{48} + \tfrac{2}{3}x$, which approximates $x^{1/2}$ on $[\tfrac{1}{4}, 1]$ with error in modulus not greater than $\tfrac{1}{48}$. As an aside, we point out that this approximation may be used to estimate the square root of any positive number. We need to multiply it by an appropriate power of 4 to put it into the range $[\tfrac{1}{4}, 1]$. See Problem 5.28. $\square$

To provide further insight into minimax approximations, recall the minimax problem which we solved in Chapter 4, where we wished to minimize

$$\max_{-1 \leqslant x \leqslant 1} |(x - x_0)\cdots(x - x_n)|$$

over all choices of interpolating points $x_0, x_1, \ldots, x_n$. In fact, we considered the more general minimax problem

$$\min \max_{-1 \leqslant x \leqslant 1} |p_{n+1}(x)|, \tag{5.45}$$

where $p_{n+1} \in P_{n+1}$ with leading term x^{n+1}. We found that the solution is given by

$$p_{n+1}(x) = T_{n+1}(x)/2^n.$$

We rephrase this result in the terminology of (5.40), with $a = -1$, $b = 1$ and $f(x) = x^{n+1}$. Thus (5.45) is replaced by

$$\min_{p \in P_n} \max_{-1 \leqslant x \leqslant 1} |x^{n+1} - p(x)|$$

and the minimum is attained when

$$x^{n+1} - p(x) = T_{n+1}(x)/2^n.$$

This is the error when p is considered as an approximation to x^{n+1}. Because of the oscillatory property of T_{n+1}, this error oscillates between its $n + 2$ points of maximum modulus on $[-1, 1]$.

Definition 5.4 A continuous function E is said to *equioscillate* on n points of $[a, b]$ if there exist n points x_i, with $a \leqslant x_1 < x_2 < \cdots < x_n \leqslant b$, such that

$$|E(x_i)| = \max_{a \leqslant x \leqslant b} |E(x)|, \qquad i = 1, \ldots, n,$$

and

$$E(x_i) = -E(x_{i+1}), \qquad i = 1, \ldots, n - 1. \quad \square$$

Thus the error function in Example 5.14 equioscillates on three points, while the error function $T_{n+1}/2^n$, which occurs in the solution of (5.45), equioscillates on $n + 2$ points. This property holds generally for minimax approximations.

Theorem 5.2 If f is continuous on $[a, b]$ and p denotes a minimax approximation $\in P_n$ to f on $[a, b]$, then $f - p$ equioscillates on at least $n + 2$ points of $[a, b]$.

Proof Write $E(x) = f(x) - p(x)$ and suppose that E equioscillates on s points. If $s = 1$, as in Fig. 5.4, we could add $\frac{1}{2}C$ to p to give *two* equioscillations and reduce the maximum modulus of the error E. Therefore $s > 1$.

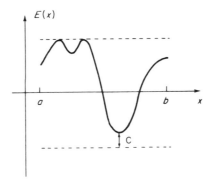

Fig. 5.4. The error must equioscillate at more than one point (Theorem 5.2).

We assume now that $s < n + 2$ and construct an appropriate *polynomial* which, when added to p, reduces all s "humps" in the error function. We divide $[a, b]$ into closed sub-intervals, each of width δ, so that on each sub-interval I, $x, x' \in I$ implies that

$$|E(x) - E(x')| < \max_{a \leqslant x \leqslant b} |E(x)|.$$

This is possible, due to the continuity and therefore uniform continuity† of E. Let $X_1, X_2, \ldots, X_m$ denote those sub-intervals (of width δ) on which the maximum modulus of E is attained. (See Fig. 5.5.) The sub-division of $[a, b]$ is arranged, as we now see, so that E is of constant sign on each X_i. We will assume that E has opposite signs on consecutive X_i (we have $+$, $-$, $+$ in Fig. 5.5), so that we have just s intervals X_i. We will also assume that E is

† See Chapter 2, Definition 2.10.

positive on X_1. These assumptions do not seriously affect the proof, but save introducing unimportant detail. One consequence of these assumptions is that $E(x)$ cannot attain its maximum modulus at an end point of any X_i (except possibly at $x = a$ or $x = b$) as then $E(x)$ would have the same sign on two consecutive X_i. We select points $x_1, x_1, \ldots, x_{s-1}$ so that x_i lies between intervals X_i and X_{i+1}, but belongs to neither.

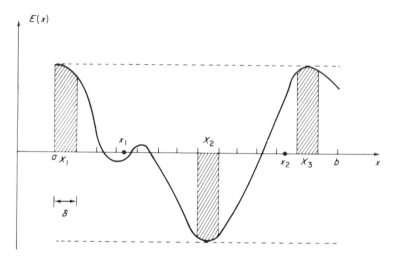

Fig. 5.5. The equioscillation theorem 5.2.

We now seek a number ε so that, with

$$q(x) = (x_1 - x)\cdots(x_{s-1} - x),$$

the polynomial

$$p^*(x) = p(x) + \varepsilon q(x) \tag{5.46}$$

is a *better* approximation to f, contrary to our assumption about p. Note that $s - 1 \leqslant n$ and thus $p^* \in P_n$. We write

$$E^*(x) = f(x) - p^*(x) = f(x) - p(x) - \varepsilon q(x),$$

so that

$$E^*(x) = E(x) - \varepsilon q(x). \tag{5.47}$$

Observe that q changes sign at, and only at, the x_i. Also, q is positive on X_1 and therefore takes the same sign as E on each X_i. Let us restrict ε so that $\varepsilon > 0$ and

$$\varepsilon|q(x)| < |E(x)|, \quad \text{for all } x \in X_i, \ i = 1, \ldots, s. \tag{5.48}$$

We can always choose ε small enough to satisfy (5.48), since $E(x)$ cannot be zero on any X_i. Therefore $\max|E^*(x)| < \max|E(x)|$ on each X_i, that is, we can reduce all s humps of the error curve. However, we must take care not to create larger humps elsewhere, as we reduce those on the X_i.

Let R denote the union of the sub-intervals of width δ, other than the X_i, into which we have divided $[a, b]$. Then from (5.47)

$$\max_{x \in R}|E^*(x)| \leqslant \max_{x \in R}|E(x)| + \varepsilon \max_{x \in R}|q(x)|. \tag{5.49}$$

Note that we have the *strict inequality*

$$\max_{x \in R}|E(x)| < \max_{a \leqslant x \leqslant b}|E(x)|,$$

as $E(x)$ attains its maximum modulus only at interior points of the X_i (except possibly at $x = a$ or $x = b$). Thus it is possible to restrict ε further, if necessary, to make the second term on the right of the inequality (5.49) small enough so that

$$\max_{x \in R}|E^*(x)| < \max_{a \leqslant x \leqslant b}|E(x)|.$$

With such a choice of ε, $E^* = f - p^*$ is of smaller modulus than E on the *whole* of $[a, b]$. Therefore $p^* \in P_n$ is a better approximation to f than p, which gives the required contradiction. Notice that, if $s \geqslant n + 2$, we can no longer assert that $p^* \in P_n$ and so cannot force the contradiction in this case. $\square$

Knowing the equioscillation property, it is not difficult to show the uniqueness of the minimax polynomial.

Theorem 5.3 If f is continuous on $[a, b]$, there is a unique minimax polynomial approximation of degree at most n for f on $[a, b]$.

Proof Suppose p and p^* are both best approximations for f and let e denote the common maximum error on $[a, b]$. Let $\xi_0, \xi_1, \ldots, \xi_{n+1}$ denote $n + 2$ equioscillation points of the error $f - p$. Then at any ξ_r

$$p(\xi_r) - p^*(\xi_r) = (f(\xi_r) - p^*(\xi_r)) - (f(\xi_r) - p(\xi_r))$$
$$= (f(\xi_r) - p^*(\xi_r)) \mp e,$$

where the $\mp$ signs alternate as $r = 0, 1, \ldots, n + 1$. Since $|f(\xi_r) - p^*(\xi_r)| \leqslant e$, the polynomial $p - p^*$ is alternately $\geqslant 0$ and $\leqslant 0$ on the $n + 2$ points ξ_r. We deduce from Problems 2.3 and 2.4 that $p - p^*$, which belongs to P_n, has at least $n + 1$ zeros and so must be identically zero. This proves the uniqueness of the minimax approximation. $\square$

We can now prove a converse of Theorem 5.2, which then enables us to state that the error equioscillation property *characterizes* minimax approximations.

Theorem 5.4   Let f be continuous on $[a, b]$ and suppose that, for a certain polynomial $p \in P_n$, $f - p$ equioscillates on $n + 2$ points of $[a, b]$. It follows that p is the minimax approximation from P_n for f on $[a, b]$.

Proof   Suppose, on the contrary, that $p^* \in P_n$ is the best approximation. Let e and e^* denote the maximum errors obtained from p and p^* respectively. From our supposition, we have $e^* < e$. If $\xi_0, \ldots, \xi_{n+1}$ denote points on which $f(x) - p(x)$ equioscillates, we have, exactly as in the proof of Theorem 5.3, that

$$p(\xi_r) - p^*(\xi_r) = f(\xi_r) - p^*(\xi_r) \mp e.$$

Since $|f(\xi_r) - p^*(\xi_r)| \leqslant e^* < e$, the right side of this last equation is positive and negative alternately. Again we deduce that $p - p^*$, which belongs to P_n, has at least $n + 1$ zeros and so is identically zero. $\square$

5.7 Chebyshev series

Having studied the equioscillation property, we can now appreciate the relevance of Chebyshev series. Intuitively, if the Chebyshev series $\sum'^{\infty}_{r=0} a_r T_r(x)$ Theorem 5.4 we see that if the error were *exactly* $a_{n+1} T_{n+1}(x)$ then the partial sum

$$f(x) - \sum_{r=0}^{n} {}' a_r T_r(x) = \sum_{r=n+1}^{\infty} a_r T_r(x)$$

will be dominated by the leading term $a_{n+1} T_{n+1}(x)$, which equioscillates on $n + 2$ points. (We assume here that $a_{n+1} \neq 0$, although if $a_{n+1} = 0$, the first non-zero term in the error will still equioscillate on $n + 2$ points.) From Theorem 5.4 we see that if the error were *exactly* $a_{n+1} T_{n+1}(x)$, then

$$s_n(x) = \sum_{r=0}^{n} {}' a_r T_r(x)$$

would be precisely the minimax approximation for $f(x)$. If the series converges rapidly, so that $a_{n+2}, a_{n+3}, \ldots$ are small compared with a_{n+1}, we can expect s_n to be close to the minimax approximation.

There are algorithms for computing minimax approximations. However, these algorithms are not easy to use and hence the emphasis on Chebyshev series, which are much easier to handle.

Example 5.15   Consider the function $x^{1/2}$ on $[\frac{1}{4}, 1]$, for which we found the minimax straight line in Example 5.14. We make a linear change of variable, so that $x^{1/2}$ for $\frac{1}{4} \leqslant x \leqslant 1$ becomes $(\frac{3}{8}u + \frac{5}{8})^{1/2}$ for $-1 \leqslant u \leqslant 1$. To four decimal places, the first few Chebyshev coefficients a_r for this function are

r	0	1	2	3	4	5
a_r	1.5420	0.2465	−0.0202	0.0033	−0.0007	0.0002

The remaining coefficients are zero, to four decimal places. Knowing these coefficients, we construct partial sums of the Chebyshev series. In Table 5.1 these are compared with the minimax polynomials of the same degree. To make it easy to compare the polynomials themselves, we have expressed the Chebyshev partial sums explicitly in powers of u. $\square$

Table 5.1   Comparison of Chebyshev series and minimax approximations of degree 0, 1, 2 for the function $(\frac{3}{8}u + \frac{5}{8})^{1/2}$ on $[-1, 1]$, which is $x^{1/2}$ on $[\frac{1}{4}, 1]$.

Degree (n) of poly-nomial	Minimax polynomial	Maximum error	Chebyshev partial sum	Maximum error	$\lvert a_{n+1} \rvert$
0	0.75	0.25	0.7710	0.2710	0.2465
1	$0.7708 + 0.2500u$	0.0208	$0.7710 + 0.2465u$	0.0245	0.0202
2	$0.7920 + 0.2465u$ $- 0.0420u^2$	0.0035	$0.7912 + 0.2465u$ $- 0.0404u^2$	0.0043	0.0033

The Chebyshev coefficients, as we saw in (5.30), are given by

$$a_r = \frac{2}{\pi} \int_0^\pi f(\cos \theta) \cos r\theta \, d\theta. \tag{5.50}$$

It is not always possible to evaluate these integrals exactly, as we were able to do in Example 5.11. We have to find some method for approximating to these coefficients. The most obvious approach is to approximate to the integral in (5.50) by some numerical integration rule. On using the trapezoidal rule with N equal sub-intervals (see Chapter 6), we obtain from (5.50), as an approximation for a_r,

$$\alpha_r = \frac{2}{N} \sum_{j=0}^{N}{}'' f(x_j) T_r(x_j), \tag{5.51}$$

where $x_j = \cos(\pi j / N)$. In Example 5.13 we saw that these same numbers α_r are coefficients in a finite orthogonal series for f defined on the point set $\{x_0, \ldots, x_N\}$, whose members are the extreme points of T_N. This series has partial sums

$$\sum_{r=0}^{n}{}' \alpha_r T_r(x) \tag{5.52}$$

which we can now think of as approximations to the partial sums of the Chebyshev series. Of course each α_r depends on N; we expect that, in general, as N is increased in (5.51), we will obtain closer approximations α_r to the Chebyshev coefficient a_r, given by (5.50).

We now derive an explicit relation between the α_r and the a_r. First we extend the orthogonality relation (5.36) to polynomials T_r with $r > N$ by noting that

$$T_{2kN \pm r}(x_j) = \cos\left(\frac{(2kN \pm r)\pi j}{N}\right) = \cos\left(2k\pi j \pm \frac{r\pi j}{N}\right).$$

From the periodicity of the cosine function, we have

$$T_{2kN \pm r}(x_j) = T_r(x_j). \tag{5.53}$$

Let us now assume that the Chebyshev series for f converges uniformly to f on $[-1, 1]$. Then it follows from (5.51) that

$$\alpha_r = \frac{2}{N} \sum_{j=0}^{N}{}'' \left(\sum_{s=0}^{\infty}{}' a_s T_s(x_j)\right) T_r(x_j)$$

$$= \frac{2}{N} \sum_{s=0}^{\infty}{}' a_s \sum_{j=0}^{N}{}'' T_s(x_j) T_r(x_j). \tag{5.54}$$

From (5.36) and (5.53), if $r \neq 0$ or N, the only summations over j which are non-zero in (5.54) are those for which $s = r$, $2N - r$, $2N + r$, $4N - r$, $4N + r$, and so on. So, if $r \neq 0$ or N,

$$\alpha_r = a_r + \sum_{k=1}^{\infty} (a_{2kN-r} + a_{2kN+r}). \tag{5.55}$$

We find that (5.55) holds for $r = 0$ and N also, on examining these cases separately. If we keep r fixed and take N large enough, we expect from (5.55) that α_r will approximate closely to a_r.

We now consider an alternative method for approximating to the Chebyshev coefficients a_r, which is based on another orthogonality property of the Chebyshev polynomials. This time, the orthogonality is on the *zeros* of T_N, instead of on the extreme points as discussed above. We have (see Problem 5.29) that for $0 \leqslant r, s \leqslant N$,

$$\sum_{j=1}^{N} T_r(x_j^*) T_s(x_j^*) = \begin{cases} 0, & r \neq s \quad \text{or} \quad r = s = N \\ N/2, & r = s \neq 0 \quad \text{or} \quad N \\ N, & r = s = 0 \end{cases} \tag{5.56}$$

where the x_j^* are the zeros of T_N. Following the same procedure as in Example 5.13, we find that for $n \leqslant N - 1$

$$\sum_{r=0}^{n}{}' \alpha_r^* T_r(x), \tag{5.57}$$

with

$$\alpha_r^* = \frac{2}{N} \sum_{j=1}^{N} f(x_j^*) T_r(x_j^*),\qquad(5.58)$$

is the least squares approximation to f on the point set $\{x_1^*, \ldots, x_N^*\}$. The weights are all equal in this case. Thus the polynomial (5.57) is another approximation, easily computed, to the partial Chebyshev series. If we put $n = N$ in (5.57) the least squares approximation must coincide with the interpolating polynomial for f constructed at the x_j^*, that is, the zeros of $T_N(x)$, which we discussed at the end of Chapter 4. Since then, we have in this chapter followed a full circle from interpolation via least squares and minimax approximations back to interpolation again.

We now establish two relations between α_r^* and a_r as we did for α_r and a_r. First, (5.58) may be regarded as an approximation to (5.50) obtained by using a quadrature rule. In this case (see Chapter 6), it is the *mid-point rule*. Second, we replace $f(x_j^*)$ in (5.58) by its Chebyshev series (again assumed to be uniformly convergent) and extend the orthogonality relation (5.56) by the relation

$$T_{2kN \pm r}(x_j^*) = (-1)^k T_r(x_j^*).\qquad(5.59)$$

Thus we obtain

$$\alpha_r^* = a_r + \sum_{k=1}^{\infty} (-1)^k (a_{2kN-r} + a_{2kN+r})\qquad(5.60)$$

for $0 \leqslant r < N$.

The above analysis suggests a simple, reliable method for estimating Chebyshev coefficients a_r, for $0 \leqslant r \leqslant n$. First we select a value of $N > n$ and compute α_r and α_r^* $(0 \leqslant r \leqslant n)$ from (5.51) and (5.58). If $|\alpha_r - \alpha_r^*|$ is sufficiently small, we use $(\alpha_r + \alpha_r^*)/2$ to approximate to a_r. Otherwise, we increase N and re-calculate α_r and α_r^*. Finally, we observe from (5.55) and (5.60) that, if the Chebyshev coefficients tend to zero rapidly, we have

$$a_r \simeq \alpha_r - (a_{2N-r} + a_{2N+r})$$

and

$$a_r \simeq \alpha_r^* + (a_{2N-r} + a_{2N+r}).$$

Thus we expect that a_r will usually lie between α_r and α_r^* and will be more closely estimated by $(\alpha_r + \alpha_r^*)/2$.

5.8 Economization of power series

Suppose we wish to obtain a polynomial approximation for $\sin x$ on $[-1, 1]$, with error not greater than 0.5×10^{-2} in modulus. Forgetting all knowledge of interpolating polynomials, minimax approximations and Chebyshev series, we might use the Taylor polynomial

$$p_5(x) = x - \frac{x^3}{3!} + \frac{x^5}{5!} \tag{5.61}$$

with error

$$\frac{x^6}{6!} \left[\frac{d^6}{dx^6} \sin x \right]_{x=\xi_x}$$

for some ξ_x. Thus the modulus of the error on $[-1, 1]$ is not greater than $\frac{1}{720}$.† We now express the powers of x which appear in (5.61) in terms of the Chebyshev polynomials:

$$x = T_1(x)$$

$$x^3 = (3T_1(x) + T_3(x))/4$$

$$x^5 = (10T_1(x) + 5T_3(x) + T_5(x))/16.$$

Substituting these into (5.61) and collecting like terms, we obtain

$$p_5(x) = \tfrac{169}{192}T_1(x) - \tfrac{5}{128}T_3(x) + \tfrac{1}{1920}T_5(x), \tag{5.62}$$

and note the small coefficient of $T_5(x)$. (More generally, in a series of powers of x which ends with $a_n x^n$, the series in terms of Chebyshev polynomials will end with the term $a_n T_n(x)/2^{n-1}$, since $T_n(x)$ has leading term $2^{n-1}x^n$.) If in (5.62) we neglect the term in $T_5(x)$, we incur an additional error in modulus not greater than $\frac{1}{1920}$. Therefore, on $[-1, 1]$ we can approximate to $\sin x$ by the *cubic* polynomial

$$p_3(x) = \tfrac{169}{192}T_1(x) - \tfrac{5}{128}T_3(x) = \tfrac{383}{384}x - \tfrac{5}{32}x^3$$

with an error whose modulus cannot exceed $\frac{1}{720} + \frac{1}{1920}$, which is within the required accuracy of 0.5×10^{-2}. This simple process is known as *economization* of power series. It is sometimes possible to neglect further terms besides the last one in the transformed series of Chebyshev polynomials.

5.9 Convergence of minimax polynomials

As we remarked in Chapter 3, the main justification for using polynomials to approximate to a function f which is continuous on a finite interval $[a, b]$ is Weierstrass' theorem, which we restate more rigorously here.

Theorem 5.5 (Weierstrass' theorem.)   If f is continuous on $[a, b]$ then, given any $\varepsilon > 0$, there exists a polynomial p such that $|f(x) - p(x)| < \varepsilon$ for all $x \in [a, b]$.

---

† By considering the Taylor polynomial p_6 ($= p_5$) we can obtain the error bound $1/7!$. The bound given here is sufficient for the purposes of this example.

There are several quite different proofs of this fundamental theorem. One proof uses Bernstein polynomials†. ☐

Definition 5.5 The nth Bernstein polynomial for f on $[0, 1]$ is

$$B_n(f; x) = \sum_{j=0}^{n} f\left(\frac{j}{n}\right)\binom{n}{j} x^j (1 - x)^{n-j}. \quad \square$$

By making a linear change of variable, we may construct similar polynomials on any finite interval $[a, b]$. It can be shown rigorously that, if f is continuous on $[0, 1]$, the sequence $\{B_n(f; x)\}$ converges uniformly to $f(x)$ on $[0, 1]$. In addition, derivatives of the Bernstein polynomials converge to derivatives of f (if these exist). Also, if f is convex (Definition 2.6), so is each Bernstein polynomial. Unfortunately, such good behaviour has to be paid for: the Bernstein polynomials converge *very slowly* to f, and thus are of little direct use for approximating to f. However, the uniform convergence of the Bernstein polynomials to f gives us a constructive proof of Weierstrass' theorem and shows us that the minimax polynomials must also converge uniformly to f.

5.10 Further types of approximation

If we wish to approximate to a function f on $[a, b]$ (or on some finite point set whose points belong to $[a, b]$) we may divide $[a, b]$ into N sub-intervals, by introducing points of sub-division

$$a = x_0 < x_1 < \cdots < x_N = b, \tag{5.63}$$

and use some form of approximating or interpolating polynomial on each sub-interval $[x_{i-1}, x_i]$, $1 \leqslant i \leqslant N$. This is called *piecewise* polynomial approximation. In general, the approximating polynomials used on neighbouring intervals $[x_{i-1}, x_i]$ and $[x_i, x_{i+1}]$ will *not* take the same value at the common point x_i. It is natural to seek piecewise approximations which are continuous and whose first few derivatives are also continuous at the nodes of sub-division, so as to give smooth joins. These are called *spline* approximations.

The commonest are *cubic splines*, which use polynomials of degree three on each sub-interval, chosen so that the function values and first two derivatives are continuous. More precisely, a function s is a cubic spline approximation for f on the given sub-division of $[a, b]$ if it is a polynomial of degree at most three on each sub-interval $[x_{i-1}, x_i]$, $1 \leqslant i \leqslant N$, with $s(x_i) = f(x_i)$, $0 \leqslant i \leqslant N$, and with the first two derivatives of s continuous at x_i, $1 \leqslant i \leqslant N - 1$. Therefore s and its first two derivatives are continuous on $[a, b]$. Note that s *interpolates* f at $x = x_0, \ldots, x_N$, but that the derivatives s' and s''

† Named after the Russian mathematician S. Bernstein, whose proof of Weierstrass' theorem was published in 1912.

do not necessarily take the values of f' and f'' (if these exist) at the sub-dividing nodes. We need to supply two more conditions to give a unique s. One possibility is to fix (arbitrarily) the values of s' at the end-points $x = x_0$ and $x = x_N$. For further reading on cubic splines, see Rivlin.

Another type of piecewise cubic approximation is the cubic Hermite approximation. The function f is again matched by a cubic polynomial on each sub-interval $[x_{i-1}, x_i]$. In this case we find polynomials ϕ_i such that ϕ_i and ϕ_i' take the same values as f and f' at both end points x_{i-1} and x_i. To simplify the notation, we illustrate this for the sub-interval $[0, 1]$. If

$$\phi(x) = (1 - x)^2(1 + 2x)f(0) + x^2(3 - 2x)f(1)$$
$$+ (1 - x)^2 xf'(0) + x^2(x - 1)f'(1), \tag{5.64}$$

we may easily verify that ϕ and ϕ' match f and f' at $x = 0$ and 1. As a generalization of this, we have the order $2n + 1$ piecewise Hermite approximation, in which a polynomial $\phi_i \in P_{2n+1}$ is used on each sub-interval. We choose ϕ_i such that $\phi_i, \phi_i', \ldots, \phi_i^{(n)}$ match $f, f', \ldots, f^{(n)}$ respectively at the end points of $[x_{i-1}, x_i]$.

In our discussion of least squares approximations, we went from polynomials to a more general class of functions constructed from Chebyshev sets. There is a more obvious generalization of polynomials. If we include the arithmetical operation of division, along with the operations of addition and multiplication which enable us to construct polynomials, we obtain the *rational functions*. Any rational (polynomial) function may be written in the form

$$r_{m,n}(x) = p_m(x)/q_n(x), \tag{5.65}$$

where p_m and q_n are polynomials of degree m and n respectively. To give a unique representation in the form (5.65), it is necessary to make the leading coefficient of $q_n(x)$ unity or apply some other suitable condition. The equioscillation property of minimax polynomial approximations generalizes to rational approximations. See Ralston.

Problems

Section 5.2

5.1   Find the least squares straight line for $f(x) = x^{1/2}$ on $[0, 1]$.

5.2   Find the least squares straight line approximation for the following data.

x	0.0	0.1	0.2	0.3	0.4	0.5	0.6
$f(x)$	2.9	2.8	2.7	2.3	2.1	2.1	1.7

5.3 Show that, if the functions $\psi_0, \ldots, \psi_n$ form a Chebyshev set on some interval $[a, b]$, they are also linearly independent on $[a, b]$.

5.4 A set of functions $\{\psi_0, \ldots, \psi_n\}$ is said to be linearly dependent on a point set $\{x_0, \ldots, x_N\}$ if there exist numbers $a_0, \ldots, a_n$, not all zero, such that

$$\sum_{r=0}^{n} a_r \psi_r(x_i) = 0, \qquad i = 0, 1, \ldots, N.$$

Show that the functions x, x^3, x^5 are linearly dependent on the point set $\{-2, -1, 0, 1, 2\}$.

5.5 If a_1 and a_2 are not both zero, find a number α so that

$$a_0 + a_1 \cos x + a_2 \sin x = a_0 + (a_1{}^2 + a_2{}^2)^{1/2} \cos(x - \alpha).$$

Then argue that the right side of the equation can be zero at no more than two points of $[0, 2\pi)$ and so $\{1, \cos x, \sin x\}$ is a Chebyshev set on $[0, 2\pi)$. *Hint:* use the identity

$$\cos(x - \alpha) = \cos x \cos \alpha + \sin x \sin \alpha.$$

5.6 For the data of Problem 5.2, make a linear transformation so that $\sum x_i = \sum x_i{}^3 = 0$. Hence find the least squares quadratic polynomial approximation for this data.

5.7 Find the least squares planar approximation, that is, of the form $z = a_0 + a_1 x + a_2 y$, for the following five points (x, y, z): $(0, 0, 1.1)$, $(0, 1, 1.9)$, $(1, 0, 2.2)$, $(0, -1, 0.0)$, $(-1, 0, 0.1)$.

Section 5.3

5.8 Show that, in the 3-point linear smoothing formula, the smoothed value for the "centre point" may be written as $y_0 = f_0 + \frac{1}{3}\delta^2 f_0$.

5.9 Smooth the data of Problem 5.2, using the 3-point linear smoothing formula.

5.10 To obtain a 5-point planar smoothing formula, which may be applied to Problem 5.7, let $f_{i,j}$ denote the value of $f(x, y)$ at $x = i, y = j$ for $(i, j) = (0, 0)$, $(\pm 1, 0)$, $(0, \pm 1)$ and let $z_{i,j}$ denote the smoothed value. By solving the appropriate normal equations, show that $z_{0,0}$ is simply the average of the five values $f_{i,j}$ and that

$$z_{1,0} = z_{0,0} + (f_{1,0} - f_{-1,0})/2.$$

Section 5.4

5.11   Show that a set of orthogonal functions is necessarily linearly independent. (*Hint:* show that

$$\sum_{r=0}^{n} a_r^2 \int_a^b [\psi_r(x)]^2 \, dx = \int_a^b \left(\sum_{r=0}^{n} a_r \psi_r(x) \right)^2 dx = 0,$$

if $\sum_{r=0}^{n} a_r \psi_r = 0$.)

5.12   By using the well known trigonometrical identities, like for example
(2.7) and (2.8), show that the set of functions $\{1, \cos x, \sin x, \ldots, \cos kx, \sin kx\}$
is orthogonal on $[-\pi, \pi]$ and verify the formulae (5.16) for the Fourier
coefficients.

5.13   Obtain the classical Fourier series for $f(x) = |x|$ on $[-\pi, \pi]$.

5.14   Find the classical Fourier series for $f(x) = x^2$ on $[-\pi, \pi]$.

Section 5.5

5.15   The set of polynomials $\{p_0, \ldots, p_n\}$ with $p_r \in P_r$ is orthogonal with
respect to a weight function ω on $[a, b]$. Show that, given any $q \in P_n$, there is
a set of coefficients $\{a_0, \ldots, a_n\}$ such that

$$q(x) = \sum_{r=0}^{n} a_r p_r(x).$$

Obtain an expression for the a_r. (*Hint:* the weighted least squares polynomial
approximation to q out of P_n is unique and must be q.)

5.16   Show that the Chebyshev polynomials of the second kind U_r, defined
by (5.27), are orthogonal on $[-1, 1]$ with respect to weight function $(1 - x^2)^{1/2}$.

5.17   From (5.27) deduce that the orthogonal polynomials U_r satisfy the
same recurrence relation as the T_r.

5.18   Show that the zeros of U_n are $x = \cos(j\pi/(n + 1))$, $1 \leq j \leq n$, and
show from the recurrence formula that U_n has leading term $2^n x^n$.

5.19   Verify that the conditions of Theorem 5.1 apply to the function
$f(x) = x^{1/2}$ on $[0, 1]$. Hence verify that the best L_1 straight line approximation for $x^{1/2}$ on $[0, 1]$ is that given in Example 5.1.

5.20   Suppose that the polynomials p_r are orthogonal on $[a, b]$ with respect to
weight function ω. By making the change of variable $u = (2x - b - a)/(b - a)$,
find polynomials which are orthogonal on $[-1, 1]$. What is the weight
function in this case?

5.21   Find the Chebyshev series for $f(x) = \cos^{-1} x$.

5.22   Derive the Chebyshev series for $f(x) = (1 - x^2)^{1/2}$.

5.23   By calculating the coefficients β_k, γ_k which are used in the recurrence formula (5.31), construct polynomials of degree r, $0 \leqslant r \leqslant 3$, which are orthogonal on the point set $\{-2, -1, 1, 2\}$.

5.24   To verify (5.36), one may proceed as follows. Express each $T_r(x_j)$ as $\cos(\pi rj/N)$ and apply the trigonometrical identity (2.8) to give the left side of (5.36) as a *sum* of cosines. Find the sum by using the identity

$$\cos k\theta = \frac{\sin(k + \tfrac{1}{2})\theta - \sin(k - \tfrac{1}{2})\theta}{2 \sin \tfrac{1}{2}\theta}.$$

Section 5.6

5.25   Obtain the minimax first degree polynomial for $f(x) = x^{1/2}$ on $[0, 1]$.

5.26   Obtain the minimax first degree polynomial for $f(x) = 1/(1 + x)$ on $[0, 1]$.

5.27   Let m and M denote the minimum and maximum values of a continuous function f on $[a, b]$. Show that the minimax polynomial of degree zero for f on $[a, b]$ is $\tfrac{1}{2}(m + M)$ and find the maximum error.

5.28   Given any $a > 0$, show that there is a unique integer k such that $\tfrac{1}{4} \leqslant 4^k a < 1$. If $x = 4^k a$, we saw in Example 5.14 that, for $\tfrac{1}{4} \leqslant x \leqslant 1$, $x^{1/2} \simeq \tfrac{17}{48} + \tfrac{2}{3}x$. Deduce that $a^{1/2} \simeq \tfrac{17}{48}2^{-k} + \tfrac{2}{3}2^k a$ and find the maximum error incurred by using this approximation. Thus estimate $\sqrt{2}$, $\sqrt{3}$ and $\sqrt{10}$ and estimate the error in each case.

Section 5.7

5.29   Apply the method described in Problem 5.24 to verify (5.56).

5.30   For the error in p_N, the interpolating polynomial defined by (5.52) with $n = N$ (which may be regarded as an approximation to the Chebyshev partial series), we have

$$f(x) - p_N(x) = (x - x_0)\cdots(x - x_N)f^{(N+1)}(\xi_x)/(N + 1)!$$

from (4.16), where $x_j = \cos(\pi j/N)$. We assume that the $(N + 1)$th derivative of f exists and is continuous on $[-1, 1]$. Applying the results of Problem 5.18, show that the above product of factors $x - x_r$ is

$$(x^2 - 1)U_{N-1}(x)/2^{N-1} = -(\sin \theta \sin N\theta)/2^{N-1},$$

on substituting $x = \cos \theta$. Deduce that

$$\max_{-1 \leqslant x \leqslant 1} |f(x) - p_N(x)| \leqslant \frac{1}{2^{N-1}.(N + 1)!} \max_{-1 \leqslant x \leqslant 1} |f^{(N+1)}(x)|.$$

5.31   Deduce from Theorem 5.2 that $p(x)$, the minimax polynomial $\in P_n$ for $f(x)$ on $[-1, 1]$, is the *interpolating polynomial* for $f(x)$ constructed at certain points $\tilde{x}_0, \ldots, \tilde{x}_n$ belonging to $[-1, 1]$.

5.32   With the notation and result of Problem 5.31, deduce from the error formula (4.16) that

$$E_n(f) = \max_{-1 \leqslant x \leqslant 1} \left(|(x - \tilde{x}_0)\cdots(x - \tilde{x}_n)| \cdot |f^{(n+1)}(\xi_x)|\right)/(n + 1)!$$

and hence that

$$E_n(f) \geqslant \max_{-1 \leqslant x \leqslant 1} |(x - \tilde{x}_0)\cdots(x - \tilde{x}_n)| \cdot \min_{-1 \leqslant x \leqslant 1} |f^{(n+1)}(x)|/(n + 1)!,$$

assuming that $f^{(n+1)}$ is continuous on $[-1, 1]$. From this and Theorem 4.3, deduce that

$$E_n(f) \geqslant \frac{1}{2^n(n + 1)!} \min_{-1 \leqslant x \leqslant 1} |f^{(n+1)}(x)|.$$

5.33   If $f^{(n+1)}$ is continuous on $[-1, 1]$, use the result of Problem 5.32 and (4.67) with

$$M_{n+1} = \max_{-1 \leqslant x \leqslant 1} |f^{(n+1)}(x)|$$

to show that there exists a number $\xi \in (-1, 1)$ such that

$$E_n(f) = \frac{1}{2^n(n + 1)!} |f^{(n+1)}(\xi)|.$$

5.34   Write

$$f(x) = f(\tfrac{1}{2}(b - a)t + \tfrac{1}{2}(b + a)) = g(t),$$

so that

$$\inf_{p \in P_n} \max_{a \leqslant x \leqslant b} |f(x) - p(x)| = \inf_{p \in P_n} \max_{-1 \leqslant t \leqslant 1} |g(t) - p(t)|.$$

Deduce from the result of Problem 5.33 that, if $f^{(n+1)}$ is continuous on $[a, b]$, there exists a number $\xi \in (a, b)$ such that

$$\inf_{p \in P_n} \max_{a \leqslant x \leqslant b} |f(x) - p(x)| = \frac{2}{(n + 1)!} \left(\frac{b - a}{4}\right)^{n+1} |f^{(n+1)}(\xi)|.$$

Apply this result to find the smallest degree of polynomial which will approximate to e^x with a maximum error of at most 10^{-6} on $[0, 1]$. (See also Problems 3.4, 4.27 and 4.29.)

Section 5.8

5.35 Show that, to approximate to e^x on $[-1, 1]$ to an accuracy within 10^{-6} by a Taylor polynomial constructed at $x = 0$, we require the polynomial of degree $n = 9$, or more (cf. Problem 3.4). Verify that if $n = 9$ the maximum error is less than 0.75×10^{-6}. Given that

$$x^9 = T_9(x)/256 + \text{(terms involving lower } odd \text{ order Chebyshev polynomials)},$$

$$x^8 = T_8(x)/128 + \text{(terms involving lower } even \text{ order Chebyshev polynomials)},$$

find how many terms of the above Taylor polynomial can be "economized" so that the maximum error remains less than 10^{-6}.

Section 5.10

5.36 By making the appropriate change of variable in (5.64), obtain the cubic polynomial associated with the sub-interval $[x_{i-1}, x_i]$ which is part of the piecewise cubic Hermite approximation.

Chapter 6

Numerical Differentiation and Integration

6.1 Numerical differentiation

We will consider methods for approximating to f', given the values of $f(x)$ at certain points. If p is some polynomial approximation to f, we might use p' as an approximation to f'. However, we need to be careful: the maximum modulus of $f'(x) - p'(x)$ on a given interval $[a, b]$ can be much larger than the maximum modulus of $f(x) - p(x)$, as the following example shows.

Example 6.1 Suppose that $f(x) - p(x) = 10^{-2}T_n(x)$. Putting $x = \cos\theta$, we find as in (4.65) that

$$T_n'(x) = n\frac{\sin n\theta}{\sin\theta} = n^2\left(\frac{\sin n\theta}{n\theta}\right)\left(\frac{\theta}{\sin\theta}\right).$$

Since (see §2.2)

$$\lim_{\theta \to 0} \frac{\sin\theta}{\theta} = 1,$$

we find that $T_n'(1) = n^2$. It can be shown that this is the maximum modulus of T_n' on $[-1, 1]$. If $n = 10$, say, the maximum modulus of $f - p$ is 10^{-2} on $[-1, 1]$, whereas that of $f' - p'$ is unity. $\square$

As an approximation to f, we now take the polynomial p_n which interpolates f at distinct points $x_0, \ldots, x_n$. The error formula (4.16) is

$$f(x) - p_n(x) = \pi_{n+1}(x)\frac{f^{(n+1)}(\xi_x)}{(n+1)!}, \tag{6.1}$$

where

$$\pi_{n+1}(x) = (x - x_0)\cdots(x - x_n). \tag{6.2}$$

This is valid (see Theorem 4.2) if $f^{(n+1)}$ exists on some interval $[a, b]$ which contains $x, x_0, \ldots, x_n$. The number ξ_x (depending on x) belongs to (a, b). Differentiating (6.1), we obtain

$$f'(x) - p_n'(x) = \pi_{n+1}'(x)\frac{f^{(n+1)}(\xi_x)}{(n+1)!} + \frac{\pi_{n+1}(x)}{(n+1)!}\frac{d}{dx}f^{(n+1)}(\xi_x). \tag{6.3}$$

In general, we can say nothing further about the second term on the right of (6.3). We cannot perform the differentiation *with respect to x* of $f^{(n+1)}(\xi_x)$, since ξ_x is an unknown function of x. Thus, for general values of x, (6.3) is useless for determining the accuracy with which p_n' approximates to f'. However, if we restrict x to one of the values $x_0, \ldots, x_n$, then $\pi_{n+1}(x) = 0$ and the unknown second term on the right of (6.3) becomes zero. Putting $x = x_r$, say, we have

$$f'(x_r) - p_n'(x_r) = \pi_{n+1}'(x_r)\frac{f^{(n+1)}(\xi_r)}{(n+1)!}, \tag{6.4}$$

where we have written ξ_r for the value of ξ_x when $x = x_r$. For equally spaced x_r, we write p_n in the forward difference form (4.40):

$$p_n(x) = p_n(x_0 + sh) = f_0 + \binom{s}{1}\Delta f_0 + \binom{s}{2}\Delta^2 f_0 + \cdots + \binom{s}{n}\Delta^n f_0. \tag{6.5}$$

From $x = x_0 + sh$, we have $dx/ds = h$ and therefore

$$p_n'(x) = \frac{ds}{dx}\cdot\frac{d}{ds}p_n(x_0 + sh) = \frac{1}{h}\frac{d}{ds}p_n(x_0 + sh).$$

Thus

$$p_n'(x) = \frac{1}{h}\left[\Delta f_0 + \tfrac{1}{2}(2s - 1)\Delta^2 f_0 + \cdots + \frac{d}{ds}\binom{s}{n}\cdot\Delta^n f_0\right]. \tag{6.6}$$

To calculate $\pi_{n+1}'(x_r)$, we write

$$\pi_{n+1}(x) = (x - x_r)\cdot\prod_{j \neq r}(x - x_j) \tag{6.7}$$

and obtain π_{n+1}' by differentiating the product on the right of (6.7). We have

$$\pi_{n+1}'(x) = \left(\frac{d}{dx}(x - x_r)\right)\prod_{j \neq r}(x - x_j) + (x - x_r)\cdot\frac{d}{dx}\prod_{j \neq r}(x - x_j).$$

On putting $x = x_r$, the second term becomes zero and we obtain

$$\pi_{n+1}'(x_r) = \prod_{j \neq r}(x_r - x_j) = (-1)^{n-r}h^n r! \, (n - r)!, \tag{6.8}$$

since $x_r - x_j = (r - j)h$. We therefore now write (6.4) as

$$f'(x_r) - p_n'(x_r) = (-1)^{n-r}h^n\frac{r!(n - r)!}{(n + 1)!}f^{(n+1)}(\xi_r). \tag{6.9}$$

We now give some differentiation rules obtained from (6.6), with error estimates provided by (6.9), for different values of n and r.

Putting $n = 1$ in (6.6), we obtain

$$p_1'(x) = \frac{1}{h}\Delta f_0 = \frac{f_1 - f_0}{h}$$

and with $r = 0$ in (6.9) we have

$$f'(x_0) = \frac{f_1 - f_0}{h} - \tfrac{1}{2}hf''(\xi_0), \tag{6.10}$$

where $\xi_0 \in (x_0, x_1)$. If we use $(f_1 - f_0)/h$ as an approximation to $f'(x_0)$, (6.10) provides an estimate for the error.

We now choose $n = 2$ in (6.6) and choose $r = 1$, since we are then obtaining the derivative at x_1, which is the centre of the interpolating points x_0, x_1 and x_2. If $x = x_1$ then $s = 1$ and (6.6) becomes

$$p_2'(x_1) = \frac{1}{h}[\Delta f_0 + \tfrac{1}{2}\Delta^2 f_0] = \frac{f_2 - f_0}{2h}. \tag{6.11}$$

This is an approximation to $f'(x_1)$ with an error (from (6.9) with $n = 2$, $r = 1$)

$$f'(x_1) - \frac{f_2 - f_0}{2h} = -\tfrac{1}{6}h^2 f^{(3)}(\xi_1). \tag{6.12}$$

Both differentiation rules obtained above,

$$f'(x) \simeq \frac{f(x + h) - f(x)}{h} \tag{6.13}$$

and

$$f'(x) \simeq \frac{f(x + h) - f(x - h)}{2h} \tag{6.14}$$

are in common use. The former is the more obvious rule, since $f'(x)$ is defined as the limit of the right side of (6.13), as $h \to 0$. However, the midpoint rule (6.14), which uses symmetrically placed interpolating points, is usually preferred because the error (6.12) is of order h^2. For some applications, it is convenient to use the approximation

$$f'(x) \simeq \frac{f(x) - f(x - h)}{h}. \tag{6.15}$$

(See Problem 6.4.) Note that all three differentiation rules have a simple geometrical interpretation: the right sides of (6.13), (6.14) and (6.15) are the gradients of the chords BC, AC and AB in Fig. 6.1.

Higher derivatives of f are approximated by higher derivatives of the interpolating polynomial p_n. If we differentiate (6.1) k times ($k > 1$), we again face the insoluble problem of finding derivatives of $f^{(n+1)}(\xi_x)$ with respect to x. If $k > 1$ we cannot even estimate $f^{(k)}(x) - p_n^{(k)}(x)$ at tabulated points $x = x_r$. At the end of this section we describe another way of estimating this error. Meanwhile, we write

$$p_n''(x) = \frac{1}{h^2}\frac{d^2}{ds^2}p_n(x_0 + sh)$$

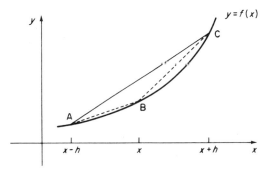

Fig. 6.1. Compare the gradients of the chords BC, AC and AB with $f'(x)$, the gradient of the tangent to $y = f(x)$ at B.

and obtain from (6.6)

$$p_n''(x) = \frac{1}{h^2}\left[\Delta^2 f_0 + \frac{d^2}{ds^2}\binom{s}{3}\cdot\Delta^3 f_0 + \cdots + \frac{d^2}{ds^2}\binom{s}{n}\cdot\Delta^n f_0 \right], \qquad (6.16)$$

where $x = x_0 + sh$. To obtain a non-zero value for $p_n''(x)$, we must take $n \geqslant 2$. Putting $n = 2$ in (6.16), we find that

$$p_2''(x) = \frac{1}{h^2}\Delta^2 f_0,$$

which is a constant. We take this as an approximation to $f''(x)$ at $x = x_1$, the central interpolating point, since we find that at this point the error is of largest order in h. Thus

$$f''(x_1) \simeq \frac{f_2 - 2f_1 + f_0}{h^2}. \qquad (6.17)$$

This is the usual approximation to the second derivative. More generally, for any $k \leqslant n$, we obtain

$$p_n^{(k)}(x) = \frac{1}{h^k}\left[\frac{d^k}{ds^k}\binom{s}{k}\cdot\Delta^k f_0 + \cdots + \frac{d^k}{ds^k}\binom{s}{n}\cdot\Delta^n f_0 \right]. \qquad (6.18)$$

If we choose $n = k$ (for k fixed), there is only one term on the right of (6.18). Since

$$\frac{d^k}{ds^k}[s(s-1)\cdots(s-k+1)] = \frac{d^k}{ds^k}[s^k] = k!,$$

we obtain from (6.18) the approximation (cf. (4.50))

$$f^{(k)}(x) \simeq \frac{\Delta^k f_0}{h^k}. \qquad (6.19)$$

We normally use (6.19) to approximate to $f^{(k)}(x)$ for x lying at the centre of the interval $[x_0, x_k]$. If $k = 2m$, we use the approximation

$$f^{(2m)}(x_m) \simeq \frac{\Delta^{2m} f_0}{h^{2m}}. \tag{6.20}$$

If $k = 2m - 1$, the interval $[x_0, x_{2m-1}]$ has no central tabulated point. In this case, generalizing the procedure adopted for estimating the first derivative, we use (6.18) with $n = k + 1 = 2m$. We find that

$$f^{(2m-1)}(x_m) \simeq \frac{1}{h^{2m-1}} (\Delta^{2m-1} f_0 + \tfrac{1}{2} \Delta^{2m} f_0). \tag{6.21}$$

(See Problem 6.5 for the details.) Since

$$\Delta^{2m} f_0 = \Delta(\Delta^{2m-1} f_0) = \Delta^{2m-1} f_1 - \Delta^{2m-1} f_0,$$

we obtain

$$f^{(2m-1)}(x_m) \simeq \frac{\Delta^{2m-1} f_0 + \Delta^{2m-1} f_1}{2h^{2m-1}}, \tag{6.22}$$

which generalizes the result for f' ($m = 1$) given by (6.14). From (6.22) and (6.20), we obtain the following approximations for the third and fourth derivatives:

$$f^{(3)}(x_2) \simeq \frac{1}{2h^3} (f_4 - 2f_3 + 2f_1 - f_0),$$

$$f^{(4)}(x_2) \simeq \frac{1}{h^4} (f_4 - 4f_3 + 6f_2 - 4f_1 + f_0).$$

Obviously we can derive other differentiation rules by retaining more terms of (6.18), that is, choosing a larger value of n for a given value of k. See, for example, Problem 6.6.

We can find error estimates for all the above rules by the use of Taylor series. As an example, we find an error estimate for the rule (6.17) for approximating to $f''(x_1)$. We assume that $f^{(4)}$ is continuous on $[x_0, x_2]$. Writing $x_0 = x_1 - h$ and $x_2 = x_1 + h$, we have

$$f(x_0) = f_1 - hf_1' + \frac{h^2}{2!} f_1'' - \frac{h^3}{3!} f_1^{(3)} + \frac{h^4}{4!} f^{(4)}(\xi_0), \tag{6.23}$$

$$f(x_2) = f_1 + hf_1' + \frac{h^2}{2!} f_1'' + \frac{h^3}{3!} f_1^{(3)} + \frac{h^4}{4!} f^{(4)}(\xi_2), \tag{6.24}$$

where $\xi_0 \in (x_0, x_1)$ and $\xi_2 \in (x_1, x_2)$. By Theorem 2.3, there exists a number $\xi \in (x_0, x_2)$ such that

$$f^{(4)}(\xi_0) + f^{(4)}(\xi_2) = 2f^{(4)}(\xi).$$

Adding (6.23) and (6.24), we find that

$$f_0 + f_2 = 2f_1 + h^2 f_1'' + \tfrac{1}{12}h^4 f^{(4)}(\xi).$$

This gives, as an error estimate for (6.17),

$$f''(x_1) = \frac{f_2 - 2f_1 + f_0}{h^2} - \frac{h^2}{12}f^{(4)}(\xi), \tag{6.25}$$

where $\xi \in (x_0, x_2)$.

6.2 Effect of errors

All numerical differentiation formulae are sensitive to rounding errors in the function values f_i. To illustrate this, we consider the relation for f' in (6.12),

$$f'(x_1) = \frac{f_2 - f_0}{2h} + E_T, \tag{6.26}$$

where

$$E_T = -\tfrac{1}{6}h^2 f^{(3)}(\xi_1) \tag{6.27}$$

denotes the truncation error. Usually we do not have the exact values f_0 and f_2, but have approximations f_0^* and f_2^*. If

$$|f_i - f_i^*| \le \tfrac{1}{2}.10^{-k}, \qquad i = 0, 2, \tag{6.28}$$

then

$$f'(x_1) = (f_2^* - f_0^*)/2h + E_R + E_T, \tag{6.29}$$

where

$$E_R = \frac{f_2 - f_0}{2h} - \frac{f_2^* - f_0^*}{2h}$$

denotes the rounding error. From (6.28) we obtain

$$|E_R| \le \frac{1}{2h}(|f_0 - f_0^*| + |f_2 - f_2^*|) \le \frac{1}{2h}.10^{-k}. \tag{6.30}$$

If $|f^{(3)}(x)| \le M_3$ for $x \in [x_0, x_2]$, we deduce from (6.29), (6.27) and (6.30) that the *total error* in evaluating f', due to truncation and rounding, is not greater than

$$|E_R| + |E_T| \le \frac{1}{2h}.10^{-k} + \frac{1}{6}h^2 M_3.$$

We write

$$E(h) = \frac{1}{2h} \cdot 10^{-k} + \frac{1}{6} h^2 M_3. \tag{6.31}$$

As h decreases, the first term on the right of (6.31) increases and the second term decreases. To find a value of h which minimizes $E(h)$, we calculate

$$E'(h) = -\frac{1}{2h^2} \cdot 10^{-k} + \frac{1}{3} h M_3.$$

We see that $E'(h) = 0$ if $h^3 = 3.10^{-k}/(2M_3)$. Thus $E(h)$ is a minimum when

$$h = \left(\frac{3}{2M_3 . 10^k} \right)^{1/3} = h_{\min}, \tag{6.32}$$

say. Substituting $h = h_{\min}$ into (6.31), we find that

$$E(h_{\min}) = \frac{1}{2} \left(\frac{3}{2 . 10^k} \right)^{2/3} M_3^{1/3}. \tag{6.33}$$

For example, if the data is accurate to six decimal places and M_3 is approximately unity, we expect from (6.33) that the differentiation rule will be accurate to about four decimal places.

Differentiation rules based on higher order interpolation are even more sensitive to rounding errors. If a formula is of the form

$$\frac{1}{h} \sum_{i=0}^{m} \alpha_i f_i,$$

the maximum error due to rounding is $\sum |\alpha_i| . 10^{-k}/(2h)$ and $\sum |\alpha_i|$ becomes large as we increase the order of the interpolation. For example, the rule in Problem 6.6 has a truncation error $O(h^4)$ and the maximum error due to rounding is $3.10^{-k}/(4h)$, which is larger than (6.30).

Example 6.2   To illustrate the effect of different choices of h on the accuracy of the differentiation rule (6.26), we list the moduli of the errors of approximating to the derivative of e^x at $x = 1$, assuming that values of e^x correct to four decimal places are used. In this case, $k = 4$, M_3 is approximately e and, from (6.32), $h_{\min} \simeq 0.04$. The results in Table 6.1 support this choice of h. □

Table 6.1   Accuracy of approximations to the derivative of e^x at $x = 1$, using (6.26) with different choices of h.

h	0.002	0.004	0.01	0.02	0.04	0.1	0.2	0.4
Modulus of error	0.0183	0.0067	0.0017	0.0008	0.0005	0.0047	0.0182	0.0731

We now turn to the truncation error. There is a simple procedure which usually improves the estimate $(f_2 - f_0)/2h$ for $f'(x_1)$. First we require a refined version of the truncation error. If we extend the Taylor series in (6.23) and (6.24) by one further term and subtract, we obtain

$$f_2 - f_0 = 2hf'(x_1) + \frac{h^3}{3}f^{(3)}(x_1) + \frac{h^5}{120}(f^{(5)}(\xi_0') + f^{(5)}(\xi_2')). \quad (6.34)$$

If $f^{(5)}$ is continuous on $[x_0, x_2]$, then by Theorem 2.3, the coefficients of h^5 may be simplified to $f^{(5)}(\xi)/60$, for some $\xi \in (x_0, x_2)$. Thus (6.34) becomes

$$f'(x_1) = \frac{(f_2 - f_0)}{2h} - \frac{h^2}{6}f^{(3)}(x_1) - \frac{h^4}{120}f^{(5)}(\xi). \quad (6.35)$$

To emphasize the dependence on h, we rewrite (6.35) as

$$f'(x_1) = G(x_1; h) - \frac{h^2}{6}f^{(3)}(x_1) - \frac{h^4}{120}f^{(5)}(\xi_h), \quad (6.36)$$

where

$$G(x; h) = \frac{f(x + h) - f(x - h)}{2h}. \quad (6.37)$$

Replacing h by $h/2$ in (6.36), we obtain

$$f'(x_1) = G(x_1; h/2) - \frac{h^2}{24}f^{(3)}(x_1) - \frac{h^4}{1920}f^{(5)}(\xi_{h/2}). \quad (6.38)$$

The elimination of the term in h^2 between (6.36) and (6.38) yields

$$f'(x_1) = \frac{4G(x_1; h/2) - G(x_1; h)}{3} + \frac{h^4}{1440}(4f^{(5)}(\xi_h) - f^{(5)}(\xi_{h/2}). \quad (6.39)$$

When considered as an approximation to $f'(x_1)$, the first term on the right of (6.39) is of higher order accuracy than either $G(x; h)$ or $G(x; h/2)$. This is an example of a process called *extrapolation to the limit*.

More generally, suppose that some quantity $g(x)$ is approximated by $G(x; h)$, where h is a step size which may be chosen arbitrarily. Thus above, $g(x) = f'(x)$ and $G(x; h)$ is given by (6.37). For any given x, we suppose that $G(x; h) \to g(x)$ as $h \to 0$. Suppose also that, for h sufficiently small, the error between g and G can be expressed as

$$g(x) = G(x; h) + h^2 E_2(x) + h^4 E_4(x) + h^6 E_6(x) + \cdots, \quad (6.40)$$

where $E_2, E_4, \ldots$ are *independent* of h. It is perhaps not obvious in the foregoing example, where we approximate to f', that we do not need to know the functions $E_2, E_4, \ldots$ which appear in (6.40). For, on replacing h by $h/2$ in (6.40) and eliminating the term in h^2, we obtain

$$g(x) = G^{(1)}(x; h) + h^4 E_4^{(1)}(x) + h^6 E_6^{(1)}(x) + \cdots, \quad (6.41)$$

where

$$G^{(1)}(x; h) = \frac{4G(x; h/2) - G(x; h)}{3} \tag{6.42}$$

and $E_4^{(1)}$, $E_6^{(1)}$ are multiples of E_4, E_6 which do not concern us. (Note that the superfix 1 in (6.41) does not denote differentiation.) Replacing h by $h/2$ in (6.41) and eliminating the term in h^4, we obtain

$$g(x) = G^{(2)}(x; h) + h^6 E_6^{(2)}(x) + \cdots,$$

where

$$G^{(2)}(x; h) = \frac{16G^{(1)}(x; h/2) - G^{(1)}(x; h)}{15} \tag{6.43}$$

and $E_6^{(2)}$ is a multiple of $E_6^{(1)}$ (and therefore of E_6). Note that $G^{(2)}(x; h)$ involves $G^{(1)}(x; h/2)$ which, from (6.42), is seen to involve $G(x; h/4)$. Thus, to calculate $G^{(2)}(x; h)$, which approximates to $g(x)$ with order h^6 accuracy, we require $G(x; h)$, $G(x; h/2)$ and $G(x; h/4)$. In general, we may compute

$$G^{(m)}(x; h) = \frac{4^m G^{(m-1)}(x; h/2) - G^{(m-1)}(x; h)}{4^m - 1}. \tag{6.44}$$

If $G^{(0)}$ is identified with G, (6.44) holds for $m \geqslant 1$. By induction, we see that $G^{(m)}(x; h)$ requires the evaluation of $G(x; h)$, $G(x; h/2)$, ..., $G(x; h/2^m)$ and approximates to $g(x)$ with an error of order $h^{2(m+1)}$. The calculation of the $G^{(m)}$ is shown schematically in Table 6.2. Each entry in columns 2, 3 and 4 is

Table 6.2   Repeated extrapolation to the limit.

$G(x; h)$			
	$G^{(1)}(x; h)$		
$G(x; h/2)$		$G^{(2)}(x; h)$	
	$G^{(1)}(x; h/2)$		$G^{(3)}(x; h)$
$G(x; h/4)$		$G^{(2)}(x; h/2)$	
	$G^{(1)}(x; h/4)$		
$G(x; h/8)$			

computed from the two nearest entries in the column to the left, using (6.44). The evaluation of $G^{(m)}$ for $m > 1$ is called *repeated* extrapolation to the limit.

Example 6.3   The numbers in the first column of Table 6.3 are the results of applying the differentiation rule $(f_2 - f_0)/2h$ for e^x at $x = 1$ with $h = 0.8$, 0.4 and 0.2. The remaining numbers of Table 6.3 are the result of repeated extrapolation to the limit. The last number, 2.7183, is the correct derivative to four decimal places. □

It is essential that there is a power series in h of the form (6.40) if repeated extrapolation to the limit is to be used. For numerical differentiation using

Table 6.3 Extrapolation to the limit (Example 6.3).

3.0176		
	2.7160	
2.7914		2.7183
	2.7182	
2.7365		

(6.14), it can be shown that (6.40) exists provided all derivatives of f exist. The effects of using extrapolation to the limit when no such power series in h exists will be illustrated in §6.4 for numerical integration.

6.3 Numerical integration

We now consider the inverse problem of differentiation, namely that of integration. Given a function f defined on a finite interval $[a, b]$, we wish to evaluate the definite integral

$$\int_a^b f(x)\, dx, \tag{6.45}$$

assuming that f is integrable. (See §2.4.) If $f(x) \geqslant 0$ on $[a, b]$ the integral (6.45) has the value of the area bounded by the curve $y = f(x)$, the x-axis and the ordinates $x = a$ and $x = b$. If we can find a function F such that $F' = f$, then (Theorem 2.11) we can evaluate the integral using the relation

$$\int_a^b f(x)\, dx = F(b) - F(a).$$

Sometimes considerable skill is required to obtain F, perhaps by making a change of variable or integrating by parts. Even if F can be found, it may still be more convenient to use a numerical method to estimate (6.45) if the evaluation of F requires a great deal of computation. If we cannot find F or if f is known only for certain values of x, we use a numerical method for evaluating (6.45).

An obvious approach is to replace the integrand f in (6.45) by an approximating polynomial and integrate the polynomial. We discard the Taylor polynomial, since it requires the evaluation of derivatives of f. First we use an interpolating polynomial constructed at equally spaced points $x_r = x_0 + rh$, $0 \leqslant r \leqslant n$. If $f^{(n+1)}$ is continuous on $[x_0, x_n]$, we have from (4.41) that

$$f(x_0 + sh) = f_0 + \binom{s}{1} \Delta f_0 + \cdots + \binom{s}{n} \Delta^n f_0 + h^{n+1}\binom{s}{n+1} f^{(n+1)}(\xi_s),$$

(6.46)

where $\xi_s \in (x_0, x_n)$. (We need to replace $[x_0, x_n]$ by an appropriate larger interval if s is outside the interval $0 \leqslant s \leqslant n$.) We can construct different integration rules by choosing different values of n in (6.46). Thus, with $n = 0$,

$$f(x_0 + sh) = f_0 + h\binom{s}{1} f'(\xi_s).$$

Integrating $f(x)$ over $[x_0, x_1]$, that is, integrating with respect to s over $[0, 1]$, we obtain

$$\int_{x_0}^{x_1} f(x)\, dx = hf_0 + h^2 \int_0^1 sf'(\xi_s)\, ds.$$

(6.47)

We replaced dx by $h\,.ds$, as $x = x_0 + sh$. In the integrand on the right of (6.47), s does not change sign on $[0, 1]$ and $f'(\xi_s)$ is a continuous function of s. Thus by the mean value theorem for integrals, Theorem 2.12, there is a number $s = \bar{s}$ with $\xi_{\bar{s}} \in (x_0, x_1)$ such that

$$\int_{x_0}^{x_1} f(x)\, dx = hf_0 + h^2 f'(\xi_{\bar{s}}) \int_0^1 s\, ds.$$

Writing ξ in place of $\xi_{\bar{s}}$, we have

$$\int_{x_0}^{x_1} f(x)\, dx = hf_0 + \tfrac{1}{2}h^2 f'(\xi).$$

(6.48)

The first term on the right of (6.48) gives the integration rule (also called a *quadrature* rule); the second term is the error term. This is called the rectangular quadrature rule, since the integral (the area under the curve $y = f(x)$ in Fig. 6.2) is approximated by the rectangle of width $h = x_1 - x_0$ and height f_0.

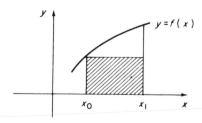

Fig. 6.2. The rectangular quadrature rule.

Putting $n = 1$ and integrating $f(x)$ over $[x_0, x_1]$, we similarly obtain

$$\int_{x_0}^{x_1} f(x)\, dx = hf_0 + \tfrac{1}{2}h\,\Delta f_0 + h^3 f''(\xi) \int_0^1 \binom{s}{2}\, ds,$$

where f'' is assumed continuous on $[x_0, x_1]$. We have again been able to apply Theorem 2.12, since $\binom{s}{2}$ does not change sign on $[0, 1]$. This gives

$$\int_{x_0}^{x_1} f(x)\, dx = \frac{h}{2}(f_0 + f_1) - \frac{h^3}{12} f''(\xi), \qquad (6.49)$$

where $\xi \in (x_0, x_1)$ and is usually distinct from the ξ appearing in (6.48). This is the *trapezoidal* rule plus error term. The integral is approximated by $\tfrac{1}{2}h(f_0 + f_1)$, the area of the trapezium which appears shaded in Fig. 6.3.

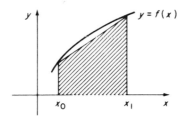

Fig. 6.3. The trapezoidal rule.

The trapezoidal rule is usually applied in a *composite form*. To estimate the integral of f over $[a, b]$, we divide $[a, b]$ into N sub-intervals of equal length $h = (b - a)/N$. The end points of the sub-intervals are $x_i = a + ih$, $i = 0$, $1, \ldots, N$, so that $x_0 = a$ and $x_N = b$. We now apply the trapezoidal rule to each sub-interval $[x_{i-1}, x_i]$, $i = 1, 2, \ldots, N$. (See Fig. 6.4). Thus we have from (6.49), on distinguishing the numbers ξ occurring in the error terms,

$$\int_{x_0}^{x_N} f(x)\, dx = \frac{h}{2}(f_0 + f_1) - \frac{h^3}{12} f''(\xi_1) + \cdots + \frac{h}{2}(f_{N-1} + f_N) - \frac{h^3}{12} f''(\xi_N).$$

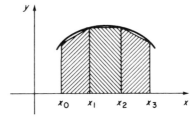

Fig. 6.4. Composite form of the trapezoidal rule.

Maintaining the assumption that f'' is continuous on $[a, b]$, we can combine the error terms (see Theorem 2.3) to give

$$\int_a^b f(x)\, dx = \frac{h}{2}(f_0 + 2f_1 + 2f_2 + \cdots + 2f_{N-1} + f_N) - \frac{Nh^3}{12} f''(\xi), \quad (6.50)$$

for some $\xi \in (a, b)$. This may be rewritten as

$$\int_a^b f(x)\, dx = h(f_0 + f_1 + \cdots + f_{N-1}) + \frac{h}{2}(f_N - f_0) - (x_N - x_0)\frac{h^2}{12} f''(\xi), \quad (6.51)$$

since $x_N - x_0 = Nh$. Incidentally, we note from (6.51) that the composite trapezoidal rule consists of the composite rectangular rule plus the *correction term* $h(f_N - f_0)/2$. We note that the error in (6.51) is $O(h^2)$ or, equivalently, we can say that the rule is exact for every polynomial belonging to P_1. (The second derivative in the error term is zero for such polynomials.)

Now suppose that, due to rounding, the f_i are in error by at most $\frac{1}{2}.10^{-k}$. Then we see from (6.50) that the error in the trapezoidal rule due to rounding is not greater than

$$\frac{h}{2}(1 + 2 + 2 + \cdots + 2 + 1).\frac{1}{2}.10^{-k} = Nh.\frac{1}{2}.10^{-k} = (b - a).\frac{1}{2}.10^{-k}.$$

Thus, rounding errors do not seriously affect the accuracy of the quadrature rule. This is generally true of numerical integration, unlike numerical differentiation. (See Problem 6.26.)

There is a generalization of the composite trapezoidal rule called Gregory's formula:

$$\int_{x_0}^{x_N} f(x)\, dx \simeq h(f_0 + \cdots + f_N) + h \sum_{j=0}^{m} c_{j+1}(\nabla^j f_N + (-1)^j \Delta^j f_0), \quad (6.52)$$

where

$$c_j = (-1)^j \int_{-1}^0 \binom{-s}{j} ds. \quad (6.53)$$

If $m \le N$, the rule uses values $f(x_i)$ with every $x_i \in [x_0, x_N]$. With $m = 0$, since $c_1 = -\frac{1}{2}$, (6.52) is simply the trapezoidal rule. (The next few values of the c_j are $-\frac{1}{12}$, $-\frac{1}{24}$, $-\frac{19}{720}$. We see from (6.53) that $c_j < 0$ for all j.) We state without proof that if $m = 2k$ or $m = 2k + 1$, then the rule (6.52) has error $O(h^{2k+1})$ and integrates exactly every polynomial belonging to P_{2k+1}. Thus it is desirable to use (6.52) with an even value of m. For example, if $m = 2$ and $N \ge 2$, we have the integration rule

$$\int_{x_0}^{x_N} f(x)\, dx \simeq h(f_0 + \cdots + f_N) - \frac{5h}{8}(f_0 + f_N)$$

$$+ \frac{h}{6}(f_1 + f_{N-1}) - \frac{h}{24}(f_2 + f_{N-2}), \quad (6.54)$$

which is exact if $f \in P_3$. For the derivation of Gregory's formula, see Ralston.

Returning to (6.46), we now choose $n = 3$ and integrate $f(x)$ over $[x_0, x_2]$. This time the error term contains the binomial coefficient $\binom{s}{4}$ which *does* change sign over $[x_0, x_2]$. However, it can be shown that, in this case, the error term can still be written in the form

$$h^5 f^{(4)}(\xi) \int_0^2 \binom{s}{4} ds = -\frac{h^5}{90} f^{(4)}(\xi),$$

as if Theorem 2.12 were applicable. (See Isaacson and Keller.) We obtain

$$\int_{x_0}^{x_2} f(x) \, dx = \frac{h}{3} (f_0 + 4f_1 + f_2) - \frac{h^5}{90} f^{(4)}(\xi), \tag{6.55}$$

where $\xi \in (x_0, x_2)$. This is called Simpson's rule. Like the trapezoidal rule, Simpson's rule is usually applied in composite form. Combining the error terms as we did for the trapezoidal rule, we obtain

$$\int_{x_0}^{x_{2N}} f(x) \, dx = \frac{h}{3} (f_0 + 4f_1 + 2f_2 + 4f_3 + \cdots + 2f_{2N-2} + 4f_{2N-1} + f_{2N})$$

$$- (x_{2N} - x_0) \frac{h^4}{180} f^{(4)}(\xi), \tag{6.56}$$

where $\xi \in (x_0, x_{2N})$. Note that the composite form of Simpson's rule requires an even number of sub-intervals. For a rule of comparable accuracy which permits an odd number of sub-intervals, we have the Gregory formula (6.54). Note that both (6.54) and (6.56) are exact if the integrand $f \in P_3$.

The basic trapezoidal and Simpson rules (6.49) and (6.55) are special cases of the form

$$\int_{x_0}^{x_n} f(x) \, dx \simeq \sum_{i=0}^{n} w_i f(x_0 + ih) \tag{6.57}$$

where the w_i are chosen so that the rule is exact if $f \in P_n$. These are called the closed Newton–Cotes formulae. For $n = 3$ we have $w_0 = w_3 = 3h/8$ and $w_1 = w_2 = 9h/8$. For *even* values of n the rules turn out to be exact for $f \in P_{n+1}$ (as for Simpson's rule, with $n = 2$). In any integration rule, such as (6.57), the numbers w_i are called the *weights* since the rule consists of a weighted sum of certain function values. There are also *open* Newton–Cotes formulae, which are of the form

$$\int_{x_0}^{x_n} f(x) \, dx \simeq \sum_{i=1}^{n-1} w_i f(x_0 + ih).$$

The terms "closed" and "open" refer to whether the end-points x_0 and x_n are or are not included in the rule.

To obtain the simplest open Newton–Cotes rule, we integrate (6.46), with $n = 1$, over $[x_0 - h, x_0 + h]$. This gives

$$\int_{x_0 - h}^{x_0 + h} f(x)\, dx = 2hf_0 + h^3 \int_{-1}^{1} \binom{s}{2} f''(\xi_s)\, ds.$$

As in the derivation of Simpson's rule, the error term here is troublesome, since $\binom{s}{2}$ changes sign on $[-1, 1]$ and Theorem 2.12 is inapplicable. However, again it is possible to write the error in the form

$$h^3 f''(\xi) \int_{-1}^{1} \binom{s}{2}\, ds = \tfrac{1}{3} h^3 f''(\xi),$$

as we will see shortly. Applying this rule to f on the interval $[x_0, x_1]$, we obtain

$$\int_{x_0}^{x_1} f(x)\, dx = hf(x_0 + \tfrac{1}{2}h) + \frac{h^3}{24} f''(\xi), \qquad (6.58)$$

where $\xi \in (x_0, x_1)$. This is called the *mid-point* rule. At first sight this rule appears to be more economical than the trapezoidal rule on $[x_0, x_1]$, since both have accuracy of order h^3 but (6.58) requires only one evaluation of f instead of two evaluations for the trapezoidal rule. This is misleading, since the composite form of (6.58) is

$$\int_{x_0}^{x_N} f(x)\, dx = h \sum_{r=0}^{N-1} f(x_r + \tfrac{1}{2}h) + (x_N - x_0) \frac{h^2}{24} f''(\xi), \qquad (6.59)$$

where $\xi \in (x_0, x_N)$. This requires N evaluations of f, which is only one fewer than the $N + 1$ evaluations required by the composite trapezoidal rule (6.51). Note the opposite signs taken by the error terms in (6.51) and (6.59). If f'' has constant sign on $[x_0, x_N]$, the value of the integral must therefore lie *between* the estimates provided by the mid-point and trapezoidal rules.

Let T_N and M_N denote† the approximations to $\int_a^b f(x)\, dx$ obtained by using respectively the trapezoidal and mid-point rules with N sub-intervals. If f'' is constant, we can eliminate the error terms between (6.51) and (6.59) and obtain the quadrature rule

$$\int_a^b f(x)\, dx \simeq (2M_N + T_N)/3. \qquad (6.60)$$

From (6.51) and (6.59), we see that this rule involves values of $f(x)$ at the $2N + 1$ equally spaced points $x = x_0 + \tfrac{1}{2}rh$, $r = 0, 1, \ldots, 2N$. Indeed,

$$(2M_N + T_N)/3 = S_{2N} \qquad (6.61)$$

---

† No confusion should arise from this use of T_N, which we use elsewhere to denote the Chebyshev polynomial.

where S_{2N} is the result of applying Simpson's rule with $2N$ sub-intervals on $[a, b]$. Thus the elimination of the h^2 error term in this way gives Simpson's rule.

Example 6.4 In Table 6.4 are listed the results of approximating to $\int_0^1 (1 + x)^{-1} dx$ by the trapezoidal, mid-point and Simpson rules, for different numbers (N) of sub-intervals. As the second derivative of $(1 + x)^{-1}$ is positive on [0, 1], the numbers in columns 1 and 2 of the table provide bounds for the integral, whose value is $\ln 2 = 0.6931$, to 4 decimal places. □

Table 6.4   Approximations to $\int_0^1 (1 + x)^{-1} dx$.

N	Trapezoidal rule	Mid-point rule	Simpson's rule
1	0.7500	0.6667	—
2	0.7083	0.6857	0.6944
3	0.7000	0.6898	—
4	0.6970	0.6912	0.6933

We found in §6.1 that truncation errors for differentiation rules could be estimated by using Taylor series. The same is true of integration rules. For example, if $F'(x) = f(x)$ and we write $u = x_0 + \frac{1}{2}h$, we have

$$\int_{x_0}^{x_1} f(x) \, dx = F(u + \tfrac{1}{2}h) - F(u - \tfrac{1}{2}h)$$
$$= F(u) + \tfrac{1}{2}hF'(u) + \tfrac{1}{8}h^2 F''(u) + \tfrac{1}{48}h^3 F^{(3)}(\xi_1)$$
$$- [F(u) - \tfrac{1}{2}hF'(u) + \tfrac{1}{8}h^2 F''(u) - \tfrac{1}{48}h^3 F^{(3)}(\xi_0)], \quad (6.62)$$

where ξ_0 and $\xi_1 \in (x_0, x_1)$. Using Theorem 2.3 to combine the two error terms, we see that (6.62) yields the mid-point rule plus error, as given by (6.58).

Each quadrature rule studied so far consists of a weighted sum of values $f(x_i)$, where the abscissae x_i are *within* the range of integration. It is not always possible to use such rules. For example, suppose we wish to integrate a function f over $[x_n, x_{n+1}]$, where $x_n = x_0 + nh$, and the only known values of f are $f(x_i)$, $0 \leqslant i \leqslant n$. This type of problem arises in the numerical solution of ordinary differential equations (see Chapter 11) and it is convenient to use the backward difference form of the interpolating polynomial, constructed at $x = x_n, x_{n-1}, \ldots, x_{n-m}$, where $m \leqslant n$. From (4.16) and (4.43), we have for any $s > 0$

$$f(x_n + sh) = \sum_{j=0}^{m} (-1)^j \binom{-s}{j} \nabla^j f_n + (-1)^{m+1} h^{m+1} \binom{-s}{m+1} f^{(m+1)}(\xi_s).$$
$$(6.63)$$

In (6.63), $\xi_s \in (x_{n-m}, x_n + sh)$ and $f^{(m+1)}$ is assumed continuous on this interval. We now integrate (6.63) over the interval $x_n \leqslant x \leqslant x_{n+1}$. Because of the change of variable $x = x_n + sh$, we must replace dx by $h \cdot ds$ to give

$$\int_{x_n}^{x_{n+1}} f(x)\, dx = h \sum_{j=0}^{m} (-1)^j \nabla^j f_n \int_0^1 \binom{-s}{j} ds$$

$$+ (-1)^{m+1} h^{m+2} \int_0^1 \binom{-s}{m+1} f^{(m+1)}(\xi_s)\, ds. \qquad (6.64)$$

Again, Theorem 2.12 is applicable and we obtain

$$\int_{x_n}^{x_{n+1}} f(x)\, dx = h \sum_{j=0}^{m} b_j \nabla^j f_n + h^{m+2} b_{m+1} f^{(m+1)}(\xi_n), \qquad (6.65)$$

where $\xi_n \in (x_{n-m}, x_{n+1})$ and

$$b_j = (-1)^j \int_0^1 \binom{-s}{j} ds.$$

It is sometimes appropriate to make use of $f(x_{n+1})$ also and construct the interpolating polynomial at $x = x_{n+1}, x_n, \ldots, x_{n-m}$. We have

$$f(x_{n+1} + sh) = \sum_{j=0}^{m+1} (-1)^j \binom{-s}{j} \nabla^j f_{n+1} + (-1)^{m+2} h^{m+2} \binom{-s}{m+2} f^{(m+2)}(\xi'_s),$$

$$\qquad (6.66)$$

which is just (6.63) with n and m increased by 1. In this case, $x_n \leqslant x \leqslant x_{n+1}$ corresponds to $-1 \leqslant s \leqslant 0$. We apply Theorem 2.12 and obtain

$$\int_{x_n}^{x_{n+1}} f(x)\, dx = h \sum_{j=0}^{m+1} c_j \nabla^j f_{n+1} + h^{m+3} c_{m+2} f^{(m+2)}(\xi'_n), \qquad (6.67)$$

where $\xi'_n \in (x_{n-m}, x_{n+1})$ and the c_j (see (6.53)) are the same numbers as occur in Gregory's formula.

6.4 Romberg integration

The elimination of error terms in numerical quadrature can be achieved by extrapolation to the limit, which we discussed in §6.2. If this process is to be valid for, say, the trapezoidal rule, we need to be able to express the error as a series of even powers of h. Fortunately, there is such a series if f is sufficiently differentiable. We have the Euler–Maclaurin formula,

$$\int_{x_0}^{x_N} f(x)\, dx - T_N = h^2 E_2 + h^4 E_4 + h^6 E_6 + \cdots, \qquad (6.68)$$

where

$$E_{2r} = -\frac{B_{2r}}{(2r)!} (f^{(2r-1)}(x_N) - f^{(2r-1)}(x_0))$$

and the coefficients B_{2r} are the Bernouilli numbers, defined by

$$\frac{x}{e^x - 1} = \sum_{r=0}^{\infty} B_r x^r / r!.$$

See Ralston for the derivation of (6.68). (Note the similarity between (6.68) and Gregory's formula (6.52). The first correction term in the latter yields the trapezoidal rule and the remaining terms consist of finite *differences* in place of the *derivatives* in (6.68).) The coefficients E_{2r} in (6.68) are independent of h, as required for the extrapolation process. Repeated extrapolation to the limit, when applied to the trapezoidal rule, is called Romberg integration. It is easily verified that the result of the first extrapolation to the limit is Simpson's rule, that is,

$$S_{2N} = (4T_{2N} - T_N)/3. \tag{6.69}$$

Example 6.5 In Table 6.5 are given the results of applying Romberg integration to the integral

$$\int_0^1 \frac{dx}{1 + x}.$$

The first column gives T_1, T_2, T_4 and T_8 and the remaining columns are set out as in Table 6.2. For comparison, the result to six decimal places is 0.693147. Simpson's rule with 8 sub-intervals gives 0.693155, as in the second

Table 6.5 Results of Romberg integration with $N = 1$, 2, 4 and 8 sub-intervals for $\int_0^1 (1 + x)^{-1} dx$ (Example 6.5).

0.750000			
	0.694444		
0.708333		0.693175	
	0.693254		0.693148
0.697024		0.693148	
	0.693155		
0.694122			

column of the table, which is consistent with (6.69). Thus Romberg integration has estimated the integral with an error not greater than 1.5×10^{-6}, using only nine function evaluations. From the error term of (6.51), we can estimate how many evaluations are required to achieve such accuracy, using the trapezoidal rule. Since in this case $\frac{1}{4} \leqslant f''(x) \leqslant 2$, for $0 \leqslant x \leqslant 1$, we see that over one hundred and forty evaluations are needed. $\square$

If there is no series of the form (6.68) because one of the derivatives of f does not exist at some points in the range of integration, the results of applying the formulae can be misleading, as the next example illustrates.

Example 6.6 We try repeated extrapolation to the limit for the integral

$$\int_0^{0.8} x^{1/2} \, dx.$$

The integrand is not differentiable at $x = 0$. Table 6.6 shows results taking $N = 1, 2, 4, 8, 16$. The two underlined entries are quite inaccurate, as the result should be 0.47703 to five decimal places. □

Table 6.6   Erroneous Romberg table for $\int_0^{0.8} x^{1/2} \, dx$.

0.35777				
	0.45657			
0.43187		0.47066		
	0.46978		0.47484	
0.46030		0.47477		0.47626
	0.47446		0.47625	
0.47092		0.47623		
	0.47612			
0.47482				

In looking at a table of extrapolated results, we should look for signs of convergence as we progress down each *column*. Looking at just the last entries of the columns is misleading, as we found in Example 6.6.

6.5 Gaussian integration

We now consider quadrature formulae of the form

$$\int_a^b f(x) \, dx \simeq \sum_{i=0}^n w_i f(x_i), \tag{6.70}$$

where the abscissae are not necessarily equally spaced, as hitherto in this chapter. We again replace f by p_n, the interpolating polynomial constructed at $x = x_0, \ldots, x_n$. We write p_n in the Lagrange form (4.9),

$$p_n(x) = L_0(x).f(x_0) + \cdots + L_n(x).f(x_n)$$

and integrate p_n over $[a, b]$ to give the right side of (6.70). We thus obtain the weights,

$$w_i = \int_a^b L_i(x) \, dx, \qquad 0 \leqslant i \leqslant n. \tag{6.71}$$

An explicit expression for $L_i(x)$ is given in (4.13). If we take the x_i as equally spaced on $[a, b]$, we have an alternative derivation of the Newton–Cotes rules discussed in §6.3. We now study the error of the quadrature rule (6.70) for arbitrarily spaced $x_i \in [a, b]$. Integrating the error of the interpolating polynomial (4.16), we obtain

$$\int_a^b f(x)\,dx - \sum_{i=0}^n w_i f(x_i) = \frac{1}{(n+1)!} \int_a^b \pi_{n+1}(x) f^{(n+1)}(\xi_x)\,dx, \quad (6.72)$$

where $f^{(n+1)}$ is assumed to exist on $[a, b]$ and

$$\pi_{n+1}(x) = (x - x_0) \cdots (x - x_n). \quad (6.73)$$

Thus the integration rule is exact if $f \in P_n$, since then $f^{(n+1)}(x) \equiv 0$ and the right side of (6.72) is zero.

We can regard (6.70) in another way. First we note that if the rule (6.70) is exact for functions f and g, it is also exact for any function $\alpha f + \beta g$, where α and β are arbitrary real numbers. For

$$\int_a^b (\alpha f(x) + \beta g(x))\,dx = \alpha \int_a^b f(x)\,dx + \beta \int_a^b g(x)\,dx$$

$$= \sum_{i=0}^n w_i(\alpha f(x_i) + \beta g(x_i)).$$

Hence, the rule (6.70) is exact for $f \in P_n$ if, and only if, it is exact for the monomials $f(x) = 1, x, \ldots, x^n$. If (6.70) is to be exact for $f(x) = x^j$, we require

$$\int_a^b x^j\,dx = \sum_{i=0}^n w_i x_i^j. \quad (6.74)$$

The left side of (6.74) is known. This suggests that, by taking $j = 0, 1, \ldots, 2n + 1$, we set up $2n + 2$ equations to solve for the $2n + 2$ unknowns w_i and x_i, $i = 0, 1, \ldots, n$. If these equations have a solution, the resulting integration rule will be exact for $f \in P_{2n+1}$. We will not pursue the direct solution of these equations, although this is not as difficult as might at first appear. (See Davis and Rabinowitz.)

Instead, we return to (6.72). It is convenient to use the divided difference form of the error $f - p_n$, as in (4.47),

$$f(x) - p_n(x) = \pi_{n+1}(x) f[x, x_0, \ldots, x_n].$$

Thus we have

$$\int_a^b f(x) - \sum_{i=0}^n w_i f(x_i) = \int_a^b \pi_{n+1}(x) f[x, x_0, \ldots, x_n]\,dx. \quad (6.75)$$

We now prove:

Lemma 6.1 If $f[x, x_0, \ldots, x_k]$ is a polynomial (in x) of degree $m > 0$, then $f[x, x_0, \ldots, x_{k+1}]$ is a polynomial of degree $m - 1$.

Proof From (4.36),

$$f[x, x_0, \ldots, x_{k+1}] = \frac{f[x_0, \ldots, x_{k+1}] - f[x, x_0, \ldots, x_k]}{x_{k+1} - x}. \tag{6.76}$$

Now $x - x_{k+1}$ is a factor of the numerator on the right of (6.76), since

$$f[x_0, \ldots, x_{k+1}] - f[x_{k+1}, x_0, \ldots, x_k] = 0.$$

This follows from the fact that the order of the arguments in a divided difference is irrelevant, which is apparent from the symmetric form of divided differences, (4.32). Since the numerator on the right of (6.76) is a polynomial of degree m and $x - x_{k+1}$ is a factor, it follows that $f[x, x_0, \ldots, x_{k+1}]$ is a polynomial of degree $m - 1$. $\square$

Returning to (6.75), if $f \in P_{2n+1}$, an induction argument using Lemma 6.1 shows that $f[x, x_0, \ldots, x_n] \in P_n$. Suppose that $\{p_0, p_1, \ldots, p_{n+1}\}$ is a set of orthogonal polynomials on $[a, b]$, that is,

$$\int_a^b p_r(x)p_s(x)\, dx \begin{cases} = 0, & r \neq s \\ \neq 0, & r = s \end{cases}$$

and $p_r \in P_r$, $r = 0, 1, \ldots, n + 1$. For some set of real numbers $\alpha_0, \alpha_1, \ldots, \alpha_n$,

$$f[x, x_0, \ldots, x_n] = \alpha_0 p_0(x) + \alpha_1 p_1(x) + \cdots + \alpha_n p_n(x)$$

(see Problem 5.15) and, from the orthogonality relation, the right side of (6.75) will be zero if $\pi_{n+1}(x) = \alpha p_{n+1}(x)$ for some real number $\alpha \neq 0$.

On $[-1, 1]$, these orthogonal polynomials p_n are the Legendre polynomials. (See Example 5.10.) Thus, if the x_i are chosen as the zeros of the Legendre polynomial of degree $n + 1$ and the weights w_i are determined from (6.74) with $[-1, 1]$ for $[a, b]$, the integration rule

$$\int_{-1}^1 f(x)\, dx \simeq \sum_{i=0}^n w_i f(x_i) \tag{6.77}$$

is exact for $f \in P_{2n+1}$. These are called Gaussian integration rules, after K. F. Gauss.

The first few Legendre polynomials (see Example 5.10) are $1, x, (3x^2 - 1)/2$, $(5x^3 - 3x)/2$. With $n = 0$ in (6.77) we have a rule with one abscissa, $x = 0$. This gives

$$\int_{-1}^1 f(x)\, dx \simeq 2f(0) \tag{6.78}$$

which is exact for $f \in P_1$. This is the mid-point rule, which we derived earlier as (6.58). With $n = 1$, the abscissae are the zeros of $(3x^2 - 1)/2$, which are

$x = \pm 1/\sqrt{3}$. The weights w_0 and w_1 are easily found from the requirement that (6.77), with $n = 1$, must integrate the monomials 1 and x exactly. Thus we obtain the Gaussian rule

$$\int_{-1}^{1} f(x)\, dx \simeq f(-1/\sqrt{3}) + f(1/\sqrt{3}), \qquad (6.79)$$

which is exact for $f \in P_3$ and so is comparable with Simpson's rule. With $n = 2$, the abscissae are the zeroes of $(5x^3 - 3x)/2$, which are $x = 0, \pm\sqrt{\frac{3}{5}}$. Since the rule (6.77) with $n = 2$ has to integrate 1, x and x^2 exactly, we have:

$$2 = \quad w_0 + w_1 + w_2$$
$$0 = -\sqrt{\tfrac{3}{5}}w_0 + \sqrt{\tfrac{3}{5}}w_2$$
$$\tfrac{2}{3} = \quad \tfrac{3}{5}w_0 + \tfrac{3}{5}w_2.$$

The integration rule is therefore

$$\int_{-1}^{1} f(x)\, dx \simeq \left(5f(-\sqrt{\tfrac{3}{5}}) + 8f(0) + 5f(\sqrt{\tfrac{3}{5}})\right)/9, \qquad (6.80)$$

which is exact for $f \in P_5$.

Gaussian rules are also used in composite forms, similar to the trapezoidal, mid-point and Simpson rules. It can be shown that if $f^{(2n+2)}$ is continuous on $[-1, 1]$, the error of the $(n + 1)$-point Gaussian rule (6.77) is of the form $d_{n+1}f^{(2n+2)}(\xi)$, where $\xi \in (-1, 1)$ and

$$d_n = \frac{2^{2n+1}(n!)^4}{(2n + 1)[(2n)!]^3}, \qquad (6.81)$$

which decreases rapidly as n increases. (See Ralston.) To apply a Gaussian rule to the integral

$$\int_{a}^{a+h} g(t)\, dt, \qquad (6.82)$$

we make the change of variable $t = a + \frac{1}{2}h(x + 1)$, which maps $a \leqslant t \leqslant a + h$ onto $-1 \leqslant x \leqslant 1$. We have

$$g(t) = g\left(a + \tfrac{1}{2}h(x + 1)\right) = f(x),$$

say, and on differentiating k times, we have

$$\frac{d^k}{dx^k} f(x) = \left(\frac{h}{2}\right)^k \frac{d^k}{dt^k} g(t).$$

Thus, the error in using the Gaussian $(n + 1)$-point rule to estimate (6.82) is

$$\left(\frac{h}{2}\right)^{2n+2} d_{n+1} g^{(2n+2)}(\eta), \qquad (6.83)$$

where $\eta \in (a, a + h)$. The high power of $h/2$ in (6.83) makes the error very small for small values of h.

Example 6.7 To illustrate the accuracy of Gaussian rules, we apply the 3-point rule to

$$\int_0^1 \frac{dx}{1 + x}$$

to give the result 0.693122. The same rule applied in composite form on $[0, \frac{1}{2}]$ and $[\frac{1}{2}, 1]$, that is with six function evaluations, gives 0.693146. The last error is not greater than 1.5×10^{-6}, since the correct result is $\ln 2 = 0.693147$ to six decimal places. The accuracy is thus comparable with that achieved by Romberg integration, with nine function evaluations, in Example 6.5. $\square$

There also exist quadrature formulae of the form

$$\int_a^b \omega(x)f(x) \, dx \simeq \sum_{i=0}^n w_i f(x_i), \tag{6.84}$$

where ω is some fixed weight function. Weights w_i and abscissae x_i can be found so that (6.84) is exact for all $f \in P_{2n+1}$. A similar argument to that used for the derivation of the classical Gaussian rules shows that the x_i in (6.84) must be chosen as the zeros of the polynomial of degree $n + 1$ belonging to the sequence of polynomials which are orthogonal on $[a, b]$ with respect to the weight function ω. The resulting formulae (6.84) are all referred to as Gaussian rules. In particular, if $[a, b]$ is taken as $[-1, 1]$ and $\omega(x) = (1 - x^2)^{-1/2}$, the orthogonal polynomials are the Chebyshev polynomials. (The resulting formulae are also called the Chebyshev quadrature rules.) In this case, the interpolating polynomial is

$$p_n(x) = \sum_{i=0}^n L_i(x)f(x_i),$$

where the x_i are the zeros of T_{n+1} and (see Problem 6.23) we may write

$$L_i(x) = \frac{T_{n+1}(x)}{(x - x_i)T'_{n+1}(x_i)}.$$

Thus from (6.84) the weights are given by

$$w_i = \int_{-1}^1 \frac{(1 - x^2)^{-1/2}T_{n+1}(x)}{(x - x_i)T'_{n+1}(x_i)} \, dx. \tag{6.85}$$

We now use the formula†

$$\frac{1}{2}\big(T_{n+1}(x)T_n(y) - T_{n+1}(y)T_n(x)\big) = (x - y) \sum_{r=0}^n{}' T_r(x)T_r(y), \tag{6.86}$$

† Σ' denotes summation with the first term halved.

which is derived in Problem 6.24. (This is a special case of the Christoffel–Darboux formula, which holds for any set of orthogonal polynomials. See Davis.) Putting $y = x_i$ in (6.86) we have $T_{n+1}(y) = 0$ and we divide both sides of (6.86) by $\frac{1}{2}(x - x_i)T_n(x_i)T'_{n+1}(x_i)$ to give for (6.85)

$$w_i = \frac{2}{T_n(x_i)T'_{n+1}(x_i)} \sum_{r=0}^{n}{}' T_r(x_i) \int_{-1}^{1} (1 - x^2)^{-1/2}T_r(x) \, dx.$$

From the orthogonality of $T_0 = 1$ and T_r, $r \neq 0$, only the first term of the summation is non-zero, so that

$$w_i = \frac{\pi}{T_n(x_i)T'_{n+1}(x_i)}. \tag{6.87}$$

Putting $x = \cos\theta$ we obtain, as in (4.65),

$$T_n(x)T'_{n+1}(x) = (n + 1) \sin(n + 1)\theta \cos n\theta \,/ \sin\theta. \tag{6.88}$$

We now write $x_i = \cos\theta_i$ and use the identity

$$\cos n\theta = \cos((n + 1)\theta - \theta) = \cos(n + 1)\theta \cos\theta + \sin(n + 1)\theta \sin\theta.$$

Since $\cos(n + 1)\theta_i = 0$, we have

$$\cos n\theta_i = \sin(n + 1)\theta_i \sin\theta_i$$

and, from (6.88),

$$\begin{aligned} T_n(x_i)T'_{n+1}(x_i) &= (n + 1) \sin^2(n + 1)\theta_i \\ &= (n + 1)(1 - \cos^2(n + 1)\theta_i) = n + 1. \end{aligned}$$

Thus from (6.87) we have

$$w_i = \pi/(n + 1)$$

and therefore, for the Chebyshev rules, all the weights are equal.

Note that the Gaussian rules with weight functions may be used to integrate $\int_a^b g(x) \, dx$ by writing

$$g(x) = \omega(x).[g(x)/\omega(x)]$$

and applying the rule (6.84) to the function $f(x) = g(x)/\omega(x)$.

An obvious disadvantage of Gaussian rules is that their weights and abscissae are a little awkward to handle: we need to compute them with appropriate accuracy and store them in the computer. Romberg integration, whose weights and abscissae are very simple, is used more often than Gaussian rules.

6.6 Indefinite integrals

Suppose we wish to evaluate the *indefinite integral*

$$F(x) = \int_a^x f(t)\, dt \tag{6.89}$$

for $a < x \leqslant b$, rather than the definite integral (6.45). We can choose an appropriate step size h and approximate to the right side of (6.89) for $x = a + h, a + 2h, \ldots$ in turn. Each integral can be estimated by some quadrature rule, for example the trapezoidal rule. Intermediate values of $F(x)$ can be estimated by using interpolation.

Another method is to replace f by some approximating polynomial p and approximate to F by integrating p. If $|f(x) - p(x)| < \varepsilon$ for $a \leqslant x \leqslant b$, then

$$\left| F(x) - \int_a^x p(t)\, dt \right| < \int_a^x \varepsilon\, dx \leqslant (b - a)\varepsilon$$

for $a \leqslant x \leqslant b$. Taking $[a, b]$ as $[-1, 1]$, a suitable choice of p (see §5.7) is the modified Chebyshev series

$$\sideset{}{'}\sum_{r=0}^{n} \alpha_r T_r(x), \tag{6.90}$$

where

$$\alpha_r = \frac{2}{N} \sideset{}{''}\sum_{j=0}^{N} f(x_j) T_r(x_j) \tag{6.91}$$

and $x_j = \cos(\pi j/N)$, with $N > n$. To integrate (6.90) we need to integrate $T_r(x)$, whose indefinite integral is clearly a polynomial of degree $r + 1$ and is thus expressible as a sum of multiples of $T_0, T_1, \ldots, T_{r+1}$. In fact, we find that the indefinite integral is quite simple for, putting $t = \cos\theta$, we have

$$\int T_r(t)\, dt = -\int \cos r\theta \sin\theta\, d\theta$$

$$= \frac{1}{2} \int [\sin(r - 1)\theta - \sin(r + 1)\theta]\, d\theta$$

$$= \frac{1}{2} \left(\frac{\cos(r + 1)\theta}{r + 1} - \frac{\cos(r - 1)\theta}{r - 1} \right) + C$$

if $r \neq 1$, C being an arbitrary constant. Choosing C in agreement with the condition that $T_r(-1) = (-1)^r$, we deduce that

$$\int_{-1}^x T_r(t)\, dt = \frac{1}{2} \left(\frac{T_{r+1}(x)}{r + 1} - \frac{T_{r-1}(x)}{r - 1} \right) + \frac{(-1)^{r+1}}{r^2 - 1}, \tag{6.92}$$

where $r > 1$. We also find that

$$\int_{-1}^{x} T_1(t)\, dt = \tfrac{1}{4}T_2(x) - \tfrac{1}{4},$$

$$\int_{-1}^{x} T_0(t)\, dt = T_1(x) + 1.$$

Integrating (6.90) we have

$$\int_{-1}^{x} \left(\sum_{r=0}^{n}{}' \alpha_r T_r(t) \right) dt = \sum_{j=0}^{n+1} \beta_j T_j(x),$$

where, from (6.92) and the following relations, we find that

$$\beta_j = \frac{1}{2j}(\alpha_{j-1} - \alpha_{j+1}), \qquad 1 \leqslant j \leqslant n - 1 \qquad (6.93)$$

and

$$\beta_0 = \tfrac{1}{2}\alpha_0 - \tfrac{1}{4}\alpha_1 + \sum_{r=2}^{n} (-1)^{r+1}\alpha_r/(r^2 - 1).$$

If we define $\alpha_{n+1} = \alpha_{n+2} = 0$, (6.93) holds also for $j = n$ and $n + 1$.

6.7 Improper integrals

Hitherto we have restricted our attention to so-called proper integrals, whose range of integration is finite and whose integrands are bounded. We now consider the following two problems.

(i) Estimate $\int_a^b f(x)\, dx$, where f has a singularity at $x = b$, but is defined on any interval $[a, b - \varepsilon]$, where $0 < \varepsilon < b - a$. We suppose that

$$\lim_{\varepsilon \to 0} \int_a^{b-\varepsilon} f(x)\, dx$$

exists, write $\int_a^b f(x)\, dx$ to denote this limit and say that the integral converges. An example of this type of integral is $\int_0^1 (1 - x)^{-1/2}\, dx$.

(ii) Estimate $\int_a^\infty f(x)\, dx$, where f is defined on any finite interval $[a, b]$ and

$$\lim_{b \to \infty} \int_a^b f(x)\, dx$$

exists. We write $\int_a^\infty f(x)\, dx$ to denote the value of this limit and say that the integral converges. As an example, we have $\int_1^\infty x^{-2}\, dx$ which, like our previous example, can be evaluated without recourse to numerical methods.

A technique which is sometimes useful in both cases (i) and (ii) (and also for proper integrals) is to make a suitable change of variable. Thus the substitution $t = (1 - x)^{1/2}$ changes the improper integral (case (i) above) $\int_0^1 (1 - x)^{-1/2} \sin x \, dx$ into the proper integral $2 \int_0^1 \sin(1 - t^2) \, dt$, which may be estimated by using a standard quadrature rule.

Another useful technique is that of "subtracting out" the singularity. Consider the improper integral

$$I = \int_0^1 x^{-p} \cos x \, dx, \qquad 0 < p < 1,$$

with a singularity at $x = 0$. We can approximate to $\cos x$ by the first term of its Taylor series about $x = 0$, to give

$$I = \int_0^1 x^{-p} \, dx + \int_0^1 x^{-p}(\cos x - 1) \, dx.$$

The first integral has the value $(1 - p)^{-1}$. A numerical method may be used for the second integral, whose integrand has no singularity, since $\cos x - 1$ behaves like $-\frac{1}{2}x^2$ at $x = 0$. However, $x^{-p}(\cos x - 1)$ behaves like $-\frac{1}{2}x^{2-p}$, whose second and higher derivatives are singular at $x = 0$. We therefore could not guarantee that numerical integration would give very accurate results and could subtract out a little more. For example, we have

$$I = \int_0^1 x^{-p} p_4(x) \, dx + \int_0^1 x^{-p}(\cos x - p_4(x)) \, dx, \qquad (6.94)$$

where

$$p_4(x) = 1 - \frac{x^2}{2!} + \frac{x^4}{4!}.$$

At $x = 0$, the second integrand in (6.94) is zero and its first five derivatives exist on $[0, 1]$. We could therefore safely use Simpson's rule (whose error term involves the fourth derivative) to estimate the second integral. The first integral in (6.94) may be evaluated explicitly.

Infinite integrals can be estimated by truncating the interval at a suitable point. We replace $\int_a^\infty f(x) \, dx$ by $\int_a^b f(x) \, dx$, where b is chosen large enough to give a sufficiently good approximation. We then use a quadrature rule to estimate the latter integral. For example,

$$\left| \int_b^\infty \frac{\sin^2 x}{1 + e^x} \, dx \right| < \int_b^\infty e^{-x} \, dx = e^{-b}$$

and $e^{-b} < \frac{1}{2}.10^{-6}$ for $b \geqslant 15$. Thus we can estimate $\int_0^\infty \sin^2 x / (1 + e^x) \, dx$ correct to six decimal places by calculating a sufficiently accurate approximation to the integral over the range $[0, 15]$.

It is sometimes convenient to use Gaussian rules to estimate improper integrals. Thus to evaluate $\int_a^b \omega(x)f(x)\,dx$, where f is "well-behaved" on $[a, b]$, but $\omega(x) \geqslant 0$ has a singularity, we could use a Gaussian rule based on the polynomials which are orthogonal with respect to ω on $[a, b]$. The obvious disadvantage is the need to compute the required abscissae and weights. For infinite integrals we have Gaussian rules of the form

$$\int_0^\infty \omega(x)f(x)\,dx \simeq \sum_{i=0}^n w_i f(x_i) \tag{6.95}$$

and

$$\int_{-\infty}^\infty \omega(x)f(x)\,dx \simeq \sum_{i=0}^n w_i f(x_i) \tag{6.96}$$

which are exact for $f \in P_{2n+1}$. The choice of $\omega(x) = \exp(-x)$ in (6.95) gives the Gauss–Laguerre rules. With $\omega(x) = \exp(-x^2)$ in (6.96), we obtain the Gauss–Hermite rules. The names of Laguerre and Hermite are given to the respective sets of orthogonal polynomials associated with these formulae. Values of the weights and abscissae for these two formulae are tabulated in Ralston for $2 \leqslant n \leqslant 5$.

6.8 Multiple integrals

Domains of integration can become rather complicated when the dimension is greater than one. Here we shall discuss only two-dimensional integrals over a rectangular region. By means of a linear change of variable, the rectangle may be transformed into a square. Now, using any one-dimensional quadrature rule, we can write

$$\int_a^b \left(\int_a^b f(x, y)\,dx \right) dy \simeq \int_a^b \left(\sum_{i=0}^N w_i f(x_i, y) \right) dy$$

$$= \sum_{i=0}^N w_i \int_a^b f(x_i, y)\,dy$$

$$\simeq \sum_{i=0}^N w_i \left(\sum_{j=0}^N w_j f(x_i, y_j) \right),$$

where, in fact, $x_i = y_i$. The numbers x_i and w_i are the abscissae and weights of the one-dimensional quadrature rule. Thus we have the two-dimensional rule

$$\int_a^b \int_a^b f(x, y)\,dx\,dy \simeq \sum_{i,j=0}^N w_i w_j f(x_i, y_j). \tag{6.97}$$

This is called a *product* integration rule. It generalizes obviously to higher dimensional multiple integrals over rectangular regions. For a product composite Simpson rule in two dimensions with, say, 4 sub-intervals in each direction, the pattern of relative weights 1, 4, 2, 4, 1 is "multiplied" to give the array in Table 6.7. The sum of the numbers in the table is $(1+4+2+4+1)^2 = 144$. Since the rule has to integrate the function $f(x, y) \equiv 1$ exactly over the region with area $(b - a)^2$, each number in the table must be multiplied by $(b - a)^2/144$ to give the appropriate weight $w_i w_j$ for (6.97).

Table 6.7   Relative weights of a
product Simpson rule.

1	4	2	4	1
4	16	8	16	4
2	8	4	8	2
4	16	8	16	4
1	4	2	4	1

It is easy to derive error estimates for product rules, given an error estimate for the one-dimensional rule on which the product formula is based. We will obtain an error bound for (6.97) based on Simpson's rule, with step-size h. Let R denote the square $a \leqslant x \leqslant b$, $a \leqslant y \leqslant b$ and write

$$\int_a^b f(x, y) \, dx = \sum_{i=0}^N w_i f(x_i, y) - (b - a) \frac{h^4}{180} f_{4x}(\xi_y, y), \qquad (6.98)$$

using (6.56). We have written f_{4x} for $\partial^4 f/\partial x^4$, assumed continuous on R. Then

$$\int_a^b \left(\int_a^b f(x, y) \, dx \right) dy = \sum_{i=0}^N w_i \int_a^b f(x_i, y) \, dy + E, \qquad (6.99)$$

where

$$E = -(b - a) \frac{h^4}{180} \int_a^b f_{4x}(\xi_y, y) \, dy.$$

Replacing $\int_a^b f(x_i, y) \, dy$, for each i, by Simpson's rule with error term, we obtain the total error in the product rule,

$$-(b - a) \frac{h^4}{180} \sum_{i=0}^N w_i f_{4y}(x_i, \eta_i) + E, \qquad (6.100)$$

assuming that $\partial^4 f/\partial y^4$ is continuous on R. If the moduli of f_{4x} and f_{4y} are bounded by M_x and M_y respectively, we deduce from (6.100) that the modulus of the error of the product rule is not greater than

$$(b - a)^2 \frac{h^4}{180} (M_x + M_y). \qquad (6.101)$$

Problems

Section 6.1

6.1 Write down the forward difference formula up to the second differences and differentiate to obtain the approximation

$$f'(x_0) \simeq (-3f_0 + 4f_1 - f_2)/2h.$$

6.2 Obtain the error term for the approximation in Problem 6.1.

6.3 Show that a differentiation rule of the form

$$f'(x_0) \simeq \alpha_0 f_0 + \alpha_1 f_1 + \alpha_2 f_2$$

is exact for all $f \in P_2$ if, and only if, it is exact for $f(x) = 1$, x and x^2. Hence find values of α_0, α_1 and α_2 so that the rule is exact for $f \in P_2$.

6.4 Use the backward difference formula with error term,

$$f(x_1 + sh) = f_1 + s\nabla f_1 + h^2 \binom{-s}{2} f''(\xi_s),$$

to obtain

$$f'(x_1) = \frac{f_1 - f_0}{h} + \tfrac{1}{2}hf''(\xi).$$

6.5 Show that the first two leading terms of $s(s - 1) \cdots (s - k)$, regarded as a polynomial in s, are $s^{k+1} - \tfrac{1}{2}k(k + 1)s^k$. Deduce that

$$\frac{d^k}{ds^k}\binom{s}{k + 1} = s - \tfrac{1}{2}k.$$

Put $k = 2m - 1$, $n = 2m$ and $s = m$ in (6.18), and so obtain the differentiation rule (6.22).

6.6 Put $n = 4$ in (6.6) and then put $s = 2$ to obtain

$$f'(x_2) \simeq \frac{1}{12h}(f_0 - 8f_1 + 8f_3 - f_4),$$

which is exact for $f \in P_4$.

6.7 By differentiating the best first degree polynomial, in the least squares sense, for f at $x = x_0$, $x_0 \pm h$, $x_0 \pm 2h$, obtain the differentiation rule

$$f'(x_0) \simeq [2f(x_0 + 2h) + f(x_0 + h) - f(x_0 - h) - 2f(x_0 - 2h)]/10h.$$

(*Hint:* it is sufficient to consider the case where $x_0 = 0$.)

6.8 If we repeat Problem 6.7, but construct the least squares straight line at $x = x_0$, $x_0 \pm h$ only, what approximation is obtained for $f'(x_0)$?

Section 6.2

6.9 Consider the effect of rounding errors and the truncation error in the rule (6.25) for estimating f''. If $|f^{(4)}(x)| \leq M_4$ and the f_i are in error by at most $\frac{1}{2} . 10^{-k}$, show that a choice of h near to

$$h = \left(\frac{24 . 10^{-k}}{M_4}\right)^{1/4}$$

may be expected to minimize the total error.

6.10 Suppose that $|f^{(4)}(x)| \leq 1$ for the function f tabulated below. From these tabulated values, estimate $f''(1.8)$ using the rule (6.25) with all three possible choices of step size h. Given that $f(x) = \sin x$, see how the best value of h (that is $h = 0.1, 0.2$ or 0.3) compares with that predicted in the last problem.

x	1.5	1.6	1.7	1.8	1.9	2.0	2.1
$f(x)$	0.9975	0.9996	0.9917	0.9738	0.9463	0.9093	0.8632

6.11 Assuming that f is sufficiently differentiable, use Taylor series to show that the "half-way" interpolation formula

$$f(x_0) \simeq \tfrac{1}{2}(f(x_0 + \tfrac{1}{2}h) + f(x_0 - \tfrac{1}{2}h)) = H(x_0; h)$$

has an error of a type which makes extrapolation to the limit valid.

6.12 Show that, if the interpolation formula in Problem 6.11 is used to estimate $f(x_0)$, knowing the two values $f(x_0 \pm \tfrac{1}{2}h)$, the approximation $[9H(x_0; h/3) - H(x_0; h)]/8$ has a leading error term of order h^4. Apply this with $h = 0.3$ to the data of Problem 6.10 to estimate $\sin(1.85)$. (Note that we have used $h/3$-extrapolation instead of the usual $h/2$-extrapolation because the former is more convenient for this application.)

Section 6.3

6.13 Verify that Simpson's rule is exact for all $f \in P_3$ by verifying it holds exactly for $f(x) = 1, x, x^2$ and x^3.

6.14 Derive an integration rule

$$\int_{x_0}^{x_3} f(x)\, dx \simeq h(a_0 f_0 + a_1 f_1 + a_2 f_2 + a_3 f_3)$$

which is exact for $f \in P_3$. (*Hint:* set the mid-point $x_0 + 3h/2$ to zero, put $h = 2$ and make the rule exact for $f(x) = 1, x, x^2$ and x^3.)

6.15 Obtain the open Newton–Cotes formula of the form

$$\int_{x_0}^{x_3} f(x)\, dx \simeq h(b_1 f_1 + b_2 f_2).$$

6.16 Show that the integration rule

$$\int_{x_0}^{x_1} f(x)\, dx \simeq \frac{h}{2}(f(x_0) + f(x_1)) - \frac{h^2}{12}(f'(x_1) - f'(x_0))$$

is exact for $f \in P_3$.

6.17 Deduce from the result of Problem 6.16 that

$$\int_{x_0}^{x_N} f(x)\, dx \simeq T_N - \frac{h^2}{12}(f'(x_N) - f'(x_0)),$$

where T_N denotes the result of applying the composite trapezoidal rule, is exact for $f \in P_3$. (This is called the trapezoidal rule with end correction and is obtained by truncating the right side of the Euler–Maclaurin formula (6.68) after one correction term.)

6.18 Verify that the special case of Gregory's formula given by (6.54) is exact for all $f \in P_3$. (*Hint:* put $x_0 = 0$, $h = 1$ and use the identities $\sum_{r=1}^{N} r = \frac{1}{2}N(N+1)$; $\sum_{r=1}^{N} r^2 = \frac{1}{6}N(N+1)(2N+1)$; $\sum_{r=1}^{N} r^3 = \frac{1}{4}N^2(N+1)^2$.)

6.19 Verify (6.61), that $(2M_N + T_N)/3 = S_{2N}$.

Section 6.4

6.20 Verify that one step of Romberg integration, that is the elimination of the h^2 error term between T_N and T_{2N}, is equivalent to the composite Simpson's rule.

Section 6.5

6.21 Verify directly that the 2-point Gaussian quadrature rule (6.79) is exact for all $f \in P_3$.

6.22 Verify directly that the 3-point Gaussian quadrature rule (6.80) is exact for all $f \in P_5$.

6.23 With π_{n+1} defined by (6.73), show that

$$L_r(x) = \frac{(x - x_0) \cdots (x - x_{r-1})(x - x_{r+1}) \cdots (x - x_n)}{(x_r - x_0) \cdots (x_r - x_{r-1})(x_r - x_{r+1}) \cdots (x_r - x_n)}$$

$$= \frac{\pi_{n+1}(x)}{(x - x_r)\pi'_{n+1}(x_r)}.$$

(*Hint:* use (6.8).)

6.24 If $\chi_n(x, y) = T_{n+1}(x)T_n(y) - T_{n+1}(y)T_n(x)$, where T_n denotes the Chebyshev polynomial, show that

$$\chi_n(x, y) = 2(x - y)T_n(x)T_n(y) + \chi_{n-1}(x, y)$$

for $n \geqslant 1$ and deduce that

$$\tfrac{1}{2}\chi_n(x, y) = (x - y) \sum_{r=0}^{n}{}' T_r(x)T_r(y).$$

6.25 For a given function f, the Chebyshev coefficient a_r (see (5.29)) is

$$a_r = \frac{2}{\pi} \int_{-1}^{1} (1 - x^2)^{-1/2} f(x)T_r(x) \, dx.$$

Show that, if the N-point Chebyshev quadrature formula is used to estimate the integral, we obtain the approximation $a_r \simeq \alpha_r^*$, where α_r^* is defined by (5.58).

6.26 Consider the quadrature formula

$$\int_a^b f(x) \simeq \sum_{i=0}^{n} w_i f(x_i),$$

where each $w_i > 0$ and the rule is exact for the function $f(x) \equiv 1$. If the $f(x_i)$ are in error by at most $\tfrac{1}{2}.10^{-k}$, show that the error in the quadrature formula due to this is not greater than $(b - a)\tfrac{1}{2}.10^{-k}$.

Section 6.6

6.27 Obtain a quadratic polynomial approximation for

$$F(x) = \int_0^x e^{-t^2} \, dt$$

by integrating the interpolating polynomial for e^{-t^2} constructed at $t = 0$ and 0.1. Estimate the error in the resulting approximation to F on $[0, 0.1]$.

6.28 One method of obtaining a polynomial approximation to

$$F(x) = \int_a^x f(t) \, dt$$

is to use the Taylor polynomial approximation of degree $n - 1$ to $f(t)$ at $t = a$. Show that the resulting approximation to $F(x)$ is the Taylor polynomial of degree n for $F(x)$ at $x = a$.

Section 6.7

6.29   Deduce from the inequality $e^{-x^2} < e^{-Mx}$ for $x > M$ that

$$\int_M^\infty e^{-x^2}\, dx < e^{-M^2}/M.$$

Find a value of M so that this last number is smaller than 10^{-4} and hence estimate $\int_0^\infty e^{-x^2}\, dx$ to three decimal places.

Section 6.8

6.30   Obtain an error estimate for the product trapezoidal rule analogous to the result (6.101) obtained for the product Simpson rule.

Chapter 7

Solution of Algebraic Equations of One Variable

7.1 Introduction

In this chapter, we discuss numerical methods for solving equations of the form

$$f(x) = 0, \tag{7.1}$$

where both x and $f(x)$ are real. Real values of x for which (7.1) holds are called *roots* of the equation. (We shall not consider any numerical methods for finding complex roots.) Examples of such equations are

$$x^2 - 5x + 2 = 0, \tag{7.2}$$

$$2^x - 5x + 2 = 0. \tag{7.3}$$

Equation (7.2) is a *quadratic equation*, whose roots are expressible in the "closed" form $x = \frac{5}{2} \pm ((\frac{5}{2})^2 - 2)^{1/2}$. It is very unusual to be able to express the roots of an equation in such a closed form. More often, we have to obtain the solution of an equation numerically, that is, we find the roots to within a given accuracy. First we *locate* the roots, usually by sketching a graph of $y = f(x)$, and then *refine* the roots. Almost all our attention here will be given to the latter process. It sometimes helps to rearrange the equation first. For instance, if we rewrite (7.3) as $2^x = 5x - 2$, we see that the required roots are the values of x where the graphs of $y = 2^x$ and $y = 5x - 2$ intersect. From Fig. 7.1 it follows that (7.3) has exactly one root (denoted by α) between 0 and 1.

We shall consider only generally applicable methods of refining the roots. Thus we shall not discuss any of the several methods available for solving *polynomial* equations. (See Ralston.) We first describe some elementary methods which have the following property in common: beginning with an interval I_0 which contains at least one root of (7.1), we construct a sequence of intervals I_n such that $I_{n+1} \subset I_n$ and therefore each I_n contains at least one root of (7.1). This will be referred to as a "bracketing" method: a root is bracketed by the end-points of each interval I_n. Obviously, we would like the length of the interval I_n to tend to zero, as $n \to \infty$, so that we may compute the root as accurately as we wish.

154

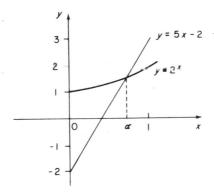

Fig. 7.1. Locating the root of $2^x = 5x - 2$.

7.2 The bisection method

Suppose that f is continuous on some interval $[x_0, x_1]$ and that $f(x_0)$ and $f(x_1)$ have opposite signs, as in Fig. 7.2. There is at least one root of $f(x) = 0$ in the interval $[x_0, x_1]$, which we shall denote by I_0. We now bisect I_0, writing $x_2 = (x_0 + x_1)/2$, and let I_1 denote the sub-interval $[x_0, x_2]$ or $[x_2, x_1]$ at whose end-points f takes opposite signs. Similarly I_1 is bisected to give an interval I_2 (half the width of I_1) at whose end points f still has opposite signs,

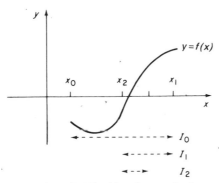

Fig. 7.2. The bisection method.

and so on. See Fig. 7.2. This bracketing method, called the bisection method, is set out as a flow diagram in Fig. 7.3. In the flow diagram, y_0 and y_1 denote $f(x_0)$ and $f(x_1)$ respectively and the process is terminated when the root is pinned down in an interval of at most 10^{-6}. We may select the mid-point of this last interval to approximate the root, with an error not greater than $\frac{1}{2} . 10^{-6}$. Note that, by appropriate re-naming of the x_i and y_i in the flow diagram, the interval under scrutiny at any time is always denoted by $[x_0, x_1]$. Advantages of the bisection method are its simplicity and the fact that we may predict, in advance, how many iterations of the main "loop" of calcu-

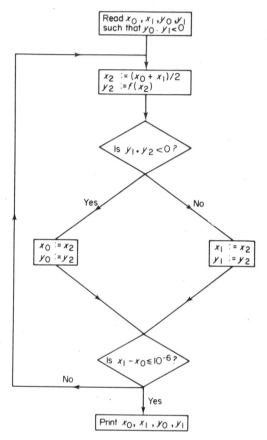

Fig. 7.3. A flow diagram for the bisection method.

lations will be required since the interval is halved during each iteration. For example, if initially in Fig. 7.3, $x_1 - x_0 = 1$, the number of iterations or bisections necessary will be 20, since $2^{-20} < 10^{-6} < 2^{-19}$.

Example 7.1 For equation (7.3), $f(0) = 3$, $f(1) = -1$ and the bisection algorithm can be used on the interval $[0, 1]$. The reader may care to verify that, after six iterations, the root is seen to lie between $\frac{46}{64}$ and $\frac{47}{64}$. ☐

7.3 Interpolation methods

It can sometimes be instructive to do a few calculations. In particular, this may help us discover more efficient algorithms. In working through Example 7.1, we soon discover one inefficiency. It seems unsatisfactory simply to bisect an interval $[x_0, x_1]$ for which $f(x_0)$ and $f(x_1)$ have opposite signs, since this bisection takes no account of the *magnitudes* of the numbers $f(x_0)$ and $f(x_1)$. An obvious alternative, as displayed in Fig. 7.4, is to join the points

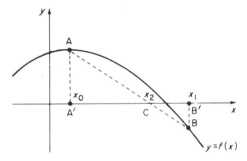

Fig. 7.4. The regula falsi method.

$(x_0, f(x_0))$ and $(x_1, f(x_1))$ by a straight line. We expect that x_2, the point where this straight line cuts the x-axis, is an improved estimate of the root. From the similar triangles AA'C and BB'C of Fig. 7.4, we notice that

$$\frac{f(x_0)}{x_2 - x_0} = \frac{-f(x_1)}{x_1 - x_2}$$

and thus

$$x_2 = x_1 - \frac{f(x_1)(x_1 - x_0)}{f(x_1) - f(x_0)}. \tag{7.4}$$

Having calculated x_2, we use it to replace x_0 or x_1, as in the bisection algorithm. Thus in Fig. 7.3 we replace the step $x_2 := (x_0 + x_1)/2$ by

$$x_2 := x_1 - \frac{y_1(x_1 - x_0)}{y_1 - y_0}, \tag{7.5}$$

as in (7.4). As a stopping criterion, it is more appropriate to ask for $|f(x_2)|$, rather than $x_1 - x_0$ to be small. The resulting bracketing method is called *the rule of false position*. This lengthy name is due to the fact that the curve $y = f(x)$ (Fig. 7.4) is replaced by the straight line AB which gives the "false position" of the root. The method is also known by the (Latin) name *regula falsi*.

Example 7.2 We apply the regula falsi algorithm to $2^x - 5x + 2 = 0$, for which the bisection method was employed in Example 7.1. With $x_0 = 0$, $x_1 = 1$ initially, the algorithm produces the results shown in Table 7.1. The numbers in the table have been rounded to three decimal places. After only three iterations, we see from the last column of Table 7.1 that the root α lies in the interval [0.732, 0.733]. This is a considerable improvement on the performance of the bisection method for this equation. □

Table 7.1   Regula falsi method for $2^x - 5x + 2 = 0$ (Example 7.2).

Iteration number	1	2	3
x_0	0	0	0
y_0	3	3	3
x_1	1	0.75	0.733
y_1	-1	-0.068	-0.003
x_2	0.75	0.733	0.732
y_2	-0.068	-0.003	0.001

In the last example, $f(x) = 2^x - 5x + 2$ and the second derivative, $f''(x) = (\ln 2)^2 . 2^x$, is positive for all x. Thus f is *convex* and, consequently, for each x_2 produced by the regula falsi method, $f(x_2)$ should always be *negative*. This is seen in Fig. 7.5. (The convexity of f is exaggerated to make the illustration clearer.) Notice what happens to the intervals $[x_0, x_1]$ on each iteration, in the case of convex f: one end-point (the left-hand one in Fig. 7.5) stays fixed throughout the operation of the regula falsi method and the length of the interval $[x_0, x_1]$ cannot shrink to zero, as we would wish. We obtain a similar effect if f is concave. It is wise to modify the method, since f will always be convex or concave in a neighbourhood of the root α unless

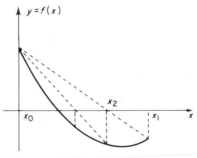

Fig. 7.5. The regula falsi method for solving $f(x) = 0$,
when f is convex.

$f''(\alpha) = 0$. One possible modification is to check whether any end-point is unchanged after two consecutive iterations. If, say, x_0 is unchanged over two iterations, then in addition to replacing the current point (x_1, y_1) by (x_2, y_2), we also replace (x_0, y_0) by $(x_0, y_0/2)$. In fact in Example 7.2, no such modification is required to obtain a small interval bracketing the root, as rounding errors cause the last value of $f(x_2)$ to be positive.

If we use the unmodified regula falsi method, we will not be able to depend on $x_1 - x_0$ becoming arbitrarily small, as occurs in the bisection method, and we need some other criterion to decide when to stop iterating. One criterion is to stop when the *residual* $f(x_2)$ is of small magnitude, that is, when $|f(x_2)| < \varepsilon$, for some $\varepsilon > 0$ of our choice. However, we do need to be careful in choosing ε, especially if $|f'(x)|$ is very small at the root so that $|f(x)|$ increases slowly as x moves away from the root. This remark applies to any iterative method for solving an equation numerically. It is desirable to write the computer program so as to stop iterating for any one of the following reasons:

(i) $|x_1 - x_0|$ is sufficiently small (for a bracketing method only),
(ii) $|f(x_n)|$ is sufficiently small, for some iterate x_n,
(iii) a certain number of iterations have been performed.

It is wise to include criterion (iii) to guard against iterating indefinitely if the method fails to converge. For a non-bracketing method which generates a sequence $x_0, x_1, \ldots$, it is common practice (cf. (i) above) to stop iterating when $|x_{n+1} - x_n|$ is sufficiently small. This is a reliable criterion for some problems (see, for example, Problem 7.26), but can be misleading.

Suppose $y = f(x)$ and that the inverse function, $x = \phi(y)$, exists. Then, in particular, $0 = f(\alpha)$ corresponds to $\alpha = \phi(0)$. Each step of the regula falsi method, as we have already seen in Example 4.6, may be regarded as inverse interpolation at the two points x_0 and x_1. We replace $\phi(y)$ by the linear interpolating polynomial $p_1(y)$ constructed at y_0 and y_1. To estimate the accuracy attained at any stage by the regula falsi method, we consider the error formula (from (4.16)):

$$\phi(y) - p_1(y) = (y - y_0)(y - y_1).\phi''(\eta_y)/2!. \tag{7.6}$$

This is valid on any interval $\alpha \leqslant y \leqslant \beta$ containing y_0 and y_1 and on which ϕ'' exists; η_y is some point of that interval. We now assume that $|\phi''(y)| \leqslant M_2$ for $y \in [\alpha, \beta]$. Putting $y = 0$ in (7.6), we obtain

$$|\alpha - x_2| \leqslant \tfrac{1}{2}M_2 y_0 y_1, \tag{7.7}$$

as by construction $p_1(0) = x_2$. We can estimate M_2 from the relation

$$\phi''(y) = -f''(x)/[f'(x)]^3, \tag{7.8}$$

assuming $f'(x) \neq 0$. (See Example 4.6, where (7.8) is derived.)

Example 7.3 Pursuing Example 7.2, where we sought the root of $2^x - 5x + 2 = 0$, we would have at the *next* iteration $x_0 = 0.732$ (with $y_0 = 0.001$) and $x_1 = 0.733$ (with $y_1 = -0.003$). We now estimate how close the next iterate x_2 would be to the root α. In this case, $f'(x) = (\ln 2).2^x - 5$ and $f''(x) = (\ln 2)^2.2^x$. From (7.8) and (7.7), a little calculation shows that the error in x_2 is smaller than $\frac{1}{2}.10^{-7}$. (To *attain* such accuracy in x_2, which is computed from (7.5), we would of course first have to re-calculate y_0 and y_1 to an accuracy of at least seven decimal places.) □

There exists a method for solving equations which is sometimes confused with the regula falsi method. This is the *secant method*, in which we begin with any two numbers $x_0 < x_1$ and calculate

$$x_{r+1} = x_r - \frac{f(x_r).(x_r - x_{r-1})}{f(x_r) - f(x_{r-1})}, \qquad r = 1, 2, \ldots, \qquad (7.9)$$

assuming that the denominator in (7.9) is never zero. Note that, unlike the regula falsi method, no attention is paid here to the signs of the numbers $f(x_r)$ and, therefore, it is *not* a bracketing method. We will return to the secant method when we discuss Newton's method in §7.6.

An obvious extension of the secant method is to use three points at a time instead of two. Suppose we begin with two approximations, x_0 and x_1, to a root of $f(x) = 0$ and that the secant method is used to compute a third approximation x_2. Instead of discarding x_0 or x_1, we may construct the unique (quadratic) interpolating polynomial p_2 for f at all three points. Suppose that $p_2(x) = 0$ has two real roots, as in Fig. 7.6, and that x_3 denotes the root nearer to x_2. We then repeat the above procedure with x_1, x_2 and x_3 in place of x_0, x_1 and x_2, and so on. This is called Muller's method.

To avoid the possibility of the polynomial equation $p_2'(x) = 0$ having complex roots, we may adopt a variant of the above method in which the root is bracketed at any stage. This time we begin with numbers $x_0 < x_1$ such that $f(x_0).f(x_1) < 0$ and select x_2 as some number in $[x_0, x_1]$. We could select x_2 as the mid-point of $[x_0, x_1]$ or as the number obtained by one iteration of the regula falsi method. The polynomial equation $p_2(x) = 0$ must then have real roots. Only *one* of these, denoted by x_3, will be inside $[x_0, x_1]$. Next, we replace either x_0 or x_1 by x_2, exactly as in the regula falsi method, and replace x_2 by x_3. For example, in Fig. 7.6 we replace x_0 by x_2 and x_2 by x_3. This completes one iteration of the method. We repeat the process with the new set $\{x_0, x_1, x_2\}$ until our stopping criterion is satisfied, as mentioned above.

It is necessary to add one further set of instructions to this algorithm. If $|x_1 - x_0|$ approaches zero, the effect of rounding error can perturb the coefficients of the polynomial p_2 sufficiently to create complex roots instead of the predicted real roots. If this occurs, we can arrange to drop the point x_2,

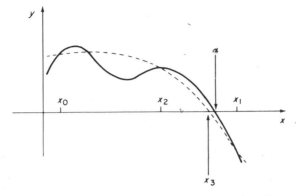

Fig. 7.6. Muller's method. The dotted line is $p_2(x)$.

and continue the search for the root by applying the bisection or regula falsi method.

Example 7.4 For the equation $2^x - 5x + 2 = 0$, with $x_0 = 0$, $x_1 = 1$ and $x_2 = 0.5$, one iteration of Muller's method gives $x_3 = 0.733$ to three decimal places. $\square$

7.4 One-point iterative methods

Hitherto we have used at least two values of x in order to predict a "better" approximation to the root, at any stage. We naturally ask whether we can devise methods which use only *one* value of x at any time. These will be called *one-point methods*. To find a root α of $f(x) = 0$, we must therefore construct a sequence $\{x_r\}$ which satisfies two criteria:

  (i) the sequence $\{x_r\}$ converges to α,
  (ii) x_{r+1} depends directly only on its predecessor x_r.

From (ii), we need to find some function g, so that the sequence $\{x_r\}$ may be computed from

$$x_{r+1} = g(x_r), \qquad r = 0, 1, \ldots, \qquad (7.10)$$

given an initial value x_0. If we happened to choose $x_0 = \alpha$, we would also want (7.10) to give $x_1 = \alpha$ and therefore g satisfies the property $\alpha = g(\alpha)$. Thus, to obtain a one-point method for finding a root α of the equation $f(x) = 0$, we rearrange the equation in a form $x = g(x)$, with $\alpha = g(\alpha)$, so that (7.10) yields a sequence $\{x_r\}$ which converges to α.

Example 7.5 Consider the equation $x = e^{-x}$, which is already of the form $x = g(x)$. The graphs of $y = x$ and $y = e^{-x}$ intersect at only one point

$x = \alpha$ (between 0 and 1), which is therefore the only root of $x = e^{-x}$. Following (7.10), let us choose $x_0 = 0.5$ and compute $x_{r+1} = e^{-x_r}$, $r = 0, 1, \ldots$ Rounding the iterates to three decimal places at each stage, we obtain the numbers displayed in Table 7.2. These numbers suggest that the process is converging, though slowly, and that the root α is 0.567, to three decimal places. Assuming that this is true, the last row of Table 7.2 suggests further that the errors $|x_r - \alpha|$ tend monotonically to zero. $\square$

Table 7.2   Solution of $x = e^{-x}$ by iterating $x_{r+1} = e^{-x_r}$.

r	0	1	2	3	4	5	6	7	8	9
x_r	0.5	0.607	0.545	0.580	0.560	0.571	0.565	0.568	0.567	0.567
$x_r - \alpha$	-0.067	0.040	-0.022	0.013	-0.007	0.004	-0.002	0.001	0.000	0.000

To investigate the behaviour of errors in the general case, we return to (7.10) and subtract from it the equation $\alpha = g(\alpha)$. Thus

$$x_{r+1} - \alpha = g(x_r) - g(\alpha) = (x_r - \alpha)g'(\xi_r), \qquad (7.11)$$

with ξ_r some point between x_r and α. This last step follows from the mean value theorem, assuming that g' exists on an interval containing x_r and α. It is easy to see how to impose conditions on g to ensure convergence of the one-point iterative method. For, if

$$|g'(x)| \leqslant L < 1 \quad \text{for all } x, \qquad (7.12)$$

we deduce from (7.11) that

$$|x_{r+1} - \alpha| \leqslant L|x_r - \alpha|, \qquad (7.13)$$

so that the errors are diminishing. Replacing r by $r - 1$ in (7.13), we obtain

$$|x_r - \alpha| \leqslant L|x_{r-1} - \alpha|, \qquad r \geqslant 1,$$

and, on applying this inequality repeatedly,

$$|x_r - \alpha| \leqslant L^r|x_0 - \alpha|, \qquad r \geqslant 0. \qquad (7.14)$$

Since $0 \leqslant L < 1$, $L^r \to 0$ as $r \to \infty$. From (7.14), $|x_r - \alpha| \to 0$ as $r \to \infty$ and the sequence $\{x_r\}$ converges to the root α. In fact, the above condition on g' allows us to show also the *existence* of the root α. The proof requires the following lemma.

Lemma 7.1   If f has continuous first derivative f' with

$$f'(x) \geqslant M > 0 \quad \text{for all real } x$$

then the equation $f(x) = 0$ has a unique root and this root lies between $x = 0$ and $x = -f(0)/M$.

Proof First we assume that $f(0) > 0$ when by the mean value theorem there is a point ξ such that

$$
\begin{aligned}
f(-f(0)/M) - f(0) &= f'(\xi)(-f(0)/M - 0) \\
&= -f'(\xi).f(0)/M \\
&\leqslant -f(0)
\end{aligned}
$$

as $-f'(\xi) \leqslant -M$. Thus

$$f(-f(0)/M) \leqslant 0,$$

showing that there must be a root between $x = 0$ and $x = -f(0)/M$. Fig. 7.7 gives a graphical representation of this argument. The slope of f must be positive and larger than that of the line AB, which has slope M. Similarly we can deal with the case $f(0) < 0$ and $f(0) = 0$ automatically gives a root.

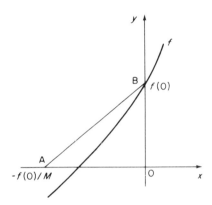

Fig. 7.7. Representation of Lemma 7.1.

It remains to show that the root is unique and this follows from Rolle's theorem 2.4. Two distinct roots would imply the existence of a point η at which $f'(\eta) = 0$ which contradicts the required property of f'. $\square$

Theorem 7.1 If g' exists, is continuous and $|g'(x)| \leqslant L < 1$ for all x, then the equation $x = g(x)$ has a unique root, say α. Further, given any x_0, the sequence $\{x_r\}$ defined by

$$x_{r+1} = g(x_r), \qquad r = 0, 1, \ldots, \tag{7.15}$$

converges to α.

Proof Since

$$\frac{d}{dx}(x - g(x)) = 1 - g'(x) \geqslant 1 - L > 0, \qquad (7.16)$$

the function $x - g(x)$ satisfies the conditions of the lemma and the existence and uniqueness of α follows. We have already shown that the sequence $\{x_r\}$ converges to α. $\square$

It is not sufficient to have $f'(x) > 0$ in Lemma 7.1 and, similarly, $|g'(x)| < 1$ in Theorem 7.1. For example, the function $f(x) = e^x$ has $f'(x) > 0$ everywhere, but there are no zeros of f. In fact the condition $|g'(x)| \leqslant L < 1$ for *all* values of x is rather restrictive. For example, this condition is not satisfied by the function $g(x) = e^{-x}$ in Example 7.5. We shall return to this point in §7.7. Meanwhile we note that the one-point method has a simple graphical interpretation, as depicted in Fig. 7.8. In the figure we start with the iterate

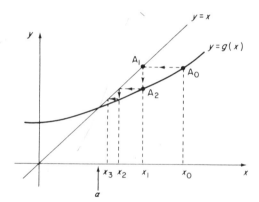

Fig. 7.8. A graphical interpretation of the
one-point method.

x_0 and determine the point A_0 which has coordinates $(x_0, g(x_0))$. We now draw the line A_0A_1 parallel to the x-axis. Since A_1 lies on the line $y = x$ and $x_1 = g(x_0)$, A_1 has coordinates (x_1, x_1). Next we compute the position of A_2 which has coordinates $(x_1, g(x_1))$. We continue along the arrowed lines. In Fig. 7.8, where $|g'(x)| < 1$, the geometrical construction suggests that the sequence $\{x_r\}$ is converging to α. The reader is urged to use the above geometrical approach to "prove" non-convergence in a case where $|g'(x)| > 1$.

7.5 Faster convergence

As we have seen, to solve an equation $f(x) = 0$ by a one-point method, we first replace $f(x) = 0$ by an equivalent equation $x = g(x)$ and then (from

(7.15)) compute a sequence $\{x_r\}$. We have to find a function g which will yield a convergent sequence. Obviously we ought to aim for *rapid* convergence, if possible. Assuming that $x_r \neq \alpha$ for all r, we deduce from (7.11) that

$$\lim_{r \to \infty} \frac{x_{r+1} - \alpha}{x_r - \alpha} = \lim_{r \to \infty} g'(\xi_r) = g'(\alpha). \qquad (7.17)$$

The last equality holds on the assumption that g' is continuous, since ξ_r lies between x_r and α. Thus the ratio of successive errors tends to the value $g'(\alpha)$. This explains the behaviour of the errors $x_r - \alpha$ in Table 7.2 for the equation $x = e^{-x}$, where $g'(\alpha) \simeq -0.6$. Since $(0.6)^4 \simeq 0.1$, we expect to gain only about one decimal place of accuracy every four iterations. This is in agreement with Table 7.2.

Now, given an equation

$$x = g(x) \qquad (7.18)$$

and any constant $\lambda \neq -1$, we may add λx to each side of the equation and divide by $1 + \lambda$. This gives a family of equations

$$x = \left(\frac{\lambda}{1 + \lambda}\right)x + \left(\frac{1}{1 + \lambda}\right)g(x), \qquad (7.19)$$

each of the form

$$x = G(x)$$

and having the same roots as (7.18). Bearing in mind (7.17), we choose λ so as to make $|G'(x)|$ small (and certainly less than unity) at a root α. From (7.19),

$$G'(x) = \left(\frac{1}{1 + \lambda}\right)(\lambda + g'(x)), \qquad (7.20)$$

so we choose a value of λ near to $-g'(\alpha)$.

Example 7.6 For the equation $x = e^{-x}$, with $g'(\alpha) \simeq -0.6$, we modify the equation as in (7.19) with $\lambda = 0.6$. This yields the iterative method

$$x_{r+1} = (3x_r + 5e^{-x_r})/8.$$

As in Example 7.5, we choose $x_0 = 0.5$ but we now obtain the iterates (rounded to five decimal places) shown in Table 7.3. Even x_1 is nearly as accurate as the x_9 of Table 7.2 and x_3 here is correct to five decimal places. $\square$

Table 7.3 Faster convergence to the root of $x = e^{-x}$.

r	0	1	2	3
x_r	0.5	0.56658	0.56713	0.56714

The above procedure for improving the rate of convergence requires an estimate of g'. We examine a method for accelerating convergence which does not require explicit knowledge of the derivative. From (7.11), if $g'(x)$ were *constant*, we could deduce that

$$\frac{x_{r+1} - \alpha}{x_r - \alpha} = \frac{x_r - \alpha}{x_{r-1} - \alpha}, \qquad (7.21)$$

with both sides of (7.21) taking the value $g'(x)$ (which is constant). We could solve (7.21) to determine the root α. Given a function g for which g' is not constant, we may still expect that the number x_{r+1}^* satisfying the equation

$$\frac{x_{r+1} - x_{r+1}^*}{x_r - x_{r+1}^*} = \frac{x_r - x_{r+1}^*}{x_{r-1} - x_{r+1}^*}$$

will be *close* to α. Thus

$$x_{r+1}^* = \frac{x_{r+1}x_{r-1} - x_r^2}{x_{r+1} - 2x_r + x_{r-1}}, \qquad (7.22)$$

assuming the denominator is non-zero. The right side of (7.22) may be re-written as x_{r+1} plus a "correction",

$$x_{r+1}^* = x_{r+1} - \frac{(x_{r+1} - x_r)^2}{x_{r+1} - 2x_r + x_{r-1}}. \qquad (7.23)$$

In terms of the backward difference operator ∇, introduced in Chapter 4, we have

$$x_{r+1}^* = x_{r+1} - \frac{(\nabla x_{r+1})^2}{\nabla^2 x_{r+1}}.$$

The calculation of a sequence $\{x_r^*\}$ in this way from a given sequence $\{x_r\}$ is called Aitken's delta-squared process or *Aitken acceleration*, after A. C. Aitken, who first suggested it in 1926.

Example 7.7 To show how rapidly the Aitken sequence $\{x_r^*\}$ can converge compared with the original sequence $\{x_r\}$, we apply the method to the sequence discussed in Example 7.5. The members of the sequence $\{x_r\}$ are given to five decimal figures (to do justice to the accuracy of the Aitken process) instead of the three figures quoted in Table 7.2. The results are shown in Table 7.4. Note from (7.23) that x_r^* is defined only for $r \geqslant 2$.

Table 7.4   Aitken's delta-squared process.

r	0	1	2	3	4	5
x_r	0.5	0.60653	0.54524	0.57970	0.56007	0.57117
x_r^*	—	—	0.56762	0.56730	0.56719	0.56716

Often a more satisfactory way of using Aitken acceleration is to choose x_0 and calculate x_1 and x_2, as usual, from $x_{r+1} = g(x_r)$, $r = 0$ and 1. Then use the acceleration technique on x_0, x_1 and x_2 (that is, (7.23) with $r = 1$) to find x_2^*. We use x_2^* as a *new* value for x_0, calculate new values $x_1 = g(x_0)$, $x_2 = g(x_1)$ and accelerate again, and so on. Thus, for the equation $x = e^{-x}$ with $x_0 = 0.5$ initially, we obtain:

$$\left. \begin{array}{l} x_0 = 0.50000 \\ x_1 = 0.60653 \\ x_2 = 0.54524 \end{array} \right\} \quad x_2^* = 0.56762$$

$$\left. \begin{array}{l} x_0 = 0.56762 \\ x_1 = 0.56687 \\ x_2 = 0.56730 \end{array} \right\} \quad x_2^* = 0.56714$$

So, by calculating only two sets of x_0, x_1 and x_2, we find the root correct to five decimal places. This is very satisfactory, considering the slowness of convergence of the original iterative process. Sometimes Aitken acceleration will not work, especially if g' fluctuates a large amount and x_0 is a long way from the root. $\square$

7.6 Higher order processes

Let $\{x_r\}$ denote a sequence defined by $x_{r+1} = g(x_r)$, for some choice of x_0, which converges to a root α of $x = g(x)$. We define

$$e_r = x_r - \alpha, \qquad r = 0, 1, \ldots, \tag{7.24}$$

to denote the error at any stage. Thus $x_{r+1} = g(x_r)$ may be written as

$$\alpha + e_{r+1} = g(\alpha + e_r).$$

If $g^{(k)}$ exists, we may replace $g(\alpha + e_r)$ by its Taylor polynomial with remainder, to give

$$\alpha + e_{r+1} = g(\alpha) + e_r g'(\alpha) + \cdots + \frac{e_r^{k-1}}{(k-1)!} g^{(k-1)}(\alpha) + \frac{e_r^k}{k!} g^{(k)}(\xi_r),$$

where ξ_r lies between α and $\alpha + e_r$. Since $\alpha = g(\alpha)$, we find that the relation between successive errors e_r and e_{r+1} has the form

$$e_{r+1} = e_r g'(\alpha) + \frac{e_r^2}{2!} g''(\alpha) + \cdots + \frac{e_r^{k-1}}{(k-1)!} g^{(k-1)}(\alpha) + \frac{e_r^k}{k!} g^{(k)}(\xi_r). \tag{7.25}$$

If $g'(\alpha) \neq 0$ and e_r is "quite small", we may neglect terms in e_r^2, e_r^3 and so on, and see that

$$e_{r+1} \simeq e_r g'(\alpha),$$

which is merely an imprecise way of saying what we knew before from (7.11). However, we see from (7.25) that if $g'(\alpha) = 0$ but $g''(\alpha) \neq 0$, e_{r+1} behaves like a multiple of $e_r{}^2$ and the process is called *second order*. More generally, if for some fixed $k \geqslant 1$,

$$g'(\alpha) = g''(\alpha) = \cdots = g^{(k-1)}(\alpha) = 0 \qquad (7.26)$$

and

$$g^{(k)}(\alpha) \neq 0,$$

then, from (7.25),

$$e_{r+1} = e_r{}^k \cdot g^{(k)}(\xi_r)/k! \qquad (7.27)$$

and the process is said to be kth order. If $g'(\alpha) \neq 0$, the process is called *first order*. If $g^{(k)}$ is continuous, we see from (7.27) that

$$\lim_{r \to \infty} e_{r+1}/e_r{}^k = g^{(k)}(\alpha)/k! \qquad (7.28)$$

This discussion may appear to be rather academic, as it may seem unlikely that the conditions (7.26) will often be fulfilled in practice, for any $k \geqslant 2$. On the contrary, however, we can easily construct second and higher order processes, since $f(x) = 0$ may be rearranged in an infinite number of ways in the form $x = g(x)$. For a given root α, different choices of g lead to processes of possibly different orders.

To derive a second order process, we may return to (7.19), where we constructed a family of equations with the same roots as $x = g(x)$. These equations are of the form $x = G(x)$, with

$$G(x) = \left(\frac{\lambda}{1 + \lambda}\right)x + \left(\frac{1}{1 + \lambda}\right)g(x)$$

with λ a constant, and, therefore,

$$G'(x) = \left(\frac{1}{1 + \lambda}\right)(\lambda + g'(x)).$$

We can make $G'(\alpha) = 0$ by choosing $\lambda = -g'(\alpha)$. To relate this to an arbitrary equation $f(x) = 0$ with a root α, we replace $g(x)$ by $x - f(x)$ so that $\lambda = f'(\alpha) - 1$. Then $x = G(x)$ becomes

$$x = x - \frac{f(x)}{f'(\alpha)},$$

suggesting the second order iterative process

$$x_{r+1} = x_r - \frac{f(x_r)}{f'(\alpha)}. \qquad (7.29)$$

This is of no practical use, since we cannot compute $f'(\alpha)$ without knowing α, which is what we are trying to find. However, let us replace $f'(\alpha)$ by $f'(x_r)$ in (7.29). This gives the method

$$x_{r+1} = x_r - \frac{f(x_r)}{f'(x_r)}, \tag{7.30}$$

which is of the form $x_{r+1} = g(x_r)$, with

$$g(x) = x - \frac{f(x)}{f'(x)}. \tag{7.31}$$

Thus $\alpha = g(\alpha)$ (since $f(\alpha) = 0$) and, on differentiating (7.31),

$$g'(x) = \frac{f(x)f''(x)}{[f'(x)]^2}, \tag{7.32}$$

provided $f'(x) \neq 0$. Thus, provided $f'(\alpha) \neq 0$, we obtain $g'(\alpha) = 0$, showing that the method (7.30) is at least second order. Differentiating (7.32) and putting $x = \alpha$, we find that

$$g''(\alpha) = f''(\alpha)/f'(\alpha),$$

and thus the method (7.30) is exactly second order if $f''(\alpha) \neq 0$. This second order process, called *Newton's method*, has a simple geometrical interpretation. The tangent to the curve $y = f(x)$ at $x = x_r$ cuts the x-axis at $x = x_{r+1}$. From Fig. 7.9 we see that

$$f'(x_r) = \frac{f(x_r)}{x_r - x_{r+1}},$$

which is in agreement with (7.30). Note that we require the condition $f'(x_r) \neq 0$ so that the tangent will cut the x-axis. This is in accordance with the restriction $f'(x) \neq 0$ made above.

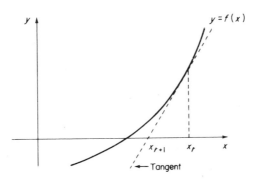

Fig. 7.9. Newton's method.

Example 7.8 For the equation $xe^x - 1 = 0$ (which we have used in earlier examples in the form $x = e^{-x}$),

$$f(x) = xe^x - 1, \qquad f'(x) = e^x(1 + x),$$

so that, for Newton's method,

$$g(x) = x - \frac{xe^x - 1}{e^x(1 + x)} = x - \frac{x - e^{-x}}{1 + x}.$$

Newton's method is

$$x_{r+1} = x_r - \frac{x_r - e^{-x_r}}{1 + x_r}.$$

The first few x_r, with $x_0 = 0.5$, are given to five decimal places in Table 7.5. □

Table 7.5   Newton's method.

r	0	1	2	3
x_r	0.5	0.57102	0.56716	0.56714

To investigate the *convergence* of Newton's method to a root α of $f(x) = 0$, we consider (7.27) with $k = 2$. Successive errors behave as

$$e_{r+1} = \tfrac{1}{2}e_r^2 \cdot g''(\xi_r), \qquad (7.33)$$

where ξ_r lies between α and x_r and g'' may be found by differentiating (7.32). We assert, without formal proof, that provided $f'(x) \neq 0$ and x_0 is chosen sufficiently close to α, we may deduce from (7.33) that the sequence $\{e_r\}$ converges to zero. Thus Newton's method converges. We also state a more useful result on convergence.

Theorem 7.2   If there exists an interval $[a, b]$ such that

(i) $f(a)$ and $f(b)$ have opposite signs,
(ii) f'' does not change sign on $[a, b]$,
(iii) the tangents to the curve $y = f(x)$ at both a and b cut the x-axis within $[a, b]$,

then $f(x) = 0$ has a unique root α in $[a, b]$ and Newton's method converges to α, for *any* $x_0 \in [a, b]$.

The proof is omitted. Condition (i) guarantees there is at least one root $\alpha \in [a, b]$ and (ii) means that f is convex or concave on $[a, b]$, which allows only one root of $f(x) = 0$ in $[a, b]$. Condition (iii) together with (ii) ensures that, if $x_r \in [a, b]$, the next iterate x_{r+1}, produced by Newton's method, is also in $[a, b]$. □

The best-known application of Newton's method is the square root algorithm. The square root of a positive number c is the positive root of the equation $x^2 - c = 0$. In this case, $f(x) = x^2 - c$ and $f'(x) = 2x$. So Newton's method is

$$x_{r+1} = x_r - \frac{x_r^2 - c}{2x_r} \quad (x_r \neq 0).$$

This simplifies to give

$$x_{r+1} = \frac{1}{2}\left(x_r + \frac{c}{x_r}\right). \tag{7.34}$$

All three conditions of Theorem 7.2 hold for the function $x^2 - c$ on an interval $[a, b]$ such that $0 < a < c^{1/2}$ and $b > \frac{1}{2}(a + c/a)$. The inequality for b ensures that both conditions (i) and (iii) hold. (See Problem 7.15.) Thus the square root algorithm converges for any $x_0 > 0$. For a more elementary proof of the convergence of the square root method (7.34), see Problem 7.16. Notice that it is easy to demonstrate the expected second order convergence of the sequence generated by (7.34). From

$$x_{r+1} - c^{1/2} = \frac{1}{2}\left(x_r + \frac{c}{x_r}\right) - c^{1/2} = \frac{1}{2x_r}(x_r^2 - 2c^{1/2}x_r + c),$$

we obtain

$$x_{r+1} - c^{1/2} = \frac{1}{2x_r}(x_r - c^{1/2})^2. \tag{7.35}$$

Note the squaring of the error $x_r - c^{1/2}$, associated with second order convergence.

Note that, given any $c > 0$, there exists an integer n such that $\frac{1}{4} \leq 4^n c < 1$ and thus $\frac{1}{2} \leq 2^n c^{1/2} < 1$. To calculate $c^{1/2}$, it therefore suffices to consider c in the range $\frac{1}{4} \leq c < 1$ and we choose

$$x_0 = \frac{2}{3}c + \frac{17}{48},$$

which is the minimax approximation to $c^{1/2}$ on $[\frac{1}{4}, 1]$. We have (see Example 5.14)

$$|x_0 - c^{1/2}| \leq \frac{1}{48}, \tag{7.36}$$

for this choice of initial iterate x_0 and after very few iterations we obtain highly accurate approximations to $c^{1/2}$ (Problem 7.18).

We cannot always use Newton's method as it is not always feasible to evaluate f'. An obvious way to modify (7.30) is to replace $f'(x_r)$ by the divided difference approximation, $(f(x_r) - f(x_{r-1}))/(x_r - x_{r-1})$. Thus (7.30) becomes

$$x_{r+1} = x_r - \frac{f(x_r) \cdot (x_r - x_{r-1})}{f(x_r) - f(x_{r-1})},$$

which we have already met as the secant method.

In Problem 7.21 an alternative method of obtaining Newton's method is described and in Problem 7.22 an extension of Newton's method to a third order process is given.

7.7 The contraction mapping theorem

Recall Theorem 7.1, which states that if g' is continuous and $|g'(x)| \leqslant L < 1$ for all values of x, then the iterative process

$$x_{r+1} = g(x_r), \qquad x_0 \text{ arbitrary},$$

converges to the unique root of $x = g(x)$. The conditions of Theorem 7.1 are not satisfied by the function $g(x) = e^{-x}$, yet convergence appeared to take place in that case (Example 7.5). It is not necessary to require $|g'(x)| \leqslant L < 1$ for *all* values of x. We will make use of a Lipschitz condition (Definition 2.12) and will require:

Definition 7.1   A function g is said to be a *contraction mapping* on an interval $[a, b]$ if

    (i) $x \in [a, b] \Rightarrow g(x) \in [a, b]$,   (7.37)
    (ii) g satisfies a Lipschitz condition on $[a, b]$ with a Lipschitz constant $L < 1$. Thus

$$x, y \in [a, b] \Rightarrow |g(x) - g(y)| \leqslant L|x - y|,$$

    where $L < 1$.
Condition (i) is sometimes referred to as the "closure condition". $\square$

We can now give a less restrictive result than Theorem 7.1.

Theorem 7.3   If there is an interval $[a, b]$ on which g is a contraction mapping, then

    (i) the equation $x = g(x)$ has a unique root (say α) in $[a, b]$,
    (ii) for any $x_0 \in [a, b]$, the sequence defined by

$$x_{r+1} = g(x_r), \qquad r = 0, 1, \ldots, \tag{7.38}$$

converges to α.

Proof   For any $x \in [a, b]$, $g(x) \in [a, b]$. In particular, for $x = a$,

$$a - g(a) \leqslant 0 \tag{7.39}$$

and, for $x = b$,

$$b - g(b) \geqslant 0. \tag{7.40}$$

The Lipschitz condition implies that g is continuous on $[a, b]$. Thus $x - g(x)$ is continuous on $[a, b]$ and, from the inequalities (7.39) and (7.40), we deduce that the equation $x - g(x) = 0$ has at least one root belonging to $[a, b]$. Let us assume that there is more than one root and let α and β $(a \leqslant \alpha, \beta \leqslant b)$ denote two distinct roots. Therefore

$$\alpha = g(\alpha), \qquad \beta = g(\beta)$$

and, subtracting these equations,

$$\alpha - \beta = g(\alpha) - g(\beta).$$

From the contraction mapping property, we deduce that

$$|\alpha - \beta| = |g(\alpha) - g(\beta)| \leqslant L|\alpha - \beta|,$$

with $0 \leqslant L < 1$. This entails $|\alpha - \beta| < |\alpha - \beta|$, which is impossible. We deduce that there is only one root, say α, belonging to $[a, b]$.

Note that, since g is a contraction mapping on $[a, b]$, $x_r \in [a, b] \Rightarrow g(x_r) \in [a, b]$, which means $x_{r+1} \in [a, b]$. By induction, every $x_r \in [a, b]$ if $x_0 \in [a, b]$. From the Lipschitz condition, it follows that

$$|x_{r+1} - \alpha| = |g(x_r) - g(\alpha)| \leqslant L|x_r - \alpha|.$$

Hence, as in the proof of Theorem 7.1, we deduce that

$$|x_r - \alpha| \leqslant L^r|x_0 - \alpha| \tag{7.41}$$

and, since $0 \leqslant L < 1$, we conclude that the sequence $\{x_r\}$ converges to α. $\square$

Looking back over the proof, we see that it is not *essential* to use a Lipschitz condition. If we prefer, we could use the mean value theorem and require instead that $|g'(x)| \leqslant L < 1$ on $[a, b]$. See Chapter 2, p. 22. It is, however, informative to use the Lipschitz condition to prepare for the extension of Theorem 7.3 to deal with iterative methods for a *system* of algebraic equations, which we discuss in Chapters 9 and 10.

From (7.41), we can derive an *a priori* estimate of the error at any stage. Since both x_0 and α belong to $[a, b]$, we obtain

$$|x_r - \alpha| \leqslant L^r(b - a). \tag{7.42}$$

See also Problem 7.25.

Example 7.9 We apply Theorem 7.3 to the equation $x = e^{-x}$ of Example 7.5. For $g(x) = e^{-x}$, $g(0.4) = 0.67$ and $g(0.7) = 0.50$, to two decimal places. Since g is also a decreasing function,

$$x \in [0.4, 0.7] \Rightarrow g(x) \in [0.4, 0.7].$$

We have $g'(x) = -e^{-x}$ and, therefore, $|g'(x)| < 0.7$ on $[0.4, 0.7]$. Thus g is a contraction mapping with $L = 0.7$ and Theorem 7.3 shows that the process

$x_{r+1} = e^{-x_r}$ converges to the unique root of $x = e^{-x}$ in $[0.4, 0.7]$, for any x_0 in that interval. This confirms the apparent convergence of Example 7.5. $\square$

As in Example 7.9, it can take some ingenuity to determine an interval $[a, b]$ on which g is a contraction mapping and therefore Theorem 7.3 is not of great *practical* importance. In particular, the closure condition is often awkward to verify. In practice, we will not often choose a first order method to solve an equation $f(x) = 0$. If f' can easily be evaluated we would probably prefer to use Newton's method because of its second order convergence. If it is difficult to evaluate f', we would opt for a "bracketing" method such as Muller's method or the modified form of the regula falsi method discussed in §7.3. In general both these methods converge faster than the first order method based on contraction mapping. On occasions, we may even wish to use the bisection method because of its simplicity, despite the relatively large number of evaluations of f which it requires.

Problems

Section 7.2

7.1   Show that the equation $1 - x - \sin x = 0$ has a root in $[0, 1]$. How many iterations of the bisection method are required to estimate this root with an error not greater than $\frac{1}{2}.10^{-4}$?

7.2   Use the bisection method to find the positive root of $x^2 - x - 1 = 0$ to one decimal place.

Section 7.3

7.3   Use the regula falsi method to estimate the root of $1 - x - \sin x = 0$, beginning with $x_0 = 0$, $x_1 = 1$ and stopping when $|1 - x_2 - \sin x_2| < \frac{1}{2}.10^{-2}$.

7.4   Find an approximation to the smallest positive root of the equation $8x^4 - 8x^2 + 1 = 0$, by using one step of Muller's method, with $x_0 = 0.3$, $x_1 = 0.5$ and $x_2 = 0.4$. (Note that the equation is $T_4(x) = 0$ and the smallest positive root is therefore $\cos(3\pi/8)$. See §4.10.)

Section 7.4

7.5   Compare the amount of calculation required to find the root of $e^x + 10x - 2 = 0$ to three decimal places by (i) the bisection method beginning with $x_0 = 0$, $x_1 = 1$ and (ii) the iterative scheme $x_{r+1} = (2 - e^{x_r})/10$, $r = 0, 1, \ldots$, with $x_0 = 0$.

7.6 A function f is such that $f'(x)$ exists and $M \geqslant f'(x) \geqslant m > 0$ for all x. Prove that the sequence $\{x_r\}$ generated by $x_{r+1} = x_r - \lambda f(x_r)$, $r = 0, 1, \ldots$, and arbitrary x_0, is convergent to the zero of f for any choice of λ in the range $0 < \lambda < 2/M$.

Section 7.5

7.7 Choose a constant λ so that the process

$$x_{r+1} = (\lambda x_r + 1 - \sin x_r)/(1 + \lambda)$$

will give a rapidly convergent method for finding the root of $1 - x - \sin x = 0$ near $x = 0.5$. Calculate the first few iterates, beginning with $x_0 = 0.5$.

7.8 Show that the number x^*_{r+1}, defined by (7.22), obtained by applying Aitken acceleration to the three successive iterates x_{r-1}, x_r and x_{r+1}, may also be expressed as

$$x^*_{r+1} = x_{r-1} - \frac{(\Delta x_{r-1})^2}{\Delta^2 x_{r-1}}.$$

7.9 If $x_r = a_1 + a_2 + \cdots + a_r$, show that the result of applying Aitken acceleration to x_{n-2}, x_{n-1} and x_n is

$$x^*_n = x_{n-2} + \frac{a^2_{n-1}}{a_{n-1} - a_n},$$

if $a_{n-1} \neq a_n$.

7.10 Working to five decimal places, apply the result of Problem 7.9, with $n = 12$, to estimate

$$\ln 2 = 1 - \tfrac{1}{2} + \tfrac{1}{3} - \tfrac{1}{4} + \cdots.$$

7.11 Let $\{x^*_r\}$ and $\{y^*_r\}$ denote sequences obtained by applying Aitken acceleration to $\{x_r\}$ and $\{y_r\}$ respectively. If $x_r = c + y_r$ for all r (where c is constant), show that $x^*_r = c + y^*_r$. (For example, if $x_0 = 2.791$, $x_1 = 2.737$, $x_2 = 2.723$, we can apply Aitken acceleration to the last two digits of each number.)

Section 7.6

7.12 Obtain a quadratic equation to which the iterative method

$$x_{r+1} = (bx_r^2 + 2cx_r)/(c - ax_r^2)$$

relates. Show that, if $c \neq 0$ and $ax^2 \neq c$ at a root, the method is *exactly* second order. What happens if $c = 0$?

7.13 Given some number $x_0 > 0$, verify that the iterative scheme $x_{r+1} = c/x_r$, $r = 0, 1, \ldots$, produces the periodic sequence $x_0, c/x_0, x_0, c/x_0, \ldots$. Show that, although this is obviously not a viable method for calculating $c^{1/2}$, Aitken acceleration on the three iterates $x_0, c/x_0, x_0$ yields the number $\frac{1}{2}(x_0 + c/x_0)$, which is equivalent to one iteration of Newton's method.

7.14 Derive an iterative process for finding the pth root of a number $c > 0$ by applying Newton's method to the equation $x^p - c = 0$.

7.15 Show that, if $c > 0$, all the conditions of Theorem 7.2 hold for the function $f(x) = x^2 - c$ on an interval $[a, b]$ with $0 < a < c^{1/2}$ and $b > \frac{1}{2}(a + c/a)$.

7.16 Deduce from (7.35) and another similar result that the sequence of iterates $\{x_r\}$, produced by the Newton square root method, satisfy

$$\frac{x_r - c^{1/2}}{x_r + c^{1/2}} = \left(\frac{x_0 - c^{1/2}}{x_0 + c^{1/2}}\right)^{2^r}.$$

Hence show that, for any $x_0 > 0$, the sequence $\{x_r\}$ converges to $c^{1/2}$.

7.17 What happens in Newton's square root process if we choose $x_0 < 0$?

7.18 If $\frac{1}{4} \leqslant c < 1$, let x_1 and x_2 denote the iterates produced by Newton's method for finding $c^{1/2}$, beginning with $x_0 = \frac{2}{3}c + \frac{17}{48}$. Noting that $x_0 > \frac{1}{2}$ as $c \geqslant \frac{1}{4}$, deduce from (7.35) that $|x_1 - c^{1/2}| < (x_0 - c^{1/2})^2$. Hence, using (7.36), show that $|x_2 - c^{1/2}| < 2.10^{-7}$.

7.19 If $f(x) = (x - \alpha)^2 h(x)$ and $h(\alpha) \neq 0$, show that $f'(\alpha) = 0$ and that, in this case, Newton's method is *not* a second order process for finding α. Find a value of λ so that the following modified version of Newton's method is at least second order for this function:

$$x_{r+1} = x_r - \lambda f(x_r)/f'(x_r).$$

7.20 Apply Newton's method to the equation $1/x - c = 0$. (Notice that this provides an algorithm for computing the reciprocal of a number c without performing the arithmetical operation of division.)

7.21 One way of obtaining iterative methods of solving $f(x) = 0$ is to use an approximating function as in Muller's method but based on Taylor polynomials. If x_n is the nth iterate, we construct the Taylor polynomial at x_n,

$$p_k(x) = f(x_n) + \frac{(x - x_n)}{1!}f'(x_n) + \cdots + \frac{(x - x_n)^k}{k!}f^{(k)}(x_n),$$

assuming the derivatives exist. We now take x_{n+1} to be a root of $p_k(x) = 0$. Show that for $k = 1$ we obtain Newton's method and that for $k = 2$

$$x_{n+1} = x_n + \frac{-f'(x_n) \pm [f'(x_n)^2 - 2f(x_n)f''(x_n)]^{1/2}}{f''(x_n)},$$

the $+$ or $-$ sign being taken according to whether $f'(x_n)$ is positive or negative. Show that the last method is at least third order if $f'(\alpha) \neq 0$ and $f''(\alpha) \neq 0$, where α is the root. The method is difficult to use as we must calculate a square root in each iteration. A better third order method is given in the next problem.

7.22 Show that the iterative method

$$x_{n+1} = x_n - \frac{f(x_n)}{f'(x_n)} - \frac{f''(x_n)}{2f'(x_n)} \left(\frac{f(x_n)}{f'(x_n)} \right)^2,$$

used to solve $f(x) = 0$, is of order 3 (at least) provided $f'(\alpha) \neq 0$ where α is the root. Obtain a third order square root process.

7.23 Suppose $y = f(x)$ has a unique inverse function $x = \phi(y)$. If ϕ is sufficiently differentiable, we have the Taylor series

$$\phi(y) = \phi(y_n) + (y - y_n)\phi'(y_n) + \frac{(y - y_n)^2}{2!} \phi''(y_n) + \cdots.$$

By expressing ϕ' and ϕ'' in terms of the derivatives of f, derive the iterative method quoted in Problem 7.22.

Section 7.7

7.24 If $x_{r+1} = g(x_r)$, $r = 0, 1, \ldots$, and g and x_0 satisfy the conditions of Theorem 7.3, deduce that

$$|x_{r+1} - x_r| \leqslant L|x_r - x_{r-1}|, \qquad r = 1, 2, \ldots,$$

and hence that

$$|x_{r+1} - x_r| \leqslant L^r|x_1 - x_0|, \qquad r = 0, 1, \ldots.$$

7.25 Applying the second inequality derived in Problem 7.24 to the inequality

$$|x_{m+n} - x_n| \leqslant |x_{m+n} - x_{m+n-1}| + \cdots + |x_{n+1} - x_n|,$$

deduce that, for $m > n \geqslant 0$,

$$|x_{m+n} - x_n| \leqslant \frac{L^n}{1 - L} |x_1 - x_0|.$$

Letting $m \to \infty$, obtain the estimate

$$|\alpha - x_n| \leqslant \frac{L^n}{1 - L} |x_1 - x_0|$$

for the error between the iterate x_n and the root α of the equation $x = g(x)$.

7.26 Deduce from Problem 7.25 that

$$|x_{n+1} - \alpha| \leqslant \frac{L}{1 - L} |x_{n+1} - x_n|.$$

(This enables us to use $x_{n+1} - x_n$ to decide when to stop iterating.)

7.27 By summing the geometric series on the right, show that the positive root of the equation

$$x^k = 1 + x + x^2 + \cdots + x^{k-1} \qquad\qquad (k \geqslant 2)$$

also satisfies $x = 2 - x^{-k}$. Show that the conditions of Theorem 7.3 apply to this last equation on the interval $[2 - 2^{1-k}, 2]$. For the case $k = 6$, find the root to three decimal places. (It may be assumed that $(1 - 2^{-k})^{-k} \leqslant 2$ for $k = 1, 2, \ldots$.)

7.28 Suppose that on some interval I, g and g' are continuous and that $x = g(x)$ has a root $\alpha \in I$ with $|g'(\alpha)| < 1$. Show that g is a contraction mapping on some sub-interval of I containing α.

7.29 Suppose that for all x, g and g' exist with $|g'(x)| \leqslant L < 1$ and that $y = g(x)$ has a unique inverse function $x = \psi(y)$ such that $\psi'(y)$ exists for all y. Show that $\psi(y)$ is not a contraction mapping, regardless of the interval chosen.

7.30 Show that we cannot expect the iterative method

$$x_{r+1} = c/x_r^{p-1}$$

to be successful in finding the pth root (p a positive integer other than $+1$) of a positive number c.

Chapter 8

Linear Equations

8.1 Introduction

In this chapter we shall be considering the solution of simultaneous *linear* algebraic equations. We shall consider separately (in Chapter 10) the more difficult problem of solving non-linear equations, as the methods for these are quite different.

If three unknowns x, y and z are related by equations of the form

$$a_1 x + b_1 y + c_1 z = d_1$$
$$a_2 x + b_2 y + c_2 z = d_2 \qquad (8.1)$$
$$a_3 x + b_3 y + c_3 z = d_3$$

where a_i, b_i, c_i and d_i ($i = 1, 2, 3$) are given constants, we say that (8.1) is a system of three linear equations. Since we have three equations in three unknowns, we shall usually expect that there is a unique solution, that is, there are unique values of x, y and z satisfying (8.1). We shall see that this is not always true and the intention in the first part of this chapter is to establish conditions which ensure the existence of solutions of linear equations. We find it convenient to express linear equations in terms of *matrices* and therefore we introduce the latter first.

8.2 Matrices

Definitions

We define an $m \times n$ matrix $\mathbf{A}$ as an ordered set of mn real (or complex) numbers a_{ij} written in the form

$$\mathbf{A} = \begin{bmatrix} a_{11} & a_{12} & \cdots & a_{1n} \\ a_{21} & a_{22} & \cdots & a_{2n} \\ \vdots & & & \vdots \\ a_{m1} & a_{m2} & \cdots & a_{mn} \end{bmatrix}.$$

The numbers a_{ij} are called the elements of $\mathbf{A}$. You will notice that $\mathbf{A}$ consists of m rows of elements arranged in n columns. We say that $\mathbf{A}$ is of *order*

179

$m \times n$ or of *dimension* $m \times n$. The element a_{ij} is in the ith row and jth column and is called the (i, j)th element of **A**. We shall usually use bold capital letters to denote matrices and italic small letters to denote elements. If $m = n$ we say that **A** is a *square* matrix of order or dimension n. We refer to the elements a_{ii}, $i = 1, 2, \ldots, n$, as the *diagonal* of a square matrix.

A *column vector* is a matrix with just one column. For example,

$$\begin{bmatrix} x_1 \\ x_2 \\ \vdots \\ x_m \end{bmatrix}$$

is of order $m \times 1$ and is called a column vector of dimension m. We usually denote such a column vector by a small bold letter, for example, **x**.

A *row vector* is a matrix with just one row. For example,

$$[x_1 \quad x_2 \quad \cdots \quad x_n]$$

is of order $1 \times n$ and is called a row vector of dimension n. We usually denote such a vector by $\mathbf{x}^T$. The superscript T indicates that the vector is a row vector obtained by writing the elements of an n-dimensional column vector **x** in a row. We also call $\mathbf{x}^T$ the *transpose* of **x**.

The set of all n-dimensional column vectors is referred to as n-dimensional space. Vector quantities in two or three dimensions, with which you are probably familiar, can be written as either column or row vectors of two or three dimensions.

Equality

Before we can discuss operations involving matrices we must define what we mean by equality for matrices. Two matrices are said to be *equal* if they are of the same order and corresponding elements are equal, that is, **A** = **B** if $a_{ij} = b_{ij}$ for all i and j.

Addition

The sum of two matrices of *equal dimensions* is defined as

$$\mathbf{C} = \mathbf{A} + \mathbf{B}$$

where $c_{ij} = a_{ij} + b_{ij}$, for all i and j. We simply add corresponding elements. Notice that the sum is not defined for two matrices of different dimensions. It can be seen that for vectors this definition of addition is consistent with the parallelogram law.

If **a** and **b** are two displacement vectors in three-dimensional space, that is, **a** represents the displacement from the origin O to the point (a_1, a_2, a_3) (see Fig. 8.1) and **b** that from O to the point (b_1, b_2, b_3), then

$$\mathbf{a} + \mathbf{b} = \begin{bmatrix} a_1 + b_1 \\ a_2 + b_2 \\ a_3 + b_3 \end{bmatrix}.$$

This is the sum of the displacement vectors given by the parallelogram law.

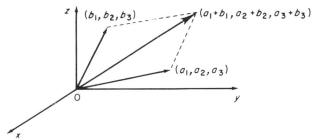

Fig. 8.1. The sum of two vectors.

The *zero matrix* of order $m \times n$, denoted by **0**, has zero for each element and, therefore,

$$\mathbf{A} + \mathbf{0} = \mathbf{A}$$

as $a_{ij} + 0 = a_{ij}$ for all i and j. From the definition of addition it is evident that the order in which addition is carried out is not important. Thus addition is commutative

$$\mathbf{A} + \mathbf{B} = \mathbf{B} + \mathbf{A}$$

and associative

$$\mathbf{A} + (\mathbf{B} + \mathbf{C}) = (\mathbf{A} + \mathbf{B}) + \mathbf{C},$$

as these properties are valid for the addition of individual elements.

Scalar multiplication

Given any real (or complex) number λ and an $m \times n$ matrix **A**, we define the $m \times n$ matrix

$$\lambda \mathbf{A} = \begin{bmatrix} \lambda a_{11} & \lambda a_{12} & \cdots & \lambda a_{1n} \\ \lambda a_{21} & \lambda a_{22} & \cdots & \lambda a_{2n} \\ \vdots & & & \vdots \\ \lambda a_{m1} & \lambda a_{m2} & \cdots & \lambda a_{mn} \end{bmatrix}.$$

We multiply each element of **A** by λ. In this context we call λ a *scalar*. In the case of vectors we obtain the usual method of scaling.

We often write $-\mathbf{A}$ instead of $(-1).\mathbf{A}$ and $\mathbf{A} - \mathbf{B}$ instead of $\mathbf{A} + (-1).\mathbf{B}$. From the definitions it is easily verified that

$$\mathbf{A} + (-\mathbf{A}) = \mathbf{0}$$

and

$$0.\mathbf{A} = \mathbf{0}.$$

Multiplication of matrices

(a) *row vector* $\times$ *column vector*

If $\mathbf{y}^T = [y_1 \quad \cdots \quad y_n]$ and

$$\mathbf{x} = \begin{bmatrix} x_1 \\ x_2 \\ \vdots \\ x_n \end{bmatrix}$$

so that $\mathbf{x}$ and $\mathbf{y}^T$ have an equal number of elements, we define

$$\mathbf{y}^T\mathbf{x} = y_1 x_1 + y_2 x_2 + \cdots + y_n x_n = \sum_{i=1}^{n} y_i x_i.$$

Note that the result consists of a single real (or complex) number since the products of corresponding elements are added together. We call $\mathbf{y}^T\mathbf{x}$ the *scalar product* (or *inner product*) of $\mathbf{y}$ and $\mathbf{x}$. This is consistent with the definition of scalar product for vectors in two- and three-dimensional space.

(b) *rectangular matrix* $\times$ *column vector*

We define the product of an $m \times n$ matrix **A** by an $n \times 1$ column vector **x** as

$$\mathbf{Ax} = \begin{bmatrix} a_{11} & a_{12} & \cdots & a_{1n} \\ a_{21} & a_{22} & \cdots & a_{2n} \\ \vdots & & & \vdots \\ a_{m1} & a_{m2} & \cdots & a_{mn} \end{bmatrix} \begin{bmatrix} x_1 \\ x_2 \\ \vdots \\ x_n \end{bmatrix} = \begin{bmatrix} a_{11}x_1 + a_{12}x_2 + \cdots + a_{1n}x_n \\ a_{21}x_1 + a_{22}x_2 + \cdots + a_{2n}x_n \\ \vdots \\ a_{m1}x_1 + a_{m2}x_2 + \cdots + a_{mn}x_n \end{bmatrix}.$$

The result is an $m \times 1$ column vector **z** with ith element

$$z_i = \sum_{j=1}^{n} a_{ij}x_j = [i\text{th row of } \mathbf{A}] \begin{bmatrix} x_1 \\ \vdots \\ x_n \end{bmatrix}.$$

This definition may seem strange at first but is in fact very useful. For example the equations (8.1) may be written as

$$\mathbf{Ax} = \mathbf{d},$$

where

$$\mathbf{A} = \begin{bmatrix} a_1 & b_1 & c_1 \\ a_2 & b_2 & c_2 \\ a_3 & b_3 & c_3 \end{bmatrix}, \quad \mathbf{x} = \begin{bmatrix} x \\ y \\ z \end{bmatrix} \quad \text{and} \quad \mathbf{d} = \begin{bmatrix} d_1 \\ d_2 \\ d_3 \end{bmatrix}.$$

(c) *rectangular matrix × rectangular matrix*

If $\mathbf{A}$ is an $m \times n$ matrix and $\mathbf{B}$ is an $n \times p$ matrix, we define their product $\mathbf{C} = \mathbf{AB}$ to be the $m \times p$ matrix with (i, j)th element

$$c_{ij} = \sum_{k=1}^{n} a_{ik}b_{kj} = [i\text{th row of } \mathbf{A}] \begin{bmatrix} j\text{th} \\ \text{column} \\ \text{of } \mathbf{B} \end{bmatrix}.$$

You will notice that the number of elements in a row of $\mathbf{A}$ must be the same as the number of elements in a column of $\mathbf{B}$ (number of columns in $\mathbf{A}$ = number of rows in $\mathbf{B}$). The multiplication rules in (a) and (b) for vectors are special cases of this more general rule.

Multiplication is not necessarily commutative even if both $\mathbf{AB}$ and $\mathbf{BA}$ are defined; that is, $\mathbf{AB} \neq \mathbf{BA}$ in general. This is true even if $\mathbf{AB}$ and $\mathbf{BA}$ are both of the same dimension. For example, if

$$\mathbf{A} = \begin{bmatrix} 3 & 4 & 2 \\ -1 & 2 & 3 \\ 2 & 1 & 4 \end{bmatrix} \quad \text{and} \quad \mathbf{B} = \begin{bmatrix} 1 & -1 & 2 \\ 2 & 1 & -2 \\ 3 & 2 & 1 \end{bmatrix}$$

$$\mathbf{AB} = \begin{bmatrix} 17 & 5 & 0 \\ 12 & 9 & -3 \\ 16 & 7 & 6 \end{bmatrix} \quad \text{and} \quad \mathbf{BA} = \begin{bmatrix} 8 & 4 & 7 \\ 1 & 8 & -1 \\ 9 & 17 & 16 \end{bmatrix}.$$

Thus the order in which two matrices are multiplied is important and we say that, for a product $\mathbf{AB}$, the matrix $\mathbf{A}$ is *postmultiplied* by $\mathbf{B}$ or that $\mathbf{B}$ is *premultiplied* by $\mathbf{A}$.

Multiplication is associative, that is, if $\mathbf{A}$, $\mathbf{B}$ and $\mathbf{C}$ are of suitable dimensions

$$\mathbf{A(BC)} = \mathbf{(AB)C}.$$

In each case the (i, j)th element is

$$\sum_k \sum_l a_{ik}b_{kl}c_{lj}.$$

Addition is distributive over multiplication, that is, for any matrices of suitable dimensions

$$\mathbf{A(B + C)} = \mathbf{AB} + \mathbf{AC}$$

and

$$\mathbf{(A + B)C} = \mathbf{AC} + \mathbf{BC}.$$

In the former the (i, j)th element of the left side is

$$\sum_k a_{ik}(b_{kj} + c_{kj}) = \sum_k a_{ik}b_{kj} + \sum_k a_{ik}c_{kj}$$

and the terms on the right are the (i, j)th elements of $\mathbf{AB}$ and $\mathbf{AC}$.

The reader is advised to familiarize himself with the multiplication rule. Self-checking examples such as that in Problem 8.1 are suggested.

(d) *unit matrix*

We define the $n \times n$ *unit matrix* $\mathbf{I}$ as

$$\mathbf{I} = \begin{bmatrix} 1 & 0 & \cdots & 0 \\ 0 & 1 & \ddots & \vdots \\ \vdots & \ddots & \ddots & 0 \\ 0 & \cdots & 0 & 1 \end{bmatrix}.$$

It may be verified from the multiplication rule that, if $\mathbf{A}$ is any $m \times n$ matrix and $\mathbf{I}$ is the $n \times n$ unit matrix, then

$$\mathbf{AI} = \mathbf{A}$$

and, if I is the $m \times m$ unit matrix,

$$\mathbf{IA} = \mathbf{A}.$$

Linear transformations

The importance of the multiplication rules is emphasized by considering matrices as linear transformations. Suppose we wish to transform from one set of coordinates (x, y, z) to another set (u, v, w) centred on the same origin. (See Fig. 8.2.) The point (x, y, z) has new coordinates of the form

$$\begin{align} u &= a_1x + b_1y + c_1z \\ v &= a_2x + b_2y + c_2z \\ w &= a_3x + b_3y + c_3z \end{align} \tag{8.2}$$

where the coefficients a_i, b_i and c_i ($i = 1, 2, 3$) are determined by the positions of the new axes. We write (8.2) in the form

$$\mathbf{u} = \mathbf{Ax} \tag{8.3}$$

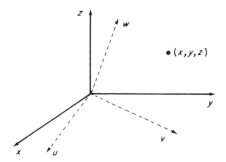

Fig. 8.2. Transformation of coordinates.

where

$$\mathbf{u} = \begin{bmatrix} u \\ v \\ w \end{bmatrix}, \quad \mathbf{A} = \begin{bmatrix} a_1 & b_1 & c_1 \\ a_2 & b_2 & c_2 \\ a_3 & b_3 & c_3 \end{bmatrix} \quad \text{and} \quad \mathbf{x} = \begin{bmatrix} x \\ y \\ z \end{bmatrix}.$$

Now suppose that we make a further linear transformation from (u, v, w) coordinates to (p, q, r) coordinates also centred on the same origin. Our general point will now have coordinates of the form

$$\begin{aligned} p &= \alpha_1 u + \beta_1 v + \gamma_1 w \\ q &= \alpha_2 u + \beta_2 v + \gamma_2 w \\ r &= \alpha_3 u + \beta_3 v + \gamma_3 w \end{aligned} \tag{8.4}$$

or, we may write,

$$\mathbf{p} = \mathbf{Bu} \tag{8.5}$$

where

$$\mathbf{p} = \begin{bmatrix} p \\ q \\ r \end{bmatrix} \quad \text{and} \quad \mathbf{B} = \begin{bmatrix} a_1 & \beta_1 & \gamma_1 \\ \alpha_2 & \beta_2 & \gamma_2 \\ \alpha_3 & \beta_3 & \gamma_3 \end{bmatrix}.$$

The question we now ask is: how can we determine the (p, q, r) coordinates of our point directly from its (x, y, z) coordinates? From (8.3) and (8.5) we would like to be able to write

$$\mathbf{p} = \mathbf{Bu} = \mathbf{B}.(\mathbf{Ax}) = (\mathbf{BA}).\mathbf{x}, \tag{8.6}$$

where $\mathbf{BA}$ is defined using the rules for matrix multiplication. It is not difficult to verify that this is, indeed, valid. For example, from (8.4),

$$p = \alpha_1 u + \beta_1 v + \gamma_1 w$$

and, on using (8.2), we obtain

$$p = \alpha_1(a_1 x + b_1 y + c_1 z) + \beta_1(a_2 x + b_2 y + c_2 z) + \gamma_1(a_3 x + b_3 y + c_3 z)$$

$$= [(\alpha_1 a_1 + \beta_1 a_2 + \gamma_1 a_3) \quad (\alpha_1 b_1 + \beta_1 b_2 + \gamma_1 b_3) \quad (\alpha_1 c_1 + \beta_1 c_2 + \gamma_1 c_3)] \begin{bmatrix} x \\ y \\ z \end{bmatrix}$$

$$= [\text{first row of } \mathbf{BA}] \begin{bmatrix} x \\ y \\ z \end{bmatrix}.$$

This is consistent with (8.6).

Block matrices

We sometimes find it convenient to write matrices in a block form. The $m \times n$ matrix $\mathbf{A}$ may be written as

$$\mathbf{A} = \begin{bmatrix} \mathbf{A}_{11} & \mathbf{A}_{12} \\ \mathbf{A}_{21} & \mathbf{A}_{22} \end{bmatrix}$$

where $\mathbf{A}_{11}$ is $r \times p$, $\mathbf{A}_{12}$ is $r \times (n - p)$, $\mathbf{A}_{21}$ is $(m - r) \times p$ and $\mathbf{A}_{22}$ is $(m - r) \times (n - p)$. The restrictions on dimensions are necessary so that, for example, $\mathbf{A}_{11}$ and $\mathbf{A}_{21}$ have the same number of columns. We call $\mathbf{A}_{11}$, $\mathbf{A}_{12}$, $\mathbf{A}_{21}$ and $\mathbf{A}_{22}$ *submatrices* of $\mathbf{A}$.

It is easily seen that if

$$\mathbf{B} = \begin{bmatrix} \mathbf{B}_{11} & \mathbf{B}_{12} \\ \mathbf{B}_{21} & \mathbf{B}_{22} \end{bmatrix}$$

is divided into blocks of the same dimensions as those of $\mathbf{A}$, then

$$\mathbf{A} + \mathbf{B} = \begin{bmatrix} \mathbf{A}_{11} + \mathbf{B}_{11} & \mathbf{A}_{12} + \mathbf{B}_{12} \\ \mathbf{A}_{21} + \mathbf{B}_{21} & \mathbf{A}_{22} + \mathbf{B}_{22} \end{bmatrix}.$$

It is not so obvious, however, that provided $\mathbf{B}$ and its blocks are of suitable dimensions

$$\mathbf{AB} = \begin{bmatrix} (\mathbf{A}_{11}\mathbf{B}_{11} + \mathbf{A}_{12}\mathbf{B}_{21}) & (\mathbf{A}_{11}\mathbf{B}_{12} + \mathbf{A}_{12}\mathbf{B}_{22}) \\ (\mathbf{A}_{21}\mathbf{B}_{11} + \mathbf{A}_{22}\mathbf{B}_{21}) & (\mathbf{A}_{21}\mathbf{B}_{12} + \mathbf{A}_{22}\mathbf{B}_{22}) \end{bmatrix}. \tag{8.7}$$

The proof of this last result is left for Problem 8.9.

8.3 Linear equations

We start by considering a simple example.

Example 8.1

Solve the set of equations

$$2x_1 + 2x_2 + 3x_3 = 3 \qquad (8.8a)$$

$$4x_1 + 7x_2 + 7x_3 = 1 \qquad (8.8b)$$

$$-2x_1 + 4x_2 + 5x_3 = -7. \qquad (8.8c)$$

We will solve these by an *elimination* process. First we use equation (8.8a) to eliminate x_1 from the other two equations. If we subtract twice (8.8a) from (8.8b) and add (8.8a) to (8.8c), we obtain the equations

$$2x_1 + 2x_2 + 3x_3 = 3$$
$$3x_2 + x_3 = -5$$
$$6x_2 + 8x_3 = -4.$$

We now eliminate x_2 from the third equation by subtracting twice the second equation. We obtain

$$2x_1 + 2x_2 + 3x_3 = 3 \qquad (8.9a)$$

$$3x_2 + x_3 = -5 \qquad (8.9b)$$

$$6x_3 = 6. \qquad (8.9c)$$

Immediately from (8.9c) we deduce that $x_3 = 1$. On substituting this value of x_3 in (8.9b) we obtain $3x_2 = -6$, so that $x_2 = -2$. Finally, from (8.9a) we have

$$2x_1 = 3 - 2x_2 - 3x_3 = 4,$$

whence $x_1 = 2$. The equations (8.9) are called *triangular* equations and the process of working in reverse order to solve them is called *back substitution*. $\square$

The method of Example 8.1 is known as *Gauss elimination* and may be extended to the m equations in n unknowns $x_1, x_2, \ldots, x_n$,

$$a_{11}x_1 + a_{12}x_2 + \cdots + a_{1n}x_n = b_1$$
$$a_{21}x_1 + a_{22}x_2 + \cdots + a_{2n}x_n = b_2 \qquad (8.10)$$
$$\vdots \qquad\qquad\qquad\qquad \vdots$$
$$a_{m1}x_1 + a_{m2}x_2 + \cdots + a_{mn}x_n = b_m$$

or

$$\mathbf{Ax} = \mathbf{b},$$

where the coefficients a_{ij} and b_j are given. We detach the coefficients and write them as the $m \times (n + 1)$ matrix

$$[\mathbf{A}\!:\!\mathbf{b}] = \begin{bmatrix} a_{11} & a_{12} & \cdots & a_{1n} \!:\! b_1 \\ a_{21} & a_{22} & \cdots & a_{2n} \!:\! b_2 \\ \vdots & & & \vdots \\ a_{m1} & a_{m2} & \cdots & a_{mn} \!:\! b_m \end{bmatrix}. \qquad (8.11)$$

The following operations may be made on the matrix (8.11) without affecting the solution of the corresponding equations.

(*a*) Rows of (8.11) may be interchanged. This merely changes the order in which the equations (8.10) are written.

(*b*) Any row of (8.11) may be multiplied by a non-zero scalar. This is equivalent to multiplying an equation in (8.10) by the scalar.

(*c*) Any row of (8.11) may be added to another. This is equivalent to adding one equation to another in (8.10).

(*d*) The first *n* columns of (8.11) may be interchanged provided corresponding unknowns are interchanged. This is equivalent to changing the order in which the unknowns are written in (8.10). For example, if we interchange the first two columns of (8.11), the first two terms in each of the equations (8.10) are interchanged and the order of the unknowns becomes $x_2, x_1, x_3, x_4, \ldots, x_n$.

We use these operations to reduce the equations to a simpler form. Firstly we ensure that the leading element a_{11} is non-zero by making row and/or column interchanges. We now subtract multiples of the first row from the other rows so as to make the rest of the coefficients in the first column zero. Thus, assuming that no interchanges were required at the start, we obtain the matrix

$$\begin{bmatrix} a_{11} & a_{12} & \cdots & a_{1n} & : & b_1 \\ 0 & a'_{22} & \cdots & a'_{2n} & : & b'_2 \\ 0 & a'_{32} & \cdots & a'_{3n} & : & b'_3 \\ \vdots & \vdots & & \vdots & & \vdots \\ 0 & a'_{m2} & \cdots & a'_{mn} & : & b'_m \end{bmatrix}$$

where, for example, the second row is obtained by subtracting $(a_{21}/a_{11}) \times$ the first row. Thus

$$a'_{22} = a_{22} - (a_{21}/a_{11})a_{12}$$

and

$$b'_m = b_m - (a_{m1}/a_{11})b_1.$$

We have eliminated x_1 from the 2nd to the *m*th equations.

We now ensure that the element a'_{22} is non-zero by making row and/or column interchanges, using rows 2 to *m* or columns 2 to *n*. Such interchanges do not affect the positions of zeros in the first column. We next subtract multiples of the second row from later rows to produce zeros in the rest of the second column, that is, we eliminate the second unknown from the 3rd to the *m*th equations.

We repeat this process and at the pth stage we have a matrix of the form:

$$
\begin{bmatrix}
\alpha_{11} & \alpha_{12} & \cdots & & \cdots & \alpha_{1n} & : & \beta_1 \\
0 & \alpha_{22} & \cdots & & \cdots & \alpha_{2n} & : & \beta_2 \\
\vdots & \ddots & \ddots & & & \vdots & & \vdots \\
\vdots & & 0 & \alpha_{pp} & \cdots & \alpha_{pn} & : & \beta_p \\
\vdots & & \vdots & \alpha_{p+1,p} & \cdots & \alpha_{p+1,n} & : & \beta_{p+1} \\
\vdots & & \vdots & \vdots & & \vdots & & \vdots \\
0 & \cdots & 0 & \alpha_{mp} & \cdots & \alpha_{mn} & : & \beta_m
\end{bmatrix}. \tag{8.12}
$$

The steps to be followed next are:

(1) interchange rows p to m or columns p to n so that α_{pp} is non-zero;

(2) subtract multiples of row p from rows $p + 1$ to m so that elements in the column below α_{pp} becomes zeros.

You will notice that neither of these steps will affect the positions of zeros in the earlier columns. At this stage α_{pp} is called the *pivot* and row p the *pivotal row*, corresponding to the *pivotal equation*.

We stop for one of three reasons: *either* it is not possible to make α_{pp} non-zero by interchanges, as all the α_{ij} in rows p to m are zero *or* $p = m$, when we have exhausted the equations *or* $p = n$, when we have exhausted the unknowns. The final form of the matrix is

$$
\begin{bmatrix}
u_{11} & u_{12} & u_{13} & \cdots & u_{1r} & u_{1,r+1} & \cdots & u_{1n} & : & c_1 \\
0 & u_{22} & u_{23} & \cdots & u_{2r} & u_{2,r+1} & \cdots & u_{2n} & : & c_2 \\
\vdots & & \ddots & \ddots & \vdots & \vdots & & \vdots & & \vdots \\
0 & \cdots & \cdots & 0 & u_{rr} & u_{r,r+1} & \cdots & u_{rn} & : & c_r \\
0 & \cdots & \cdots & \cdots & 0 & 0 & \cdots & 0 & : & c_{r+1} \\
\vdots & & & & \vdots & \vdots & & \vdots & & \vdots \\
0 & \cdots & \cdots & \cdots & 0 & 0 & \cdots & 0 & : & c_m
\end{bmatrix} \tag{8.13}
$$

which we may write in the block form

$$
\begin{bmatrix}
U & B & : & c \\
0 & 0 & : & d
\end{bmatrix}. \tag{8.14}
$$

U is the $r \times r$ *upper triangular* matrix

$$
U =
\begin{bmatrix}
u_{11} & \cdots & & \cdots & u_{1r} \\
0 & u_{22} & & & \vdots \\
\vdots & \ddots & \ddots & & \vdots \\
0 & & \cdots & 0 & u_{rr}
\end{bmatrix},
$$

B is $r \times (n - r)$ and the zero matrix below **B** is $(m - r) \times (n - r)$. The pivots u_{ii}, $i = 1, \ldots, r$, are non-zero.

We suppose that, as a result of column interchanges, the order of the unknowns $x_1, x_2, \ldots, x_n$ has been changed and that the unknowns corresponding to (8.13) are $y_1, y_2, \ldots, y_n$. If

$$\mathbf{y} = \begin{bmatrix} y_1 \\ \vdots \\ y_r \end{bmatrix} \quad \text{and} \quad \mathbf{z} = \begin{bmatrix} y_{r+1} \\ \vdots \\ y_n \end{bmatrix}$$

the equations may, from (8.14), be written in the block form

$$\begin{bmatrix} \mathbf{U} & \mathbf{B} \\ \mathbf{0} & \mathbf{0} \end{bmatrix} \begin{bmatrix} \mathbf{y} \\ \mathbf{z} \end{bmatrix} = \begin{bmatrix} \mathbf{c} \\ \mathbf{d} \end{bmatrix}. \tag{8.15}$$

Thus

$$\mathbf{U}\mathbf{y} + \mathbf{B}\mathbf{z} = \mathbf{c} \tag{8.16a}$$

and

$$\mathbf{0}\mathbf{y} + \mathbf{0}\mathbf{z} = \mathbf{d}. \tag{8.16b}$$

The original equations (8.10) and the reduced equations (8.16) are equivalent, in that a solution of (8.16) must also be a solution of (8.10) (possibly after some re-ordering of the unknowns) and *vice versa*. We therefore investigate the existence of solutions of (8.16) in detail.

The integer r in (8.13) is called the *rank* of the original coefficient matrix **A**. The existence of solutions will depend on this rank. If $r = m = n$, (8.16) is the triangular set of equations

$$\mathbf{U}\mathbf{y} = \mathbf{c}$$

where

$$\mathbf{U} = \begin{bmatrix} u_{11} & \cdots & & u_{1n} \\ 0 & \ddots & & \vdots \\ \vdots & & \ddots & \vdots \\ 0 & \cdots & 0 & u_{nn} \end{bmatrix}, \quad \mathbf{c} = \begin{bmatrix} c_1 \\ \vdots \\ c_n \end{bmatrix} \quad \text{and} \quad \mathbf{y} = \begin{bmatrix} y_1 \\ \vdots \\ y_n \end{bmatrix}.$$

with $u_{ii} \neq 0$, $i = 1, 2, \ldots, n$. We can solve these equations by back substitution. From the last equation

$$u_{nn} y_n = c_n,$$

we find

$$y_n = c_n / u_{nn}.$$

From the penultimate equation

$$u_{n-1,n-1}y_{n-1} + u_{n-1,n}y_n = c_{n-1},$$

we find

$$y_{n-1} = (c_{n-1} - u_{n-1,n}y_n)/u_{n-1,n-1}$$

and so on. We work back through the equations, introducing new unknowns one at a time and hence find all the unknowns in reverse order. Thus, for $r = m = n$, the unknowns are all uniquely determined, as we would normally expect for n equations in n unknowns. We then say that the equations (8.10) are *consistent* with *unique* solution or just simply *non-singular*. We also say that the square coefficient matrix $\mathbf{A}$ is non-singular.

If $r < m$ in (8.13) and $\mathbf{d} \neq \mathbf{0}$† in (8.16), that is, at least one of the c_{r+1}, $c_{r+2}, \ldots, c_m$ is non-zero, we cannot satisfy (8.16b) for *any* choice of $\mathbf{y}$ and $\mathbf{z}$. There is no solution of (8.16) and hence of (8.10) and we say that the equations are *inconsistent*.

If $r < m$ in (8.13) and $\mathbf{d} = \mathbf{0}$ then (8.16b) is satisfied for any $\mathbf{y}$ and $\mathbf{z}$ and (8.16a) may be written as

$$\mathbf{U}\mathbf{y} = \mathbf{c} - \mathbf{B}\mathbf{z}. \tag{8.17}$$

Similarly when $r = m < n$, (8.16b) does not apply and we obtain only (8.17). We may choose the elements of $\mathbf{z}$ arbitrarily. This fixes the right side of (8.17) and we may then solve these r triangular equations by back substitution to find $\mathbf{y}$. We say that the equations (8.10) are *consistent* with an *infinity* of *solutions*. A particular $n - r$ of the unknowns (those corresponding to $\mathbf{z}$) may be chosen arbitrarily; the remaining r unknowns are then determined uniquely.

We consider four examples which illustrate the elimination process and the different possibilities regarding the existence of solutions. For convenience we consider only examples in which all coefficients are integers; of course this will not be the case in most problems.

Example 8.2   Solve

$$x + 3x_2 - 2x_3 - 4x_4 = 3$$
$$2x_1 + 6x_2 - 7x_3 - 10x_4 = -2$$
$$-x_1 - x_2 + 5x_3 + 9x_4 = 14$$
$$-3x_1 - 5x_2 + 15x_4 = -6.$$

† By $\mathbf{d} \neq \mathbf{0}$ we mean that $\mathbf{d}$ is a vector other than the zero vector. It is possible for some components of $\mathbf{d}$ to be zero, but not all components may be zero. This is consistent with the definition of equality of matrices in §8.2.

We detach coefficients and subtract multiples of the first equation from the others to eliminate x_1.

$$\begin{bmatrix} 1 & 3 & -2 & -4 & : & 3 \\ 2 & 6 & -7 & -10 & : & -2 \\ -1 & -1 & 5 & 9 & : & 14 \\ -3 & -5 & 0 & 15 & : & -6 \end{bmatrix} \qquad \begin{bmatrix} 1 & 3 & -2 & -4 & : & 3 \\ 0 & 0 & -3 & -2 & : & -8 \\ 0 & 2 & 3 & 5 & : & 17 \\ 0 & 4 & -6 & 3 & : & 3 \end{bmatrix}$$

We interchange the second and third columns so that the second element in the second row is non-zero and use this row to eliminate the second variable.

$$\begin{bmatrix} 1 & -2 & 3 & -4 & : & 3 \\ 0 & -3 & 0 & -2 & : & -8 \\ 0 & 3 & 2 & 5 & : & 17 \\ 0 & -6 & 4 & 3 & : & 3 \end{bmatrix} \qquad \begin{bmatrix} 1 & -2 & 3 & -4 & : & 3 \\ 0 & -3 & 0 & -2 & : & -8 \\ 0 & 0 & 2 & 3 & : & 9 \\ 0 & 0 & 4 & 7 & : & 19 \end{bmatrix}$$

The last step is to eliminate the third variable from the fourth equation and we obtain the triangular set

$$\begin{bmatrix} 1 & -2 & 3 & -4 & : & 3 \\ 0 & -3 & 0 & -2 & : & -8 \\ 0 & 0 & 2 & 3 & : & 9 \\ 0 & 0 & 0 & 1 & : & 1 \end{bmatrix}.$$

Thus by back substitution

$$\begin{aligned} y_4 &= 1, \\ 2y_3 &= 9 - 3y_4, & y_3 &= 3, \\ -3y_2 &= -8 - 0 \cdot y_3 + 2y_4, & y_2 &= 2, \\ y_1 &= 3 + 2y_2 - 3y_3 + 4y_4, & y_1 &= 2. \end{aligned}$$

We have interchanged the second and third columns and, therefore,

$$x_1 = y_1, \qquad x_2 = y_3, \qquad x_3 = y_2, \qquad x_4 = y_4.$$

Hence

$$\mathbf{x} = \begin{bmatrix} 2 \\ 3 \\ 2 \\ 1 \end{bmatrix}.$$

The equations are non-singular with rank $r = n = 4$. $\square$

Example 8.3 A simple example of inconsistent equations is

$$x + 2y = 1$$
$$x + 2y = 2.$$

These equations are of rank 1. □

Example 8.4 Consider

$$\begin{bmatrix} 1 & 2 & 3 & 4 \\ 2 & 1 & -2 & 3 \\ 3 & 0 & -7 & 2 \end{bmatrix} \begin{bmatrix} x_1 \\ x_2 \\ x_3 \\ x_4 \end{bmatrix} = \begin{bmatrix} 1 \\ 1 \\ 2 \end{bmatrix}.$$

We detach coefficients and the steps of the reduction process are

$$\begin{bmatrix} 1 & 2 & 3 & 4 & : & 1 \\ 0 & -3 & -8 & -5 & : & -1 \\ 0 & -6 & -16 & -10 & : & -1 \end{bmatrix} \quad \begin{bmatrix} 1 & 2 & 3 & 4 & : & 1 \\ 0 & -3 & -8 & -5 & : & -1 \\ 0 & 0 & 0 & 0 & : & 1 \end{bmatrix}.$$

We can see from the last row that the equations are inconsistent even though we have only three equations in four unknowns. The coefficients of the unknowns in the third equation consist of twice those in the second row minus those in the first row. However, the same is not true of the right side. The equations are of rank 2. □

Example 8.5 Solve

$$\begin{bmatrix} 1 & 2 & 3 & 4 & 5 \\ 1 & 2 & 2 & 3 & 7 \\ 1 & 3 & 2 & 7 & 5 \\ 1 & 3 & 3 & 8 & 3 \end{bmatrix} \begin{bmatrix} x_1 \\ x_2 \\ x_3 \\ x_4 \\ x_5 \end{bmatrix} = \begin{bmatrix} 3 \\ 4 \\ 1 \\ 0 \end{bmatrix}.$$

The reduction process is

$$\begin{bmatrix} 1 & 2 & 3 & 4 & 5 & : & 3 \\ 0 & 0 & -1 & -1 & 2 & : & 1 \\ 0 & 1 & -1 & 3 & 0 & : & -2 \\ 0 & 1 & 0 & 4 & -2 & : & -3 \end{bmatrix} \quad \begin{bmatrix} 1 & 2 & 3 & 4 & 5 & : & 3 \\ 0 & 1 & -1 & 3 & 0 & : & -2 \\ 0 & 0 & -1 & -1 & 2 & : & 1 \\ 0 & 1 & 0 & 4 & -2 & : & -3 \end{bmatrix}$$

$$\begin{bmatrix} 1 & 2 & 3 & 4 & 5 : & 3 \\ 0 & 1 & -1 & 3 & 0 : & -2 \\ 0 & 0 & -1 & -1 & 2 : & 1 \\ 0 & 0 & 1 & 1 & -2 : & -1 \end{bmatrix} \quad \begin{bmatrix} 1 & 2 & 3 & 4 & 5 : & 3 \\ 0 & 1 & -1 & 3 & 0 : & -2 \\ 0 & 0 & -1 & -1 & 2 : & 1 \\ 0 & 0 & 0 & 0 & 0 : & 0 \end{bmatrix}.$$

Equations 2 and 3 were interchanged to make the (2, 2) element non-zero. The equations are consistent with rank $r = 3$ and we may choose x_4 and x_5 arbitrarily. By back substitution, we find

$$\begin{aligned} -x_3 &= 1 + x_4 - 2x_5, & x_3 &= -1 - x_4 + 2x_5, \\ x_2 &= -2 + x_3 - 3x_4, & x_2 &= -3 - 4x_4 + 2x_5, \\ x_1 &= 3 - 2x_2 - 3x_3 - 4x_4 - 5x_5, & x_1 &= 12 + 7x_4 - 15x_5. \;\; \square \end{aligned}$$

If in (8.13) we have rank $r < m = n$, we say that the equations (8.10) are *singular* and that the square matrix **A** is singular. Notice that singular equations have either an infinity of solutions or no solutions. You will also notice that an $n \times n$ matrix **A** is non-singular if, and only if, it is possible to choose n non-zero pivots $u_{11}, u_{22}, \ldots, u_{nn}$ in the elimination process. It is always possible to reduce an $n \times n$ matrix **A** to an upper triangular matrix **U** using the elimination process, although for $r < n$ some of the diagonal elements of **U** are zero. It can be shown that if a total of j interchanges of rows and columns are required in the process, the product $d = (-1)^j u_{11} u_{22} \cdots u_{nn}$ is independent of the order in which the pivotal equations and unknowns are chosen. We call d the *determinant* of the $n \times n$ matrix **A**. Thus **A** is non-singular if, and only if, its determinant is non-zero. (See Noble.)

We now consider the particular system of equations

$$\mathbf{Ax} = \mathbf{0}$$

where **A** is $n \times n$. These will always be consistent as the right hand side will remain zero throughout the reduction process, and, even for $r < n$, **d** = **0** in (8.16b). Thus if **A** is singular there are an infinity of solutions of this problem and, in particular, there will be a solution $\mathbf{x} \neq \mathbf{0}$. If **A** is non-singular, $r = n$ and the solution $\mathbf{x} = \mathbf{0}$ is unique. We conclude that **A** is singular if, and only if, there exists a vector $\mathbf{x} \neq \mathbf{0}$, such that $\mathbf{Ax} = \mathbf{0}$.

8.4 Vector interpretation of linear equations

We may interpret the problem of solving the equations (8.8) as that of determining x_1, x_2 and x_3 such that

$$\begin{bmatrix} 3 \\ 1 \\ -7 \end{bmatrix} = x_1 \begin{bmatrix} 2 \\ 4 \\ -2 \end{bmatrix} + x_2 \begin{bmatrix} 2 \\ 7 \\ 4 \end{bmatrix} + x_3 \begin{bmatrix} 3 \\ 7 \\ 5 \end{bmatrix}. \tag{8.18}$$

We seek x_1, x_2 and x_3 such that the vector on the left may be expressed as a linear combination of the other three vectors.

If we have any three vectors $\mathbf{a}_1$, $\mathbf{a}_2$ and $\mathbf{a}_3$ in three-dimensional space, then

$$\mathbf{v} = \alpha_1\mathbf{a}_1 + \alpha_2\mathbf{a}_2 + \alpha_3\mathbf{a}_3 \tag{8.19}$$

is also a vector in the space. By varying α_1, α_2 and α_3, we obtain all possible vectors in the space unless $\mathbf{a}_1$, $\mathbf{a}_2$ and $\mathbf{a}_3$ lie in a common plane. This result may be visualized from the parallelogram law for adding vectors. If $\mathbf{a}_1$, $\mathbf{a}_2$ and $\mathbf{a}_3$ lie in a common plane, that is, if they are coplanar, we can only generate vectors in this plane in (8.19). In Fig. 8.3, all vectors lie in the plane of the

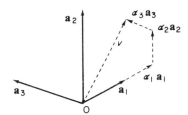

Fig. 8.3. Equation (8.19).

paper and no matter how we vary the scalars α_1, α_2 and α_3 we cannot move $\mathbf{v}$ out of that plane. If $\mathbf{a}_1$, $\mathbf{a}_2$ and $\mathbf{a}_3$ have a common direction, that is, if they are collinear, we can only generate vectors with this direction in (8.19). We call the set of all vectors generated by (8.19) the space *spanned* by $\mathbf{a}_1$, $\mathbf{a}_2$ and $\mathbf{a}_3$. In (8.18) the three vectors on the right are not coplanar or collinear and we can express any three-dimensional vector in terms of them. If the three vectors were coplanar, it would only be possible to obtain a solution if the vector on the left were in their common plane.

In the general case, we seek $x_1, \ldots, x_n$ such that

$$\mathbf{b} = x_1\mathbf{a}_1 + x_2\mathbf{a}_2 + \cdots + x_n\mathbf{a}_n$$

where $\mathbf{a}_1, \mathbf{a}_2, \ldots, \mathbf{a}_n$ are the columns of the coefficient matrix. The vector $\mathbf{b}$ must be in the space spanned by $\mathbf{a}_1, \ldots, \mathbf{a}_n$, if there is to be a solution of the equations, that is, if the equations are to be consistent.

If, for the special case $\mathbf{b} = \mathbf{0}$, there is a solution other than $\mathbf{x} = \mathbf{0}$ we say that the columns $\mathbf{a}_1, \ldots, \mathbf{a}_n$ are *linearly dependent*. If $m = n$ it follows from the argument at the end of §8.3 that the columns are linearly dependent if, and only if, $\mathbf{A}$ is singular.

Suppose that $\mathbf{a}_1$, $\mathbf{a}_2$ and $\mathbf{a}_3$ in three-dimensional space are linearly dependent and

$$\mathbf{0} = x_1\mathbf{a}_1 + x_2\mathbf{a}_2 + x_3\mathbf{a}_3$$

where at least one of x_1, x_2 and x_3 is non-zero. Assuming that $x_1 \neq 0$ we have

$$\mathbf{a}_1 = \left(\frac{-x_2}{x_1}\right)\mathbf{a}_2 + \left(\frac{-x_3}{x_1}\right)\mathbf{a}_3.$$

Thus $\mathbf{a}_1$, $\mathbf{a}_2$ and $\mathbf{a}_3$ are coplanar as $\mathbf{a}_2$ and $\mathbf{a}_3$ have at least one common plane and $\mathbf{a}_1$, a linear combination of $\mathbf{a}_2$ and $\mathbf{a}_3$, must lie in such a plane. If $x_1 = 0$, we use the same argument with $\mathbf{a}_1$ and $\mathbf{a}_2$ or $\mathbf{a}_1$ and $\mathbf{a}_3$ interchanged.

8.5 Pivoting

Example 8.6   Consider the solution of

$$10^{-12}x + y + z = 2 \qquad\qquad (8.20\text{a})$$

$$x + 2y - z = 2 \qquad\qquad (8.20\text{b})$$

$$-x + y + z = 1 \qquad\qquad (8.20\text{c})$$

if, during the calculation, numbers may be stored to only ten decimal digit accuracy in floating point form. The solution to the accuracy required is seen to be $x = y = z = 1$. However the elimination process produces

$$10^{-12}x + y + z = 2$$
$$-10^{12}y - 10^{12}z = -2.10^{12}$$
$$10^{12}y + 10^{12}z = 2.10^{12}.$$

The last two equations are obtained by adding $-10^{12} \times$ (8.20a) and $10^{12} \times$ (8.20a) with rounding. The next step yields

$$10^{-12}x + y + z = 2$$
$$-10^{12}y - 10^{12}z = -2.10^{12}$$
$$0.z = 0.$$

These suggest that z can have any arbitrary value and, by back substitution, $y = 2 - z$ and $x = 0$.

If instead we had arranged the equations as

$$x + 2y - z = 2$$
$$10^{-12}x + y + z = 2$$
$$-x + y + z = 1$$

after one step they become

$$x + 2y - z = 2$$
$$y + z = 2$$
$$3y \quad\quad = 3.$$

Because of rounding, the second equation is not changed apart from the removal of x. We obtain the solution $y = 1, z = 1, x = 1$, on working back through the equations. $\square$

The fault in the first attempt at solving Example 8.6 was that a small *pivot* (the coefficient of x in the first equation), meant that very large multiples of the first equation were added to the others, which were therefore "swamped" by the first equation, after rounding.

The trouble can be avoided if interchanges are made before each step to ensure that the pivots are as large as possible. When we have reached the stage displayed in (8.12), we look through the block of elements

$$\begin{bmatrix} \alpha_{pp} & \cdots & \alpha_{pn} \\ \vdots & & \vdots \\ \alpha_{mp} & \cdots & \alpha_{mn} \end{bmatrix}$$

and pick out one with largest modulus. This is then moved to the pivot position (p, p) by appropriate row and column interchanges. This process is called *complete pivoting*.

Sometimes elimination with *partial pivoting* is employed, when only row interchanges are permitted. At the stage of (8.12), only the column

$$\begin{bmatrix} \alpha_{pp} \\ \vdots \\ \alpha_{mp} \end{bmatrix}$$

is inspected for an element with largest modulus. Normally we are seeking the unique solution of n non-singular equations in n unknowns and, therefore, column interchanges should not be necessary to reduce the equations to triangular form. Alternatively, we could use a method in which only column interchanges are allowed. At the stage of (8.12), the row

$$[\alpha_{pp} \quad \cdots \quad \alpha_{pn}]$$

is inspected for an element with largest modulus. We shall see in §9.3, however, that partial pivoting using only row interchanges is preferable to that with only column interchanges because of the effects of rounding errors.

Example 8.6 above may seem exceptional because of one very small coefficient. Of course this was deliberately chosen as an extreme case but the lessons learned from it do hold for more reasonable equations. As a second example, we consider a more innocent looking problem.

Example 8.7 Solve

$$0.50x + 1.1y + 3.1z = 6.0$$
$$2.0x + 4.5y + 0.36z = 0.020 \qquad (8.21)$$
$$5.0x + 0.96y + 6.5z = 0.96.$$

The coefficients are given to two significant decimal digits. In an attempt to minimize rounding errors we will carry one "guarding" digit and work out multipliers and intermediate coefficients to three significant digits. Without interchanges we obtain

$$\begin{bmatrix} 0.500 & 1.10 & 3.10 & : & 6.00 \\ 0 & 0.100 & -12.0 & : & -24.0 \\ 0 & -10.0 & -24.5 & : & -59.0 \end{bmatrix} \quad \begin{bmatrix} 0.500 & 1.10 & 3.10 & : & 6.00 \\ 0 & 0.100 & -12.0 & : & -24.0 \\ 0 & 0 & -1220 & : & -2460 \end{bmatrix}.$$

Back substitution yields

$$z \simeq 2.02, \qquad y \simeq 2.40 \quad \text{and} \quad x \simeq -5.80.$$

With partial pivoting involving only row interchanges we obtain, on interchanging the first and third equations,

$$\begin{bmatrix} 5.00 & 0.960 & 6.50 & : & 0.960 \\ 0 & 4.12 & -2.24 & : & -0.364 \\ 0 & 1.00 & 2.45 & : & 5.90 \end{bmatrix} \quad \begin{bmatrix} 5.00 & 0.960 & 6.50 & : & 0.960 \\ 0 & 4.12 & -2.24 & : & -0.364 \\ 0 & 0 & 2.99 & : & 5.99 \end{bmatrix}.$$

Back substitution yields

$$z \simeq 2.00 \qquad y \simeq 1.00 \qquad x \simeq -2.60.$$

By substituting these values in equation (8.21), it may be confirmed that this is the exact solution. In Chapter 9 we shall consider in more detail the effects of rounding errors on the accuracy of the elimination process. □

8.6 Analysis of elimination method

In the rest of this chapter we shall assume that we have n equations in n unknowns

$$\mathbf{Ax} = \mathbf{b}$$

where $\mathbf{A}$ is $n \times n$. After one step of the elimination process, $\mathbf{A}$ becomes

$$\mathscr{A}_1 = \begin{bmatrix} a_{11} & a_{12} & \cdots & a_{1n} \\ 0 & a'_{22} & \cdots & a'_{2n} \\ \vdots & \vdots & & \vdots \\ 0 & a'_{n2} & \cdots & a'_{nn} \end{bmatrix},$$

assuming there are no interchanges. We have *subtracted* $(a_{i1}/a_{11}) \times$ first row from the ith row. We can restore **A** by operating on $\mathscr{A}_1$, *adding* $(a_{i1}/a_{11}) \times$ first row to the ith row for $i = 1, 2, \ldots, n$. This is equivalent to

$$\mathbf{A} = \mathscr{L}_1 \mathscr{A}_1$$

where

$$\mathscr{L}_1 = \begin{bmatrix} 1 & 0 & \cdots & & 0 \\ l_{21} & 1 & \ddots & & \vdots \\ \vdots & 0 & & & \vdots \\ \vdots & \vdots & \ddots & & 0 \\ l_{n1} & 0 & \cdots & 0 & 1 \end{bmatrix}$$

and the $l_{i1} = a_{i1}/a_{11}$, $i = 2, \ldots, n$, are the multipliers. Similarly the new right side $\mathbf{b}_1$ is related to its old value by

$$\mathbf{b} = \mathscr{L}_1 \mathbf{b}_1.$$

The second step (assuming there are no interchanges) in the elimination process is to form

$$\mathscr{A}_2 = \begin{bmatrix} a_{11} & a_{12} & \cdots & \cdots & a_{1n} \\ 0 & a'_{22} & \cdots & \cdots & a'_{2n} \\ 0 & 0 & a''_{33} & \cdots & a''_{3n} \\ \vdots & \vdots & \vdots & & \vdots \\ 0 & 0 & a''_{n3} & \cdots & a''_{nn} \end{bmatrix}$$

by subtracting $(a'_{i2}/a'_{22}) \times$ second row from the ith row for $i = 3, 4, \ldots, n$. Again we may restore $\mathscr{A}_1$ by an inverse process and

$$\mathscr{A}_1 = \mathscr{L}_2 \mathscr{A}_2$$

where

$$\mathscr{L}_2 = \begin{bmatrix} 1 & 0 & \cdot & \cdot & \cdot & 0 \\ 0 & 1 & \cdot & & & \cdot \\ \cdot & l_{32} & 1 & \cdot & & \cdot \\ \cdot & \cdot & 0 & \cdot & \cdot & \cdot \\ \cdot & \cdot & \cdot & \cdot & \cdot & 0 \\ 0 & l_{n2} & 0 & \cdot & 0 & 1 \end{bmatrix}$$

and $l_{i2} = (a'_{i2}/a'_{22})$, $i = 3, \ldots, n$, are the multipliers of the second row. Thus

$$\mathbf{A} = \mathscr{L}_1 \mathscr{A}_1 = \mathscr{L}_1 \mathscr{L}_2 \mathscr{A}_2.$$

Similarly

$$\mathbf{b} = \mathscr{L}_1\mathbf{b}_1 = \mathscr{L}_1\mathscr{L}_2\mathbf{b}_2$$

where $\mathbf{b}_2$ is the right side after two steps.

Provided there are no interchanges, by repeating the above we find that

$$\mathbf{A} = \mathscr{L}_1\mathscr{L}_2 \cdots \mathscr{L}_{n-1}\mathbf{U}$$

and

$$\mathbf{b} = \mathscr{L}_1\mathscr{L}_2 \cdots \mathscr{L}_{n-1}\mathbf{c}.$$

$\mathbf{U}$ is the final upper triangular set of equations, $\mathbf{c}$ is the final right side and

$$\mathscr{L}_j = \begin{bmatrix} 1 & & & & & & & \\ 0 & \cdot & & & & & 0 & \\ \cdot & \cdot & & \cdot & & & & \\ \cdot & & 0 & 1 & & & & \\ \cdot & & & l_{j+1,j} & 1 & & & \\ \cdot & & & & \cdot & 0 & & \cdot \\ \cdot & & & & \cdot & \cdot & \cdot & \cdot \\ 0 & \cdot & 0 & l_{nj} & 0 & \cdot & 0 & 1 \end{bmatrix} \tag{8.22}$$

where l_{ij} is the multiple of the jth row subtracted from the ith row during the reduction process.

Now

$$\mathscr{L}_1\mathscr{L}_2 \cdots \mathscr{L}_{n-1} = \begin{bmatrix} 1 & & & & & \\ l_{21} & 1 & & & 0 & \\ l_{31} & l_{32} & 1 & & & \\ \cdot & \cdot & \cdot & & \cdot & \\ \cdot & \cdot & & \cdot & & \\ \cdot & \cdot & & & \cdot & \\ l_{n1} & l_{n2} & \cdot & \cdot & \cdot\,l_{n,n-1} & 1 \end{bmatrix} = \mathbf{L}, \tag{8.23}$$

say, which is a lower triangular matrix. (See Problem 8.23.) We therefore find that, provided the equations are non-singular and no interchanges are required,

$$A = LU \tag{8.24}$$

and

$$b = Lc \tag{8.25}$$

where L is a *lower triangular* matrix with units on the diagonal and U is an *upper triangular* matrix. We say that A has been factorized into triangular factors L and U.

Example 8.8   For the equations of Example 8.1,

$$\mathcal{L}_1 = \begin{bmatrix} 1 & 0 & 0 \\ 2 & 1 & 0 \\ -1 & 0 & 1 \end{bmatrix} \quad \text{and} \quad \mathcal{L}_2 = \begin{bmatrix} 1 & 0 & 0 \\ 0 & 1 & 0 \\ 0 & 2 & 1 \end{bmatrix}.$$

Thus

$$L = \mathcal{L}_1 \mathcal{L}_2 = \begin{bmatrix} 1 & 0 & 0 \\ 2 & 1 & 0 \\ -1 & 2 & 1 \end{bmatrix}.$$

From the final set of equations we find

$$U = \begin{bmatrix} 2 & 2 & 3 \\ 0 & 3 & 1 \\ 0 & 0 & 6 \end{bmatrix} \quad \text{and} \quad c = \begin{bmatrix} 3 \\ -5 \\ 6 \end{bmatrix}.$$

It is easily verified that

$$LU = \begin{bmatrix} 2 & 2 & 3 \\ 4 & 7 & 7 \\ -2 & 4 & 5 \end{bmatrix} = A$$

and

$$Lc = \begin{bmatrix} 3 \\ 1 \\ -7 \end{bmatrix} = b. \ \square$$

If interchanges are used during the reduction process, then instead of determining triangular factors of A, we determine triangular factors of A^*

where A^* is obtained by making the necessary row and column interchanges on A.

8.7 Matrix factorization

The elimination process for n equations in n unknowns

$$Ax = b \qquad (8.26)$$

may also be described as a *factorization* method. We seek factors L (lower triangular with units on the diagonal) and U (upper triangular) such that

$$A = LU,$$

when (8.26) becomes

$$LUx = b.$$

First we find c by solving the lower triangular equations

$$Lc = b$$

by *forward substitution*. We have

$$
\begin{aligned}
1.c_1 &= b_1 \\
l_{21}c_1 + 1.c_2 &= b_2 \\
l_{31}c_1 + l_{32}c_2 + 1.c_3 &= b_3 \\
\vdots &\quad \vdots \\
l_{n1}c_1 + l_{n2}c_2 + \cdots + 1.c_n &= b_n.
\end{aligned}
$$

We find c_1 from the first equation, then c_2 from the second and so on. Secondly we find x by back substitution in the triangular equations

$$Ux = c.$$

The solution of (8.26) is this vector x, since

$$Ax = (LU)x = L(Ux) = Lc = b.$$

We can see from the above that, given any method of calculating factors L and U such that $A = LU$, we can solve the equations. One way of finding the factors L and U is to build them up from submatrices. We will use this method to describe how an $n \times n$ matrix A may be factorized in the form

$$A = LDV \qquad (8.27)$$

where

$$\mathbf{D} = \begin{bmatrix} d_1 & & & & & \\ & d_2 & & & 0 & \\ & & \cdot & & & \\ & & & \cdot & & \\ & 0 & & & \cdot & \\ & & & & & d_n \end{bmatrix}$$

is a *diagonal* matrix with $d_i \neq 0$, $i = 1, 2, \ldots, n$, $\mathbf{L}$ is a lower triangular matrix with units on the diagonal as before and $\mathbf{V}$ is an upper triangular matrix with units on the diagonal. We will show that, under certain circumstances, such a factorization exists and is unique. In this factorization we call the diagonal elements of $\mathbf{D}$ the *pivots*. We shall see later that these pivots are identical with those of the elimination method.

Let $\mathbf{A}_k$ denote the kth leading submatrix of $\mathbf{A}$. By this we mean the submatrix consisting of the first k rows and columns of $\mathbf{A}$. Note that $\mathbf{A}_1 = [a_{11}]$ and $\mathbf{A}_n = \mathbf{A}$. Suppose that for some value of k we already have a factorization

$$\mathbf{A}_k = \mathbf{L}_k \mathbf{D}_k \mathbf{V}_k$$

of the form (8.27), where $\mathbf{L}_k$, $\mathbf{D}_k$ and $\mathbf{V}_k$ are matrices of the appropriate kinds. We now seek $\mathbf{L}_{k+1}$, $\mathbf{D}_{k+1}$ and $\mathbf{V}_{k+1}$ such that

$$\mathbf{A}_{k+1} = \mathbf{L}_{k+1} \mathbf{D}_{k+1} \mathbf{V}_{k+1}.$$

We write

$$\mathbf{A}_{k+1} = \begin{bmatrix} \mathbf{A}_k & \mathbf{c}_{k+1} \\ \mathbf{r}_{k+1}^T & a_{k+1,k+1} \end{bmatrix}, \quad \text{where } \mathbf{c}_{k+1} = \begin{bmatrix} a_{1,k+1} \\ \vdots \\ a_{k,k+1} \end{bmatrix}$$

and

$$\mathbf{r}_{k+1}^T = [a_{k+1,1} \quad a_{k+1,2} \quad \cdots \quad a_{k+1,k}].$$

We seek an element d_{k+1} and vectors

$$\mathbf{l}_{k+1}^T = [l_{k+1,1} \quad l_{k+1,2} \quad \cdots \quad l_{k+1,k}], \quad \mathbf{v}_{k+1} = \begin{bmatrix} v_{1,k+1} \\ \vdots \\ v_{k,k+1} \end{bmatrix}$$

such that

$$\mathbf{A}_{k+1} = \begin{bmatrix} \mathbf{A}_k & \mathbf{c}_{k+1} \\ \mathbf{r}_{k+1}^T & a_{k+1,k+1} \end{bmatrix} = \begin{bmatrix} \mathbf{L}_k & \mathbf{0} \\ \mathbf{l}_{k+1}^T & 1 \end{bmatrix} \begin{bmatrix} \mathbf{D}_k & \mathbf{0} \\ \mathbf{0}^T & d_{k+1} \end{bmatrix} \begin{bmatrix} \mathbf{V}_k & \mathbf{v}_{k+1} \\ \mathbf{0}^T & 1 \end{bmatrix}.$$

On multiplying out the right side and equating blocks, we find that we require

$$\mathbf{A}_k = \mathbf{L}_k \mathbf{D}_k \mathbf{V}_k \tag{8.28}$$

$$\mathbf{c}_{k+1} = \mathbf{L}_k \mathbf{D}_k \mathbf{v}_{k+1} \tag{8.29}$$

$$\mathbf{r}_{k+1}^T = \mathbf{l}_{k+1}^T \mathbf{D}_k \mathbf{V}_k \tag{8.30}$$

$$a_{k+1,k+1} = \mathbf{l}_{k+1}^T \mathbf{D}_k \mathbf{v}_{k+1} + d_{k+1}. \tag{8.31}$$

Equation (8.28) is satisfied by hypothesis. We can solve (8.29) for $\mathbf{v}_{k+1}$ by forward substitution in the lower triangular equations

$$(\mathbf{L}_k \mathbf{D}_k)\mathbf{v}_{k+1} = \mathbf{c}_{k+1}. \tag{8.32}$$

We use forward substitution in the triangular equations

$$\mathbf{l}_{k+1}^T(\mathbf{D}_k \mathbf{V}_k) = \mathbf{r}_{k+1}^T \tag{8.33}$$

to find $\mathbf{l}_{k+1}^T$. Having found $\mathbf{v}_{k+1}$ and $\mathbf{l}_{k+1}^T$, we use (8.31) to determine d_{k+1}.

We have a practical method of factorizing $\mathbf{A}$ since clearly

$$\mathbf{A}_1 = [a_{11}] = [1][a_{11}][1] = \mathbf{L}_1 \mathbf{D}_1 \mathbf{V}_1$$

and the process only involves solving triangular equations. The $\mathbf{L}_k$, $\mathbf{D}_k$ and $\mathbf{V}_k$ are successively determined for $k = 1, 2, \ldots, n$ and are leading submatrices of $\mathbf{L}$, $\mathbf{D}$ and $\mathbf{V}$.

Example 8.9   Factorize

$$\mathbf{A} = \begin{bmatrix} 1 & 2 & 3 & -1 \\ 2 & -1 & 9 & -7 \\ -3 & 4 & -3 & 19 \\ 4 & -2 & 6 & -21 \end{bmatrix}.$$

$$\mathbf{A}_1 = [1], \qquad \mathbf{L}_1 = [1], \qquad \mathbf{D}_1 = [1], \qquad \mathbf{V}_1 = [1].$$

$$\mathbf{A}_2 = \begin{bmatrix} 1 & 2 \\ 2 & -1 \end{bmatrix}, \qquad \mathbf{L}_2 = \begin{bmatrix} 1 & 0 \\ 2 & 1 \end{bmatrix}, \qquad \mathbf{D}_2 = \begin{bmatrix} 1 & 0 \\ 0 & -5 \end{bmatrix}, \qquad \mathbf{V}_2 = \begin{bmatrix} 1 & 2 \\ 0 & 1 \end{bmatrix}.$$

$$\mathbf{A}_3 = \begin{bmatrix} 1 & 2 & 3 \\ 2 & -1 & 9 \\ -3 & 4 & -3 \end{bmatrix}, \qquad \mathbf{L}_3 = \begin{bmatrix} 1 & 0 & 0 \\ 2 & 1 & 0 \\ -3 & -2 & 1 \end{bmatrix},$$

$$\mathbf{D}_3 = \begin{bmatrix} 1 & 0 & 0 \\ 0 & -5 & 0 \\ 0 & 0 & 12 \end{bmatrix}, \qquad \mathbf{V}_3 = \begin{bmatrix} 1 & 2 & 3 \\ 0 & 1 & -\frac{3}{5} \\ 0 & 0 & 1 \end{bmatrix}.$$

$\mathbf{A}_4 = \mathbf{A};$

$$(\mathbf{L}_3\mathbf{D}_3)\mathbf{v}_4 = \begin{bmatrix} 1 & 0 & 0 \\ 2 & -5 & 0 \\ -3 & 10 & 12 \end{bmatrix} \mathbf{v}_4 = \mathbf{c}_4 = \begin{bmatrix} -1 \\ -7 \\ 19 \end{bmatrix},$$

$$\therefore \mathbf{v}_4 = \begin{bmatrix} -1 \\ 1 \\ \frac{1}{2} \end{bmatrix};$$

$$\mathbf{l}_4^T(\mathbf{D}_3\mathbf{V}_3) = \mathbf{l}_4^T \begin{bmatrix} 1 & 2 & 3 \\ 0 & -5 & 3 \\ 0 & 0 & 12 \end{bmatrix} = \mathbf{r}_4^T = \begin{bmatrix} 4 & -2 & 6 \end{bmatrix}$$

$$\therefore \mathbf{l}_4^T = \begin{bmatrix} 4 & 2 & -1 \end{bmatrix};$$

$$d_4 = a_{4,4} - \mathbf{l}_4^T\mathbf{D}_3\mathbf{v}_4$$

$$= -21 - \begin{bmatrix} 4 & 2 & -1 \end{bmatrix} \begin{bmatrix} -1 \\ -5 \\ 6 \end{bmatrix}$$

$$= -21 + 20 = -1.$$

Thus

$$\mathbf{A} = \begin{bmatrix} 1 & & & 0 \\ 2 & 1 & & \\ -3 & -2 & 1 & \\ 4 & 2 & -1 & 1 \end{bmatrix} \begin{bmatrix} 1 & & & 0 \\ & -5 & & \\ & & 12 & \\ 0 & & & -1 \end{bmatrix} \begin{bmatrix} 1 & 2 & 3 & -1 \\ & 1 & -\frac{3}{5} & 1 \\ 0 & & 1 & \frac{1}{2} \\ & & & 1 \end{bmatrix}. \quad \square$$

The process will fail if, and only if, any of the pivots d_k become zero as then we cannot solve (8.32) and (8.33). This can only happen if one of the leading submatrices is singular. To see this, suppose that $d_1, d_2, \ldots, d_k$ are non-zero but that when we solve (8.31) we find $d_{k+1} = 0$. We now have

$$\mathbf{A}_{k+1} = \begin{bmatrix} \mathbf{L}_k & \mathbf{0} \\ \mathbf{l}_{k+1}^T & 1 \end{bmatrix} \begin{bmatrix} \mathbf{D}_k & \mathbf{0} \\ \mathbf{0}^T & 0 \end{bmatrix} \begin{bmatrix} \mathbf{V}_k & \mathbf{v}_{k+1} \\ \mathbf{0}^T & 1 \end{bmatrix}$$

$$= \begin{bmatrix} \mathbf{L}_k & \mathbf{0} \\ \mathbf{l}_{k+1}^T & 1 \end{bmatrix} \begin{bmatrix} \mathbf{D}_k\mathbf{V}_k & \mathbf{D}_k\mathbf{v}_{k+1} \\ \mathbf{0}^T & 0 \end{bmatrix}.$$

Using a process equivalent to the elimination method of §8.3, we have reduced $\mathbf{A}_{k+1}$ to the second of these factors. This is of the form (8.14) without the right hand side vectors $\mathbf{c}$ and $\mathbf{d}$, where the upper triangular matrix

$\mathbf{D}_k \mathbf{V}_k$ is $k \times k$. Thus $\mathbf{A}_{k+1}$ is of rank k which is less than the dimension $k + 1$ and, therefore, $\mathbf{A}_{k+1}$ is singular.

The above argument and method of finding factors forms a constructive proof† of the following theorem.

Theorem 8.1 If an $n \times n$ matrix $\mathbf{A}$ is such that all n of its leading sub-matrices are non-singular, then there exists a unique factorization of the form

$$\mathbf{A} = \mathbf{LDV}, \tag{8.34}$$

where $\mathbf{L}$ is lower triangular with units on the diagonal, $\mathbf{V}$ is upper triangular with units on the diagonal and $\mathbf{D}$ is diagonal with non-zero diagonal elements.

Proof The proof of both existence and uniqueness is based on induction. The factors exist and are unique for the 1×1 leading submatrix $\mathbf{A}_1$. Given a unique factorization of the kth leading submatrix $\mathbf{A}_k$ we deduce from (8.29), (8.30) and (8.31) that $\mathbf{A}_{k+1}$ has a unique factorization. Thus the factors of $\mathbf{A} = \mathbf{A}_n$ exist and are unique. $\square$

8.8 Compact elimination methods

We do not usually bother to determine the diagonal matrix $\mathbf{D}$ of (8.34) explicitly and instead factorize $\mathbf{A}$ so that

$$\mathbf{A} = \mathbf{L(DV)} \tag{8.35}$$

or

$$\mathbf{A} = \mathbf{(LD)V}. \tag{8.36}$$

In the former we calculate $\mathbf{L}$ with units on the diagonal and $(\mathbf{DV})$. In the latter we calculate $\mathbf{V}$ with units on the diagonal and $(\mathbf{LD})$. The elimination method of §8.3 is equivalent to (8.35) and, on comparing with (8.24), we see that

$$\mathbf{U} = \mathbf{DV}.$$

From the diagonal,

$$u_{ii} = d_i, \qquad i = 1, 2, \ldots, n$$

and thus the pivots in the elimination method are the same as those for factorization.

In compact elimination methods, we seek a factorization of the form (8.35) or (8.36), but instead of building up factors from leading submatrices

---

† A proof in which an existence statement is proved by showing how to construct the objects involved. See §1.1.

as described in the last section, we find it more convenient to calculate rows and columns of **L** and **V** alternatively. For the factorization (8.35), we calculate the ith row of **(DV)** followed by the ith column of **L** for $i = 1, 2, \ldots, n$. To obtain (8.36) we calculate the ith column of **(LD)** followed by the ith row of **V**.

Example 8.10 Factorize the following matrix **A** in the form (8.35).

$$\mathbf{A} = \begin{bmatrix} 1 & 2 & 3 & -1 \\ 2 & -1 & 9 & -7 \\ -3 & 4 & -3 & 19 \\ 4 & -2 & 6 & -21 \end{bmatrix};$$

A is the matrix used in Example 8.9.

The first row of **DV** must be the same as the first row of **A**, since

$$[\text{1st row of } \mathbf{A}] = [\text{1st row of } \mathbf{L}](\mathbf{DV})$$
$$= [1 \quad 0 \quad \cdots \quad 0](\mathbf{DV})$$
$$= [\text{1st row of } \mathbf{DV}].$$

The first column of **L** is calculated from

$$\begin{bmatrix} \text{1st} \\ \text{column} \\ \text{of } \mathbf{A} \end{bmatrix} = \mathbf{L} \begin{bmatrix} \text{1st} \\ \text{column} \\ \text{of } \mathbf{DV} \end{bmatrix} = \mathbf{L} \begin{bmatrix} 1 \\ 0 \\ \vdots \\ 0 \end{bmatrix} = \begin{bmatrix} \text{1st} \\ \text{column} \\ \text{of } \mathbf{L} \end{bmatrix}.$$

We now have

$$\begin{bmatrix} 1 & 0 & 0 & 0 \\ 2 & 1 & 0 & 0 \\ -3 & l_{32} & 1 & 0 \\ 4 & l_{42} & l_{43} & 1 \end{bmatrix} \begin{bmatrix} 1 & 2 & 3 & -1 \\ 0 & u_{22} & u_{23} & u_{24} \\ 0 & 0 & u_{33} & u_{34} \\ 0 & 0 & 0 & u_{44} \end{bmatrix}.$$

The second row of **DV** is calculated using

$$[\text{2nd row of } \mathbf{A}] = [\text{2nd row of } \mathbf{L}](\mathbf{DV}).$$

For the:

(2, 2) element, $-1 =$	$4 + u_{22}$,	$\therefore u_{22} =$	-5;
(2, 3) element, $9 =$	$6 + u_{23}$,	$\therefore u_{23} =$	3;
(2, 4) element, $-7 =$	$-2 + u_{24}$,	$\therefore u_{24} =$	-5.

We now have

$$
\begin{bmatrix} 1 & 0 & 0 & 0 \\ 2 & 1 & 0 & 0 \\ -3 & l_{32} & 1 & 0 \\ 4 & l_{42} & l_{43} & 1 \end{bmatrix}
\begin{bmatrix} 1 & 2 & 3 & -1 \\ 0 & -5 & 3 & -5 \\ 0 & 0 & u_{33} & u_{34} \\ 0 & 0 & 0 & u_{44} \end{bmatrix}.
$$

The second column of **L** is calculated using

$$
\begin{bmatrix} \text{2nd} \\ \text{column} \\ \text{of } \mathbf{A} \end{bmatrix} = \mathbf{L} \begin{bmatrix} \text{2nd} \\ \text{column} \\ \text{of } \mathbf{DV} \end{bmatrix}.
$$

For the:

$$(3, 2) \text{ element,} \quad 4 = -6 - 5l_{32}, \qquad \therefore l_{32} = -2;$$
$$(4, 2) \text{ element,} \; -2 = \quad 8 - 5l_{42}, \qquad \therefore l_{42} = \quad 2.$$

We now have

$$
\begin{bmatrix} 1 & 0 & 0 & 0 \\ 2 & 1 & 0 & 0 \\ -3 & -2 & 1 & 0 \\ 4 & 2 & l_{43} & 1 \end{bmatrix}
\begin{bmatrix} 1 & 2 & 3 & -1 \\ 0 & -5 & 3 & -5 \\ 0 & 0 & u_{33} & u_{34} \\ 0 & 0 & 0 & u_{44} \end{bmatrix}.
$$

The third row of **DV** is calculated using

$$[\text{3rd row of } \mathbf{A}] = [\text{3rd row of } \mathbf{L}](\mathbf{DV}).$$

For the

$$(3, 3) \text{ element,} \; -3 = -9 - 6 + u_{33}, \qquad \therefore u_{33} = 12;$$
$$(3, 4) \text{ element,} \quad 19 = \quad 3 + 10 + u_{34}, \qquad \therefore u_{34} = \quad 6.$$

The third column of **L** and the fourth row of **DV** are similarly computed. We obtain

$$
\mathbf{A} = \begin{bmatrix} 1 & 0 & 0 & 0 \\ 2 & 1 & 0 & 0 \\ -3 & -2 & 1 & 0 \\ 4 & 2 & -1 & 1 \end{bmatrix}
\begin{bmatrix} 1 & 2 & 3 & -1 \\ 0 & -5 & 3 & -5 \\ 0 & 0 & 12 & 6 \\ 0 & 0 & 0 & -1 \end{bmatrix}
$$

which agrees with the result in Example 8.9. □

The process used in Example 8.10 is exactly equivalent to the elimination method except that intermediate values are not recorded, and hence the name, *compact elimination* method. You may see from (8.34), (8.35) and

(8.36) that there are various ways in which we may factorize **A**. As we have seen, there are also various ways in which we may order the calculations. In the literature on this topic these different variants of factorization are often given quite different names, such as Crout, Choleski, Doolittle, etc. after the names of persons who have "discovered" them.

8.9 Partial pivoting in compact elimination methods

It is possible to make the row interchanges necessary for partial pivoting in the compact elimination method. To illustrate how this is done, we consider the 4×4 case after the first row of **DV** and the first column of **L** have been determined. We have

$$
\begin{bmatrix} a_{11} & a_{12} & a_{13} & a_{14} \\ a_{21} & a_{22} & a_{23} & a_{24} \\ a_{31} & a_{32} & a_{33} & a_{34} \\ a_{41} & a_{42} & a_{43} & a_{44} \end{bmatrix} = \begin{bmatrix} 1 & 0 & 0 & 0 \\ l_{21} & ① & ⓪ & ⓪ \\ l_{31} & \times & ① & ⓪ \\ l_{41} & \times & \times & ① \end{bmatrix} \begin{bmatrix} u_{11} & u_{12} & u_{13} & u_{14} \\ 0 & \times & \times & \times \\ 0 & ⓪ & \times & \times \\ 0 & ⓪ & ⓪ & \times \end{bmatrix},
$$

where $\times$ denotes an element still to be found, ⓪ denotes a position in which the element will be zero but in the calculation so far this fact *has not been used*. Similarly ① indicates the position of an element $+1$ but again this information has not been used so far. You can see that if we now interchange any of the last three rows of **L**, we will not need to make any other changes to known elements of **L** and **DV**. However, the corresponding rows of **A** must also be interchanged, as

$$[i\text{th row of } \mathbf{A}] = [i\text{th row of } \mathbf{L}](\mathbf{DV}).$$

To decide which row interchange of **L** we should make, we first compute all the possible values of the $(2, 2)$ element of **DV** for such interchanges. Let w_j ($j = 2, 3, 4$) be the resulting $(2, 2)$ element of **DV** when the 2nd and jth rows of **L** are interchanged. We select the w_j with largest modulus to be the pivot u_{22} and make the corresponding row interchange on **L**. This is equivalent to searching through the column

$$
\begin{bmatrix} \alpha_{pp} \\ \vdots \\ \alpha_{mp} \end{bmatrix}
$$

in (8.12) for the element with largest modulus.

In the 4×4 case we have

$$l_{j1}u_{12} + 1 . w_j = a_{j2},$$

so that

$$w_j = a_{j2} - l_{j1}u_{12}.$$

Suppose now that w_2 has largest modulus, so that no row interchange is made. We choose $u_{22} = w_2$ and then, to find l_{32}, we use

$$l_{31}u_{12} + l_{32}u_{22} = a_{32}.$$

Thus

$$l_{32} = \frac{a_{32} - l_{31}u_{12}}{u_{22}} = \frac{w_3}{u_{22}}$$

and, similarly,

$$l_{42} = \frac{a_{42} - l_{41}u_{12}}{u_{22}} = \frac{w_4}{u_{22}}.$$

Hence there are no wasted calculations, as we divide the w_j, which we have already found, by the pivot u_{22} in order to obtain the second column of **L**.

It is not sensible to use complete pivoting in a compact elimination method as this would involve a large amount of unnecessary calculation.

Example 8.11 Factorize the coefficient matrix of Example 8.7 using a compact elimination method with partial pivoting and working to three significant digits.

$$\mathbf{A} = \begin{bmatrix} 0.5 & 1.1 & 3.1 \\ 2.0 & 4.5 & 0.36 \\ 5.0 & 0.96 & 6.5 \end{bmatrix}.$$

The $(1, 1)$ element of **DV** may be any of the elements in the first column of **A** and thus

$$w_1 = 0.5, \qquad w_2 = 2.0, \qquad w_3 = 5.0.$$

We choose w_3 as the pivot u_{11} and therefore interchange rows 1 and 3. For the first column of **L**, we have

$$l_{21} = \frac{w_2}{u_{11}} = \frac{2.0}{5.0} = 0.400, \qquad l_{31} = \frac{w_1}{u_{11}} = \frac{0.5}{5.0} = 0.100.$$

We obtain

$$\mathbf{A}' = \begin{bmatrix} 5.0 & 0.96 & 6.5 \\ 2.0 & 4.5 & 0.36 \\ 0.5 & 1.1 & 3.1 \end{bmatrix} = \mathbf{L(DV)}$$

$$= \begin{bmatrix} 1 & 0 & 0 \\ 0.400 & ① & ⓪ \\ 0.100 & \times & ① \end{bmatrix} \begin{bmatrix} 5.00 & 0.960 & 6.50 \\ 0 & \times & \times \\ 0 & ⓪ & \times \end{bmatrix}.$$

The possible values of the $(2, 2)$ element of $\mathbf{DV}$ are

$$w_2 = a'_{22} - l_{21}u_{12} = 4.5 - (0.400)(0.960) \simeq 4.12;$$
$$w_3 = a'_{32} - l_{31}u_{12} = 1.1 - (0.100)(0.960) \simeq 1.00.$$

We choose w_2 as pivot u_{22}. Thus no interchanges are required and

$$l_{32} = \frac{w_3}{u_{22}} = \frac{1.00}{4.12} \simeq 0.243.$$

Completing the calculations, we obtain

$$u_{23} \simeq -2.24, \qquad u_{33} \simeq 2.99. \ \square$$

8.10 Symmetric matrices

Given any $m \times n$ matrix, we call the $n \times m$ matrix obtained by interchanging the rows and columns of $\mathbf{A}$ the *tranpose* of $\mathbf{A}$ and write it as $\mathbf{A}^T$. Thus

$$\mathbf{A}^T = \begin{bmatrix} a_{11} & a_{21} & \cdots & a_{m1} \\ a_{12} & a_{22} & \cdots & a_{m2} \\ \vdots & & & \vdots \\ a_{1n} & a_{2n} & \cdots & a_{mn} \end{bmatrix}$$

and a_{ij} is the element in the jth row and ith column of $\mathbf{A}^T$. This extends the definition of the transpose of a vector, given in §8.2, to a general $m \times n$ matrix.

For any two matrices $\mathbf{A}$ and $\mathbf{B}$ whose product is defined,

$$(\mathbf{AB})^T = \mathbf{B}^T\mathbf{A}^T.$$

This follows from

$$[i\text{th row of } \mathbf{A}] \begin{bmatrix} j\text{th} \\ \text{column} \\ \text{of } \mathbf{B} \end{bmatrix} = [j\text{th row of } \mathbf{B}^T] \begin{bmatrix} i\text{th} \\ \text{column} \\ \text{of } \mathbf{A}^T \end{bmatrix}. \qquad (8.37)$$

The element on the left of (8.37) is the (i, j)th element of $\mathbf{AB}$ and is, therefore, the (j, i)th element of $(\mathbf{AB})^T$. The right side of (8.37) is the (j, i)th element of $\mathbf{B}^T\mathbf{A}^T$. We can extend this result (see Problem 8.30) to products of more than two matrices.

An $n \times n$ matrix $\mathbf{A}$ is said to be *symmetric* if

$$\mathbf{A}^T = \mathbf{A},$$

that is, $a_{ij} = a_{ji}$ for all i and j. When using a computer, we need store only approximately half the elements of a symmetric matrix as the elements below the diagonal are the same as those above. We therefore look for an elimination method which preserves symmetry when solving linear equations.

We have seen how to factorize $\mathbf{A}$ so that

$$\mathbf{A} = \mathbf{LDV} \tag{8.38}$$

where, provided the factorization exists, $\mathbf{L}$, $\mathbf{D}$ and $\mathbf{V}$ are unique. Now

$$\mathbf{A}^T = (\mathbf{LDV})^T = \mathbf{V}^T\mathbf{D}^T\mathbf{L}^T = \mathbf{V}^T\mathbf{DL}^T,$$

as any diagonal matrix $\mathbf{D}$ is symmetric. If $\mathbf{A}$ is symmetric

$$\mathbf{A} = \mathbf{A}^T = \mathbf{V}^T\mathbf{DL}^T, \tag{8.39}$$

$\mathbf{V}^T$ is a lower triangular matrix with units on the diagonal and $\mathbf{L}^T$ is an upper triangular matrix with units on the diagonal. On comparing (8.38) and (8.39) we deduce from the uniqueness of $\mathbf{L}$ and $\mathbf{V}$ that

$$\mathbf{L} = \mathbf{V}^T \quad \text{and} \quad \mathbf{L}^T = \mathbf{V}.$$

These two are equivalent and (8.39) becomes

$$\mathbf{A} = \mathbf{LDL}^T. \tag{8.40}$$

If

$$\mathbf{D} = \begin{bmatrix} d_1 & & & 0 \\ & d_2 & & \\ & & \ddots & \\ 0 & & & d_n \end{bmatrix} \quad \text{and} \quad \mathbf{E} = \begin{bmatrix} \sqrt{d_1} & & & 0 \\ & \sqrt{d_2} & & \\ & & \ddots & \\ 0 & & & \sqrt{d_n} \end{bmatrix}$$

then†

$$\mathbf{E}^2 = \mathbf{D}$$

so that

$$\mathbf{A} = \mathbf{LEEL}^T = \mathbf{LEE}^T\mathbf{L}^T \tag{8.41}$$
$$= (\mathbf{LE})(\mathbf{LE})^T.$$

$$\mathbf{A} = \mathbf{MM}^T$$

where

$$\mathbf{M} = \mathbf{LE}$$

is a lower triangular matrix and will have real elements only if all the diagonal elements of $\mathbf{D}$ are positive.

† We write $\mathbf{E}^2$ for $\mathbf{E}.\mathbf{E}$.

To calculate $\mathbf{M}$ we adapt the process described in §8.7. Equation (8.28) becomes

$$\mathbf{A}_k = \mathbf{M}_k\mathbf{M}_k^T \tag{8.42}$$

where $\mathbf{M}_k$ is the kth leading submatrix of $\mathbf{M}$. Both (8.32) and (8.33) become

$$\mathbf{M}_k\mathbf{m}_{k+1} = \mathbf{c}_{k+1} \tag{8.43}$$

where

$$\mathbf{A}_{k+1} = \begin{bmatrix} \mathbf{A}_k & \mathbf{c}_{k+1} \\ \mathbf{c}_{k+1}^T & a_{k+1,k+1} \end{bmatrix} \quad \text{and} \quad \mathbf{M}_{k+1} = \begin{bmatrix} \mathbf{M}_k & \mathbf{0} \\ \mathbf{m}_{k+1}^T & m_{k+1,k+1} \end{bmatrix}.$$

We solve (8.43) for $\mathbf{m}_{k+1}$ by forward substitution. Finally (8.31) becomes

$$(m_{k+1,k+1})^2 = a_{k+1,k+1} - \mathbf{m}_{k+1}^T\mathbf{m}_{k+1}. \tag{8.44}$$

Example 8.12 Solve

$$\begin{bmatrix} 4 & 2 & -2 \\ 2 & 2 & -3 \\ -2 & -3 & 14 \end{bmatrix} \begin{bmatrix} x_1 \\ x_2 \\ x_3 \end{bmatrix} = \begin{bmatrix} 10 \\ 5 \\ 4 \end{bmatrix}.$$

From $\mathbf{A}_1 = [4]$, $\mathbf{M}_1 = [2]$.

From $\mathbf{A}_2 = \begin{bmatrix} 4 & 2 \\ 2 & 2 \end{bmatrix}$, $\mathbf{M}_2 = \begin{bmatrix} 2 & 0 \\ 1 & 1 \end{bmatrix}$.

From $\mathbf{A}_3 = \begin{bmatrix} 4 & 2 & -2 \\ 2 & 2 & -3 \\ -2 & -3 & 14 \end{bmatrix}$, $\begin{bmatrix} 2 & 0 \\ 1 & 1 \end{bmatrix}\begin{bmatrix} m_{31} \\ m_{32} \end{bmatrix} = \begin{bmatrix} -2 \\ -3 \end{bmatrix}$.

Therefore,

$$\begin{bmatrix} m_{31} \\ m_{32} \end{bmatrix} = \begin{bmatrix} -1 \\ -2 \end{bmatrix} \quad \text{and} \quad m_{33}^2 = 14 - [-1 \quad -2]\begin{bmatrix} -1 \\ -2 \end{bmatrix} = 9.$$

Hence

$$\mathbf{M} = \begin{bmatrix} 2 & 0 & 0 \\ 1 & 1 & 0 \\ -1 & -2 & 3 \end{bmatrix}$$

and $\mathbf{A}\mathbf{x} = \mathbf{b}$ becomes $\mathbf{M}\mathbf{M}^T\mathbf{x} = \mathbf{b}$. We seek $\mathbf{c}$ such that

$$\mathbf{M}\mathbf{c} = \begin{bmatrix} 10 \\ 5 \\ 4 \end{bmatrix}.$$

By forward substitution, we obtain

$$\mathbf{c} = \begin{bmatrix} 5 \\ 0 \\ 3 \end{bmatrix}.$$

Finally, by back substitution in

$$\mathbf{M}^T\mathbf{x} = \mathbf{c},$$

we obtain

$$\mathbf{x} = \begin{bmatrix} 2 \\ 2 \\ 1 \end{bmatrix}. \quad \square$$

The above method is difficult to implement if some of the elements of the diagonal matrix $\mathbf{D}$ are negative, as this means using complex numbers in finding $\mathbf{M}$. Since each of the elements of $\mathbf{M}$ will be either real or purely imaginary, there is, however, no need to use general complex arithmetic. In such cases it is probably best to ignore symmetry and proceed as for general matrices. Another difficulty, which may arise with the method preserving symmetry, is that pivots may become small or even zero and it is difficult to make row or column interchanges without destroying the symmetry. This is only possible if, whenever rows are interchanged, the corresponding columns are also interchanged and vice versa.

There is an important class of matrices, called *positive definite* matrices, for which $\mathbf{M}$ will always exist and be real. An $n \times n$ real symmetric matrix $\mathbf{A}$ is said to be positive definite if, given any real vector $\mathbf{x} \neq \mathbf{0}$, then

$$\mathbf{x}^T\mathbf{A}\mathbf{x} > 0.$$

Note that $\mathbf{x}^T\mathbf{A}\mathbf{x}$ is a 1×1 matrix. Positive definite matrices occur in a variety of problems, for example, least squares approximation calculations. (See Problem 8.37.)

We note first that the unit matrix is positive definite, as it is symmetric and

$$\mathbf{x}^T\mathbf{I}\mathbf{x} = \mathbf{x}^T\mathbf{x} = x_1^2 + x_2^2 + \cdots + x_n^2.$$

The latter must be positive if the real vector $\mathbf{x}$ is non-zero.

Secondly we observe that, if a matrix $\mathbf{A}$ is positive definite, then it must be non-singular. For, suppose that $\mathbf{A}$ is singular so that, by the last part of §8.3, there is a vector $\mathbf{x} \neq \mathbf{0}$ for which

$$\mathbf{A}\mathbf{x} = \mathbf{0}.$$

Then clearly

$$\mathbf{x}^T\mathbf{A}\mathbf{x} = 0$$

which would contradict the positive definite property.

Theorem 8.2 An $n \times n$ real symmetric matrix $\mathbf{A}$ may be factorized in the form

$$\mathbf{A} = \mathbf{M}\mathbf{M}^T, \tag{8.45}$$

where $\mathbf{M}$ is a lower triangular real matrix with non-zero elements on the diagonal if, and only if, $\mathbf{A}$ is positive definite.

Proof If $\mathbf{A}$ can be factorized in the form (8.45), then

$$\mathbf{x}^T\mathbf{A}\mathbf{x} = \mathbf{x}^T\mathbf{M}\mathbf{M}^T\mathbf{x} = \mathbf{y}^T\mathbf{y}$$

where

$$\mathbf{y} = \mathbf{M}^T\mathbf{x}.$$

If $\mathbf{x} \neq \mathbf{0}$ we must have $\mathbf{y} \neq \mathbf{0}$ as $\mathbf{M}$ is non-singular and thus

$$\mathbf{x}^T\mathbf{A}\mathbf{x} = \mathbf{y}^T\mathbf{y} > 0,$$

showing that $\mathbf{A}$ is positive definite.

Conversely, if $\mathbf{A}$ is positive definite all the leading submatrices are positive definite and hence non-singular. (See Problem 8.35.) By Theorem 8.1, $\mathbf{A}$ may be factorized in the form

$$\mathbf{A} = \mathbf{L}\mathbf{D}\mathbf{V}$$

and since this factorization is unique and $\mathbf{A}$ is symmetric

$$\mathbf{L}^T = \mathbf{V}.$$

It remains to be shown that the diagonal elements of $\mathbf{D}$ are positive. Choose $\mathbf{x}$ so that

$$\mathbf{L}^T\mathbf{x} = \mathbf{e}_j$$

where $\mathbf{e}_j = [0 \cdots 0 \quad 1 \quad 0 \cdots 0]^T$, with the unit in the jth position, i.e. $\mathbf{e}_j$ is the jth column of $\mathbf{I}$. We can always find such an $\mathbf{x} \neq \mathbf{0}$ as $\mathbf{L}^T$ is non-singular. Now since $\mathbf{A}$ is positive definite,

$$0 < \mathbf{x}^T\mathbf{A}\mathbf{x} = \mathbf{x}^T\mathbf{L}\mathbf{D}\mathbf{L}^T\mathbf{x} = (\mathbf{L}^T\mathbf{x})^T\mathbf{D}(\mathbf{L}^T\mathbf{x})$$
$$= \mathbf{e}_j{}^T\mathbf{D}\mathbf{e}_j = d_j,$$

where d_j is the jth diagonal element of **D**. Since this holds for $j = 1, 2, \ldots, n$, it follows that all the diagonal elements of **D** are positive. $\square$

8.11 Matrix inversion

Given an $n \times n$ non-singular matrix **A**, we seek a matrix $\mathbf{A}^{-1}$ such that

$$\mathbf{A}\mathbf{A}^{-1} = \mathbf{I},$$

where **I** is the $n \times n$ unit matrix. $\mathbf{A}^{-1}$ is called the *inverse* of **A**. We usually find the columns of $\mathbf{A}^{-1}$ by solving linear equations. Let $\mathbf{x}_j$ be the jth column of $\mathbf{A}^{-1}$, so that

$$\mathbf{A}^{-1} = [\mathbf{x}_1 \quad \mathbf{x}_2 \quad \cdots \quad \mathbf{x}_n]$$

in block form. Also write

$$\mathbf{I} = [\mathbf{e}_1 \quad \mathbf{e}_2 \quad \cdots \quad \mathbf{e}_n]$$

where $\mathbf{e}_j$ is the jth column of **I**. We require

$$\mathbf{A}[\mathbf{x}_1 \quad \mathbf{x}_2 \quad \cdots \quad \mathbf{x}_n] = [\mathbf{e}_1 \quad \mathbf{e}_2 \quad \cdots \quad \mathbf{e}_n],$$

that is,

$$[\mathbf{A}\mathbf{x}_1 \quad \mathbf{A}\mathbf{x}_2 \quad \cdots \quad \mathbf{A}\mathbf{x}_n] = [\mathbf{e}_1 \quad \mathbf{e}_2 \quad \cdots \quad \mathbf{e}_n].$$

On equating corresponding columns, we obtain

$$\mathbf{A}\mathbf{x}_j = \mathbf{e}_j, \quad j = 1, 2, \ldots, n. \tag{8.46}$$

These are n sets of n linear equations, each set consisting of n equations in n unknowns. If **A** is non-singular, the solution of each set of equations exists and is unique and thus $\mathbf{A}^{-1}$ exists and is unique. We can solve the sets of equations (8.46) by a simultaneous reduction process, which we illustrate by the following example.

Example 8.13   Find the inverse of

$$\mathbf{A} = \begin{bmatrix} -1 & 8 & -2 \\ -6 & 49 & -10 \\ -4 & 34 & -5 \end{bmatrix}.$$

We write $\mathbf{A}$ and $\mathbf{I}$ side by side and carry out the usual elimination process on $\mathbf{A}$, performing the same operations on $\mathbf{I}$. We obtain

$$\begin{bmatrix} -1 & 8 & -2 : & 1 & 0 & 0 \\ -6 & 49 & -10 : & 0 & 1 & 0 \\ -4 & 34 & -5 : & 0 & 0 & 1 \end{bmatrix}$$

$$\begin{bmatrix} -1 & 8 & -2 : & 1 & 0 & 0 \\ 0 & 1 & 2 : & -6 & 1 & 0 \\ 0 & 2 & 3 : & -4 & 0 & 1 \end{bmatrix}$$

$$\begin{bmatrix} -1 & 8 & -2 : & 1 & 0 & 0 \\ 0 & 1 & 2 : & -6 & 1 & 0 \\ 0 & 0 & -1 : & 8 & -2 & 1 \end{bmatrix}. \tag{8.47}$$

We now solve by back substitution the three sets of linear equations obtained by taking the first three columns of (8.47) as an upper triangular matrix and the last three columns as three different right hand sides. For example, $\mathbf{x}_2$, the second column of $\mathbf{A}^{-1}$, is the solution of

$$\begin{bmatrix} -1 & 8 & -2 \\ 0 & 1 & 2 \\ 0 & 0 & -1 \end{bmatrix} \mathbf{x}_2 = \begin{bmatrix} 0 \\ 1 \\ -2 \end{bmatrix}.$$

We find

$$\mathbf{A}^{-1} = \begin{bmatrix} 95 & -28 & 18 \\ 10 & -3 & 2 \\ -8 & 2 & -1 \end{bmatrix}. \quad \square$$

In Example 8.13, the first three columns of (8.47) form $\mathbf{U}$ and if $\mathbf{c}_1$, $\mathbf{c}_2$ and $\mathbf{c}_3$ are the last three columns then from (8.24) and (8.25) we have

$$\mathbf{A} = \mathbf{LU}$$

and

$$\mathbf{e}_j = \mathbf{Lc}_j, \quad j = 1, 2, 3.$$

From the latter, we see that

$$\mathbf{L}[\mathbf{c}_1 \quad \mathbf{c}_2 \quad \mathbf{c}_3] = [\mathbf{e}_1 \quad \mathbf{e}_2 \quad \mathbf{e}_3]$$

and thus

$$[\mathbf{c}_1 \quad \mathbf{c}_2 \quad \mathbf{c}_3] = \mathbf{L}^{-1}.$$

The unit matrix is changed to $\mathbf{L}^{-1}$ by the elimination process.

We could consider this process as that of finding factors $\mathbf{L}$ and $\mathbf{U}$ and then calculating $\mathbf{L}^{-1}$. We finally solve

$$\mathbf{U}(\mathbf{A}^{-1}) = \mathbf{L}^{-1},$$

column by column, to obtain $\mathbf{A}^{-1}$. At each stage we need only solve triangular equations and any of the methods of factorizing $\mathbf{A}$ may be employed. Notice that the process will only succeed if $\mathbf{A}$ is non-singular as otherwise it will not be possible to solve the equations involved. There is no particular difficulty with making interchanges, but it must be remembered that if columns (rows) of $\mathbf{A}$ are interchanged the corresponding rows (columns) of $\mathbf{A}^{-1}$ are interchanged. (See Problem 8.39.)

We have defined $\mathbf{A}^{-1}$ to be such that

$$\mathbf{A}\mathbf{A}^{-1} = \mathbf{I} \tag{8.48}$$

but, provided $\mathbf{A}^{-1}$ exists, we can also show that

$$\mathbf{A}^{-1}\mathbf{A} = \mathbf{I}.$$

On postmultiplying (8.48) by $\mathbf{A}$, we have

$$\mathbf{A}\mathbf{A}^{-1}\mathbf{A} = \mathbf{I}\mathbf{A} = \mathbf{A}.$$

Now let $\mathbf{A}^{-1}\mathbf{A} = \mathbf{W}$, when

$$\mathbf{A}\mathbf{W} = \mathbf{A},$$

that is,

$$\mathbf{A}[\mathbf{w}_1 \quad \mathbf{w}_2 \quad \cdots \quad \mathbf{w}_n] = [\mathbf{a}_1 \quad \mathbf{a}_2 \quad \cdots \quad \mathbf{a}_n],$$

where $\mathbf{w}_j$ and $\mathbf{a}_j$ are the jth columns of $\mathbf{W}$ and $\mathbf{A}$ respectively. Thus

$$\mathbf{A}\mathbf{w}_j = \mathbf{a}_j, \qquad j = 1, \ldots, n.$$

For each j, this is a set of n equations with unknown vector $\mathbf{w}_j$. Since $\mathbf{A}$ is non-singular, the solution of each set is unique. It is easily verified that this solution is

$$\mathbf{w}_j = \mathbf{e}_j.$$

Thus

$$\mathbf{A}^{-1}\mathbf{A} = \mathbf{W} = [\mathbf{w}_1 \quad \cdots \quad \mathbf{w}_n] = [\mathbf{e}_1 \quad \cdots \quad \mathbf{e}_n] = \mathbf{I}.$$

If $\mathbf{A}$ and $\mathbf{B}$ are two non-singular $n \times n$ matrices,

$$(\mathbf{A}\mathbf{B})^{-1} = \mathbf{B}^{-1}\mathbf{A}^{-1}$$

since

$$\mathbf{A}\mathbf{B}(\mathbf{B}^{-1}\mathbf{A}^{-1}) = \mathbf{A}(\mathbf{B}\mathbf{B}^{-1})\mathbf{A}^{-1} = \mathbf{A}\mathbf{I}\mathbf{A}^{-1} = \mathbf{I}.$$

If we return to the problem of n non-singular equations in n unknowns

$$\mathbf{Ax} = \mathbf{b}$$

we find on premultiplication by $\mathbf{A}^{-1}$ that

$$\mathbf{x} = \mathbf{A}^{-1}\mathbf{Ax} = \mathbf{A}^{-1}\mathbf{b}.$$

Thus if $\mathbf{A}^{-1}$ is known we can easily calculate $\mathbf{x}$. In any practical problem, the work involved in calculating $\mathbf{A}^{-1}$ usually exceeds that required to find $\mathbf{x}$ by one of the elimination methods. There are, however, a few circumstances in which it is desirable to calculate $\mathbf{A}^{-1}$, especially if we have a large number of sets of equations with the same coefficient matrix, but with different right hand sides. Therefore, if $\mathbf{A}$ is $n \times n$ and

$$\mathbf{Ax}_j = \mathbf{b}_j$$

for $j = 1, 2, \ldots, p$, where $p \gg n$, it is desirable to calculate $\mathbf{A}^{-1}$, which is equivalent to solving only n sets of linear equations. We can then easily calculate

$$\mathbf{x}_j = \mathbf{A}^{-1}\mathbf{b}_j$$

for $j = 1, 2, \ldots, p$.

8.12 Tridiagonal matrices

An $n \times n$ matrix $\mathbf{A}$ is said to be *tridiagonal* (or *triple band*) if it takes the form

$$\mathbf{A} = \begin{bmatrix} b_1 & c_1 & & & & \\ a_2 & b_2 & c_2 & & \text{\Large 0} & \\ & a_3 & b_3 & c_3 & & \\ & & \ddots & \ddots & \ddots & \\ \text{\Large 0} & & & \ddots & \ddots & c_{n-1} \\ & & & & a_n & b_n \end{bmatrix} \tag{8.49}$$

Thus the (i, j)th elements are zero for $j > i + 1$ and $j < i - 1$.

The elimination method can be considerably simplified if the coefficient matrix of a linear set of equations is tridiagonal. Consider

$$\mathbf{Ax} = \mathbf{d}$$

where $\mathbf{A}$ is of the form (8.49). Assuming $b_1 \neq 0$, the first step consists of eliminating x_1 from the second equation by subtracting a_2/b_1 times the first equation from the second equation. There is no need to change the 3rd to

nth equations in the elimination of x_1. After $i - 1$ steps, assuming no interchanges are required, the equations take the form

$$
\begin{bmatrix}
\beta_1 & c_1 & & & & & \\
0 & \beta_2 & c_2 & & & \mathbf{0} & \\
 & \ddots & \ddots & \ddots & & & \\
 & 0 & \beta_i & c_i & & & \\
 & & a_{i+1} & b_{i+1} & c_{i+1} & & \\
\mathbf{0} & & & a_{i+2} & b_{i+2} & c_{i+2} & \\
 & & & & \ddots & \ddots & \ddots
\end{bmatrix}
\mathbf{x} =
\begin{bmatrix}
\delta_1 \\
\vdots \\
\delta_i \\
d_{i+1} \\
\vdots \\
d_n
\end{bmatrix} .
$$

We now eliminate x_i from the $(i+1)$th equation by subtracting a_{i+1}/β_i times the ith equation from the $(i + 1)$th equation. The other equations are not changed. The final form of the equations is

$$
\begin{bmatrix}
\beta_1 & c_1 & & & \\
 & \beta_2 & c_2 & \mathbf{0} & \\
 & & \ddots & \ddots & \\
\mathbf{0} & & \ddots & c_{n-1} & \\
 & & & & \beta_n
\end{bmatrix}
\mathbf{x} =
\begin{bmatrix}
\delta_1 \\
\vdots \\
\vdots \\
\delta_n
\end{bmatrix}
\tag{8.50}
$$

and these are easily solved by back substitution.

To reduce the equations to (8.50) we put

$$
\beta_1 = b_1
$$
$$
\delta_1 = d_1
$$

and then, for $i = 1, 2, \ldots, n - 1$, we compute

$$
m_i = a_{i+1}/\beta_i
$$
$$
\beta_{i+1} = b_{i+1} - m_i c_i
$$
$$
\delta_{i+1} = d_{i+1} - m_i \delta_i.
$$

To perform the back substitution, we compute

$$
x_n = \delta_n/\beta_n
$$

and then for $i = n - 1, n - 2, \ldots 2, 1$

$$
x_i = (\delta_i - c_i . x_{i+1})/\beta_i.
$$

If **A** is positive definite we can be certain that the method will not fail because of a zero pivot. In Problem 8.44, simple conditions on the elements a_i, b_i and c_i are given which ensure that **A** is positive definite. We shall see in Chapter 12 that tridiagonal equations occur in numerical methods of solving boundary value differential problems and that in many such applications **A** is positive definite. (See Problem 12.5.)

Problems

Section 8.2

8.1 If

$$A = \begin{bmatrix} 2 & 3 & -1 \\ -1 & 2 & 4 \end{bmatrix}, \quad B = \begin{bmatrix} 1 & 2 & 0 \\ 0 & 1 & 1 \\ 1 & -1 & 3 \end{bmatrix} \quad \text{and} \quad C = \begin{bmatrix} 2 & 1 & -1 \\ 3 & 4 & -2 \\ -1 & 0 & 1 \end{bmatrix},$$

verify by calculation that

$$A(B + C) = AB + AC.$$

8.2 Produce an example which shows that it is possible for two matrices **A** and **B** to be such that

$$AB = 0,$$

even though neither of the matrices is a zero matrix.

8.3 Produce an example which shows that it is possible for three matrices to be such that

$$AC = BC$$

even though **A** and **B** are not equal and $C \neq 0$.

8.4 An $n \times n$ matrix **A** is said to be lower triangular if $a_{ij} = 0$ for all $j > i$.

(a) Show that the product of two lower triangular matrices is lower triangular.

(b) Show that the product of two lower triangular matrices with units on the diagonal ($a_{ii} = 1$, $i = 1, 2, \ldots, n$) is a lower triangular matrix with units on the diagonal.

(c) Show that the product of two lower triangular matrices with zeros on the diagonal ($a_{ii} = 0$, $i = 1, 2, \ldots, n$) is a lower triangular matrix with zeros on the diagonal.

8.5 If the product **AB** exists, show that

$$[i\text{th row of } AB] = [i\text{th row of } A]B$$

and

$$\begin{bmatrix} j\text{th} \\ \text{column} \\ \text{of } \mathbf{AB} \end{bmatrix} = \mathbf{A} \begin{bmatrix} j\text{th} \\ \text{column} \\ \text{of } \mathbf{B} \end{bmatrix}.$$

8.6   If $\mathbf{A}$ is an $n \times n$ matrix and $\mathbf{D}$ an $n \times n$ diagonal matrix with diagonal elements $d_1, \ldots, d_n$, show that

$$[i\text{th row of } \mathbf{DA}] = d_i[i\text{th row of } \mathbf{A}].$$

What can be said about the elements of $\mathbf{AD}$?

8.7   Show that the linear transformation

$$\begin{bmatrix} u \\ v \\ w \end{bmatrix} = \begin{bmatrix} \cos\theta & \sin\theta & 0 \\ -\sin\theta & \cos\theta & 0 \\ 0 & 0 & 1 \end{bmatrix} \begin{bmatrix} x \\ y \\ z \end{bmatrix}$$

is equivalent to a rotation of the three-dimensional space about the z axis through an angle θ. (See Fig. 8.4.)

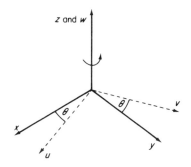

Fig. 8.4. Rotation about z-axis.

8.8   If $\mathbf{x}$ and $\mathbf{u}$ are three-dimensional vectors, show that the transformation

$$\mathbf{u} = \mathbf{A}(\mathbf{x} - \mathbf{d})$$

where $\mathbf{A}$ is a 3×3 matrix and $\mathbf{d}$ is a given vector, is equivalent to a transformation to a new system of coordinates with origin at $\mathbf{x} = \mathbf{d}$.

8.9   Verify the block multiplication rule (8.7) by considering (i, j)th elements.

Section 8.3

8.10 Show, using the elimination process, that the following equations have a unique solution and find this solution,

$$\mathbf{Ax} = \mathbf{b}$$

where

$$\mathbf{A} = \begin{bmatrix} 1 & 2 & -2 & 1 \\ 2 & 5 & -2 & 3 \\ -2 & -2 & 5 & 3 \\ 1 & 3 & 3 & 2 \end{bmatrix}, \qquad \mathbf{b} = \begin{bmatrix} 4 \\ 7 \\ -1 \\ 0 \end{bmatrix}.$$

Check your result by substituting it back in the original equations.

8.11 Use elimination with only row interchanges to show that the following equations have a unique solution and find this solution.

$$\begin{bmatrix} 1 & 1 & 3 & 2 \\ 1 & 1 & 4 & 3 \\ 2 & 1 & 1 & 2 \\ 2 & -1 & 4 & 5 \end{bmatrix} \mathbf{x} = \begin{bmatrix} 4 \\ 7 \\ 3 \\ 8 \end{bmatrix}.$$

Check your result as in Problem 8.10.

8.12 Repeat Problem 8.11 but use only column interchanges.

8.13 Show that the following equations are of rank 2 and inconsistent.

$$\begin{aligned} x + 2y + 3z &= 3 \\ x - y - 4z &= -5 \\ 3x + 3y + 2z &= 2. \end{aligned}$$

8.14 Show that the following equations are consistent if, and only if, $a = +1$ or $a = -1$.

$$\begin{aligned} x + y + z &= 1 + a^2 \\ x + 2y + 3z &= -2a \\ x + 3y + 4z &= -4a \\ x + 2y + 2z &= 2(1 - a). \end{aligned}$$

8.15 Show that the following equations are consistent, of rank 2, and find the general form of the solutions.

$$\begin{aligned} x + 2y + z &= 3 \\ 2x + 3y + 4z &= 7 \\ x + 4y - 3z &= 1. \end{aligned}$$

8.16   Given

$$A = \begin{bmatrix} 0 & 3 & 0 \\ 1 & 0 & 1 \\ 0 & 1 & 0 \end{bmatrix},$$

show that the equations $(A - \lambda I)x = 0$ admit a solution other than $x = 0$ if, and only if, $\lambda = 0$, 2 or -2. Find the general form of the solution for the three given values of λ. The three values of λ, which give a non-zero solution vector x, are called the *eigenvalues* of A. *Hint:* the equations may be written as

$$\begin{aligned} -\lambda x_1 + 3x_2 \qquad\quad &= 0 \\ x_1 - \lambda x_2 + x_3 &= 0 \\ x_2 - \lambda x_3 &= 0. \end{aligned}$$

8.17   Draw a flow diagram which describes the elimination process without interchanges for a system of n non-singular linear equations in n unknowns.

Section 8.4

8.18   Show that the vectors

$$u_1 = \begin{bmatrix} 1 \\ 2 \\ -1 \end{bmatrix}, \quad u_2 = \begin{bmatrix} 2 \\ -1 \\ 1 \end{bmatrix} \quad \text{and} \quad u_3 = \begin{bmatrix} 1 \\ 7 \\ -4 \end{bmatrix}$$

are linearly dependent and hence coplanar.

8.19   Show that $m + 1$ or more vectors in m-dimensional space are always linearly dependent. (If the vectors are $a_1, a_2, \ldots, a_n$ where $n > m$, prove that $Ax = 0$, where the a_j are columns of A, has a solution other than $x = 0$.)

8.20   If the matrix of the linear equations of (5.9) is singular, show that there exist numbers c_i, not all zero, such that

$$\sum_{i=0}^{n} c_i \int_a^b \psi_i(x)\psi_j(x)\, dx = 0, \qquad 0 \leqslant j \leqslant n.$$

Deduce that

$$\int_a^b \left[\sum_{i=0}^{n} c_i \psi_i(x) \right]^2 dx = 0,$$

which is impossible if the ψ_i are linearly independent, and hence show that the equations (5.9) have a unique solution.

Section 8.5

8.21   Repeat Example 8.6 with the same ten digit accuracy but making only column interchanges to pick suitable pivots.

8.22   Repeat Example 8.7 with the same three digit accuracy but making only column interchanges to pick suitable pivots.

Section 8.6

8.23   If

$$
\mathbf{M}_j = \begin{bmatrix}
1 & & & & & & & & \\
l_{21} & \cdot & & & & & 0 & & \\
\cdot & & \cdot & & & & & & \\
\cdot & & & 1 & & & & & \\
\cdot & & & l_{j+1,j} & 1 & & & & \\
\cdot & & & \cdot & & \cdot & & & \\
\cdot & & & \cdot & 0 & & \cdot & & \\
\cdot & & & \cdot & & \vdots & & \cdot & \\
\cdot & l_{n1} & \cdots & l_{n,j} & 0 & \cdots & 0 & & 1
\end{bmatrix}
$$

show that

$$\mathbf{M}_{j+1} = \mathbf{M}_j \mathscr{L}_{j+1}$$

where $\mathscr{L}_j$ is defined by (8.22). Hence verify (8.23) by induction.

8.24   Find the matrices $\mathscr{L}_1$, $\mathscr{L}_2$, $\mathscr{L}_3$, **L**, **U** and **c** of §8.6 for the equations of Problem 8.10. Verify that (8.24) and (8.25) are valid for your results.

Section 8.7

8.25   Use the method of §8.7 to find factors **L**, **D** and **V** of the coefficient matrix **A** of Problem 8.10.

8.26   Show that the coefficient matrix in Problem 8.11 cannot be factorized in the form **LDV**.

Section 8.8

8.27   Use the compact elimination method for the equations of Problem 8.10.

Section 8.9

8.28 Use the compact elimination method with interchanges to solve the equations of Problem 8.11.

8.29 Draw flow diagrams to describe the compact elimination method
(a) without pivoting,
(b) with partial pivoting by rows.

Section 8.10

8.30 If the product $\mathbf{ABC}$ exists, by writing $\mathbf{ABC} = \mathbf{A(BC)}$, show that $(\mathbf{ABC})^T = \mathbf{C}^T\mathbf{B}^T\mathbf{A}^T$.

8.31 If $\mathbf{A}$ and $\mathbf{B}$ are symmetric $n \times n$ matrices, show that
(a) $(\mathbf{A} + \mathbf{B})$ is symmetric,
(b) $\mathbf{AB}$ is not necessarily symmetric. (Produce a counter-example.)

8.32 An $n \times n$ matrix $\mathbf{A}$ is said to be *skew-symmetric* if

$$\mathbf{A}^T = -\mathbf{A}.$$

Show that all the diagonal elements of such a matrix are zero. Hence, by considering the $(1, 1)$ element, prove that it is not possible to factorize such a matrix in the form $\mathbf{LDV}$ where $\mathbf{D}$ has non-zero diagonal elements.

8.33 Find a matrix $\mathbf{M}$ such that (8.45) is valid for

$$\mathbf{A} = \begin{bmatrix} 4 & -2 & -4 \\ -2 & 17 & 10 \\ -4 & 10 & 9 \end{bmatrix}$$

and hence solve the equations

$$\mathbf{Ax} = \begin{bmatrix} 10 \\ 3 \\ -7 \end{bmatrix}.$$

8.34 If $\mathbf{A}$ is any non-singular symmetric matrix, show that $\mathbf{A}^2$ is a positive definite matrix. (*Hint:* $\mathbf{x}^T\mathbf{A}^2\mathbf{x} = \mathbf{x}^T\mathbf{A}^T\mathbf{Ax}$.)

8.35 Show that if $\mathbf{A}$ is a positive definite matrix, all of its leading submatrices are positive definite. (*Hint:* consider $\mathbf{x}^T\mathbf{Ax}$, where $\mathbf{x}$ is a vector whose last $n - p$ elements are zero.)

8.36 Draw a flow diagram for the symmetric matrix factorization method of §8.10 for solving linear equations with a positive definite matrix.

8.37   Show that the least squares normal equations (5.3) may be written in matrix form as $\mathbf{\Psi\Psi}^T\mathbf{a} = \mathbf{\Psi f}$, where

$$\mathbf{a}^T = [a_0 \cdots a_n], \qquad \mathbf{f}^T = [f(x_0) \cdots f(x_N)]$$

and $\mathbf{\Psi}$ is the $(n + 1) \times (N + 1)$ matrix whose (i, j)th element is $\psi_{i-1}(x_{j-1})$. Show that if $\{\psi_0, \psi_1, \ldots, \psi_n\}$ is a Chebyshev set then $\mathbf{\Psi}^T\mathbf{b} \neq \mathbf{0}$ for all $(n + 1)$-dimensional vectors $\mathbf{b} \neq \mathbf{0}$. Hence show (cf. Problem 8.34) that $\mathbf{\Psi\Psi}^T$ is positive definite and thus that the normal equations have a unique solution.

Section 8.11

8.38   Find the inverse of

$$\mathbf{A} = \begin{bmatrix} 2 & -1 & 2 \\ 4 & -4 & 6 \\ 6 & -7 & 8 \end{bmatrix}$$

by the method of §8.11. Check that $\mathbf{A}\mathbf{A}^{-1} = \mathbf{A}^{-1}\mathbf{A} = \mathbf{I}$.

8.39   Show that if columns (or rows) of $\mathbf{A}$ are interchanged the corresponding rows (or columns) of $\mathbf{A}^{-1}$ should be interchanged.

8.40   If $\mathbf{L}$ is a lower triangular matrix with units on the diagonal, show that $\mathbf{L}^{-1}$ exists and that it is a lower triangular matrix with units on the diagonal.

8.41   If $\mathbf{A}$ is the transformation matrix of Problem 8.7, show that

$$\mathbf{A}^T = \mathbf{A}^{-1}.$$

Such a matrix is called an *orthogonal* matrix.

Section 8.12

8.42   Show that the process described in §8.12 is equivalent to factorization of the tridiagonal matrix $\mathbf{A}$ into lower and upper triangular factors of the form

$$\mathbf{A} = \begin{bmatrix} 1 & & & & \\ m_1 & 1 & & \mathbf{0} & \\ & m_2 & 1 & & \\ & \mathbf{0} & & \ddots & \\ & & & m_{n-1} & 1 \end{bmatrix} \begin{bmatrix} \beta_1 & c_1 & & & \mathbf{0} \\ & \ddots & \ddots & & \\ & & \ddots & \ddots & \\ & & & \ddots & c_{n-1} \\ \mathbf{0} & & & & \beta_n \end{bmatrix}$$

8.43   Use the method of §8.12 to solve the equations

$$\begin{bmatrix} 2 & -1 & 0 & 0 & 0 \\ -1 & 2 & -1 & 0 & 0 \\ 0 & -1 & 2 & -1 & 0 \\ 0 & 0 & -1 & 2 & -1 \\ 0 & 0 & 0 & -1 & 2 \end{bmatrix} \mathbf{x} = \begin{bmatrix} 1 \\ 0 \\ 0 \\ 0 \\ 7 \end{bmatrix}.$$

Also find the triangular factors of the coefficient matrix as in Problem 8.42.

8.44   If $\mathbf{A}$ is the $n \times n$ symmetric matrix

$$\mathbf{A} = \begin{bmatrix} b_1 & c_1 & & & & \\ c_1 & b_2 & c_2 & & \text{\LARGE 0} & \\ & c_2 & b_3 & c_3 & & \\ & & \ddots & \ddots & \ddots & \\ & \text{\LARGE 0} & & \ddots & \ddots & c_{n-1} \\ & & & & c_{n-1} & b_n \end{bmatrix},$$

show that, for any n-dimensional vector $\mathbf{x}$,

$$\mathbf{x}^T \mathbf{A} \mathbf{x} = b_1 x_1{}^2 + 2c_1 x_1 x_2 + b_2 x_2{}^2 + 2c_2 x_2 x_3 + \cdots$$
$$+ b_{n-1} x_{n-1}{}^2 + 2c_{n-1} x_{n-1} x_n + b_n x_n{}^2$$
$$= (b_1 - |c_1|)x_1{}^2 + |c_1|(x_1 \pm x_2)^2 + (b_2 - |c_1| - |c_2|)x_2{}^2$$
$$+ |c_2|(x_2 \pm x_3)^2 + \cdots + (b_{n-1} - |c_{n-2}| - |c_{n-1}|)x_{n-1}^2$$
$$+ |c_{n-1}|(x_{n-1} \pm x_n)^2 + (b_n - |c_{n-1}|)x_n{}^2,$$

where the $\pm$ sign in $(x_i \pm x_{i+1})^2$ is taken to be that of c_i. Hence show that if $|c_i| > 0$, $i = 1, 2, \ldots, n - 1$, and, for $c_0 = c_n = 0$,

$$b_i \geq |c_i| + |c_{i-1}|, \qquad i = 1, 2, \ldots, n,$$

with strict inequality in the last relation for at least one i, then $\mathbf{x}^T \mathbf{A} \mathbf{x} > 0$ for all $\mathbf{x} \neq \mathbf{0}$ and thus $\mathbf{A}$ is positive definite.

Show that the coefficient matrix in Problem 8.43 is positive definite.

Chapter 9

Matrix Norms and Applications

9.1 Vector norms

We shall require some measure of the "magnitude" of a vector for satisfactory error analyses of methods of solving linear equations. You are probably familiar with the "magnitude" (or "length") of a vector in three-dimensional space. We generalize this concept to deal with n-dimensional vector spaces.

Definition 9.1 For any n-dimensional vector $\mathbf{x}$ with real (or complex) elements, we define the *norm* of $\mathbf{x}$, written as $\|\mathbf{x}\|$, to be a *real* number satisfying the following conditions.

(i) $\|\mathbf{x}\| > 0$, unless $\mathbf{x} = \mathbf{0}$, and $\|\mathbf{0}\| = 0$.
(ii) For any scalar λ, and any vector $\mathbf{x}$, $\|\lambda \mathbf{x}\| = |\lambda|\, \|\mathbf{x}\|$.
(iii) For any vectors $\mathbf{x}$ and $\mathbf{y}$,

$$\|\mathbf{x} + \mathbf{y}\| \leqslant \|\mathbf{x}\| + \|\mathbf{y}\|. \quad \square \tag{9.1}$$

The last condition is known as the *triangular inequality* and from this it may also be shown (Problem 9.1) that

$$\|\mathbf{x} - \mathbf{y}\| \geqslant \big|\, \|\mathbf{x}\| - \|\mathbf{y}\|\, \big|. \tag{9.2}$$

We may also define the norm to be a mapping of the n-dimensional space into the real numbers such that (i)–(iii) hold.

There are many ways in which one may choose norms. One of the most common is the *Euclidean* norm (or *length*) $\|\mathbf{x}\|_2$ defined by

$$\|\mathbf{x}\|_2 = (|x_1|^2 + |x_2|^2 + \cdots + |x_n|^2)^{1/2}$$

$$= \left[\sum_{i=1}^{n} |x_i|^2 \right]^{1/2}. \tag{9.3}$$

Conditions (i) and (ii) above are clearly satisfied. The verification that condition (iii) holds is more difficult and is left as Problem 9.2. In real three-dimensional space, (iii) corresponds to stating that the length of one side of a triangle cannot exceed the sum of the lengths of the other two sides.

Other possible norms include

$$\|\mathbf{x}\|_1 = |x_1| + |x_2| + \cdots + |x_n|$$

$$= \sum_{i=1}^{n} |x_i| \tag{9.4}$$

and

$$\|\mathbf{x}\|_\infty = \max_{1 \leqslant i \leqslant n} |x_i|. \tag{9.5}$$

These are particularly useful for numerical methods as the corresponding matrix norms (see §9.2) are easily computed. All of these norms may be considered special cases of

$$\|\mathbf{x}\|_p = \left[\sum_{i=1}^{n} |x_i|^p \right]^{1/p}, \tag{9.6}$$

where $p \geqslant 1$. The norms $\|\cdot\|_1$ and $\|\cdot\|_2$ correspond to $p = 1$ and $p = 2$ respectively. The *maximum* norm $\|\cdot\|_\infty$ may be considered the result of allowing p to increase indefinitely. The reader is urged to verify that (9.4) and (9.5) do define norms, by showing that conditions (i)–(iii) are satisfied. (See Problem 9.3.) It is more difficult to verify that, for a general value of $p \geqslant 1$, (9.6) satisfies condition (iii). It is easily verified that conditions (i) and (ii) are satisfied. In most of this chapter we shall use $\|\cdot\|_1$ or $\|\cdot\|_\infty$ and, where it does not matter which norm is employed, we shall simply write $\|\cdot\|$. Of course we do not mix different types of norm as otherwise (9.1) may not hold.

Example 9.1   The norms of

$$\mathbf{x} = \begin{bmatrix} 1 \\ -2 \\ 3 \end{bmatrix} \quad \text{and} \quad \mathbf{y} = \begin{bmatrix} 0 \\ 2 \\ 3 \end{bmatrix}$$

are

$$\|\mathbf{x}\|_1 = 6, \qquad \|\mathbf{y}\|_1 = 5,$$
$$\|\mathbf{x}\|_2 = \sqrt{14}, \qquad \|\mathbf{y}\|_2 = \sqrt{13},$$
$$\|\mathbf{x}\|_\infty = \|\mathbf{y}\|_\infty = 3. \quad \square$$

You will notice that two different vectors may have the same norm, so that

$$\|\mathbf{x}\| = \|\mathbf{y}\| \;\not\Rightarrow\; \mathbf{x} = \mathbf{y}. \tag{9.7}$$

However, from condition (i),

$$\|\mathbf{x} - \mathbf{y}\| = 0 \;\Rightarrow\; \mathbf{x} - \mathbf{y} = \mathbf{0} \;\Rightarrow\; \mathbf{x} = \mathbf{y}. \tag{9.8}$$

We say that a sequence of n-dimensional vectors $\{\mathbf{x}_m\}_{m=0}^{\infty}$ converges to a vector $\boldsymbol{\alpha}$ if, for $i = 1, 2, \ldots, n$, the sequence formed from the ith elements of the $\mathbf{x}_m$ converges to the ith element of $\boldsymbol{\alpha}$. We write

$$\lim_{m \to \infty} \mathbf{x}_m = \boldsymbol{\alpha}.$$

Lemma 9.1 For any type of vector norm

$$\lim_{m \to \infty} \mathbf{x}_m = \boldsymbol{\alpha}$$

if, and only if,

$$\lim_{m \to \infty} \|\mathbf{x}_m - \boldsymbol{\alpha}\| = 0.$$

Proof The lemma is clearly true for norms of the form (9.6) (including $p = \infty$) because of the continuous dependence of the norms on each element and condition (i). The proof for general norms is more difficult and may be found in Isaacson and Keller. $\square$

Note that

$$\lim_{m \to \infty} \|\mathbf{x}_m\| = \|\boldsymbol{\alpha}\| \tag{9.9}$$

is not sufficient to ensure that $\{\mathbf{x}_m\}$ converges to $\boldsymbol{\alpha}$. Indeed, such a sequence may not even converge. Consider, for example,

$$\begin{bmatrix} 1 \\ 1 \end{bmatrix}, \begin{bmatrix} -1 \\ 1 \end{bmatrix}, \begin{bmatrix} 1 \\ 1 \end{bmatrix}, \begin{bmatrix} -1 \\ 1 \end{bmatrix}, \begin{bmatrix} 1 \\ 1 \end{bmatrix}, \ldots.$$

This satisfies (9.9), with

$$\boldsymbol{\alpha} = \begin{bmatrix} 1 \\ 1 \end{bmatrix},$$

for any choice of the norms (9.6).

9.2 Matrix norms

Definition 9.2 For an $n \times n$ matrix $\mathbf{A}$, we define the *matrix norm* $\|\mathbf{A}\|$ *subordinate* to a given vector norm to be

$$\|\mathbf{A}\| = \sup_{\|\mathbf{x}\| = 1} \|\mathbf{A}\mathbf{x}\|. \quad \square$$

The supremum is taken over all n-dimensional vectors $\mathbf{x}$ with unit norm.

The supremum is attained (the proof of this is beyond the scope of this book except for the norms (9.4) and (9.5)) and, therefore, we write

$$\|\mathbf{A}\| = \max_{\|\mathbf{x}\| = 1} \|\mathbf{A}\mathbf{x}\|. \tag{9.10}$$

Lemma 9.2 A subordinate matrix norm satisfies the following properties.

 (i) $\|\mathbf{A}\| > 0$, unless $\mathbf{A} = \mathbf{0}$, and $\|\mathbf{0}\| = 0$.
 (ii) For any scalar λ and any $\mathbf{A}$, $\|\lambda \mathbf{A}\| = |\lambda| . \|\mathbf{A}\|$.
 (iii) For any two matrices $\mathbf{A}$ and $\mathbf{B}$,

$$\|\mathbf{A} + \mathbf{B}\| \leqslant \|\mathbf{A}\| + \|\mathbf{B}\|.$$

 (iv) For any n-dimensional vector $\mathbf{x}$ and any $\mathbf{A}$,

$$\|\mathbf{A}\mathbf{x}\| \leqslant \|\mathbf{A}\| . \|\mathbf{x}\|.$$

 (v) For any two matrices $\mathbf{A}$ and $\mathbf{B}$,

$$\|\mathbf{A}\mathbf{B}\| \leqslant \|\mathbf{A}\| . \|\mathbf{B}\|.$$

Proof (i) If $\mathbf{A} \neq \mathbf{0}$, there is an $\mathbf{x} \neq \mathbf{0}$ such that $\mathbf{A}\mathbf{x} \neq \mathbf{0}$. We may scale $\mathbf{x}$ so that $\|\mathbf{x}\| = 1$ and we still have $\mathbf{A}\mathbf{x} \neq \mathbf{0}$. Thus, from (9.10), $\|\mathbf{A}\| > 0$. If $\mathbf{A} = \mathbf{0}$ then $\mathbf{A}\mathbf{x} = \mathbf{0}$ for all $\mathbf{x}$ and therefore $\|\mathbf{A}\| = 0$.
 (ii) From condition (ii) for vector norms, it follows that

$$\|\lambda \mathbf{A}\| = \max_{\|\mathbf{x}\| = 1} \|\lambda \mathbf{A}\mathbf{x}\| = \max_{\|\mathbf{x}\| = 1} |\lambda| . \|\mathbf{A}\mathbf{x}\| = |\lambda| \max_{\|\mathbf{x}\| = 1} \|\mathbf{A}\mathbf{x}\|.$$

 (iii) From condition (iii) for vector norms,

$$\|\mathbf{A} + \mathbf{B}\| = \max_{\|\mathbf{x}\| = 1} \|(\mathbf{A} + \mathbf{B})\mathbf{x}\| \leqslant \max_{\|\mathbf{x}\| = 1} [\|\mathbf{A}\mathbf{x}\| + \|\mathbf{B}\mathbf{x}\|]$$

$$\leqslant \max_{\|\mathbf{x}\| = 1} \|\mathbf{A}\mathbf{x}\| + \max_{\|\mathbf{x}\| = 1} \|\mathbf{B}\mathbf{x}\| = \|\mathbf{A}\| + \|\mathbf{B}\|.$$

 (iv) For any $\mathbf{x} \neq \mathbf{0}$,

$$\|\mathbf{A}\mathbf{x}\| = \|\mathbf{x}\| . \left\| \mathbf{A} \left(\frac{1}{\|\mathbf{x}\|} \mathbf{x} \right) \right\| \leqslant \|\mathbf{x}\| . \|\mathbf{A}\|,$$

as

$$\left\| \frac{1}{\|\mathbf{x}\|} \mathbf{x} \right\| = 1$$

and $\|\mathbf{A}\|$ occurs for the maximum in (9.10). The result is trivial for $\mathbf{x} = \mathbf{0}$.

 (v) $$\|\mathbf{A}\mathbf{B}\| = \max_{\|\mathbf{x}\| = 1} \|\mathbf{A}\mathbf{B}\mathbf{x}\| = \max_{\|\mathbf{x}\| = 1} \|\mathbf{A}(\mathbf{B}\mathbf{x})\|$$

$$\leqslant \max_{\|\mathbf{x}\| = 1} \|\mathbf{A}\| . \|\mathbf{B}\mathbf{x}\|$$

by (iv) above. Thus

$$\|\mathbf{AB}\| \leqslant \|\mathbf{A}\| \max_{\|\mathbf{x}\|=1} \|\mathbf{Bx}\| = \|\mathbf{A}\| \cdot \|\mathbf{B}\|. \quad \square$$

Matrix norms other than those subordinate to vector norms may also be defined. Such matrix norms are real numbers which satisfy conditions (i), (ii), (iii) and (v) of Lemma 9.2. Throughout this chapter we shall use only subordinate matrix norms and, therefore, condition (iv) of Lemma 9.2 is valid.

We now obtain more explicit expressions for the matrix norms subordinate to the vector norms of (9.4) and (9.5). From (9.4)

$$\|\mathbf{Ax}\|_1 = \sum_{i=1}^{n} \left| \sum_{j=1}^{n} a_{ij} x_j \right|$$

$$\leqslant \sum_{i=1}^{n} \sum_{j=1}^{n} |a_{ij}| \cdot |x_j| = \sum_{j=1}^{n} \left(\sum_{i=1}^{n} |a_{ij}| \right) \cdot |x_j|$$

$$\leqslant \sum_{j=1}^{n} \left(\max_{1 \leqslant k \leqslant n} \sum_{i=1}^{n} |a_{ik}| \right) \cdot |x_j| = \left(\max_{1 \leqslant j \leqslant n} \sum_{i=1}^{n} |a_{ij}| \right) \left(\sum_{j=1}^{n} |x_j| \right).$$

The second factor of the last expression is unity if $\|\mathbf{x}\|_1 = 1$ and thus

$$\|\mathbf{A}\|_1 = \max_{\|\mathbf{x}\|_1=1} \|\mathbf{Ax}\|_1 \leqslant \max_{1 \leqslant j \leqslant n} \sum_{i=1}^{n} |a_{ij}|. \qquad (9.11)$$

Suppose the maximum in the last expression is attained for $j = p$. We now choose $\mathbf{x}$ such that

$$x_p = 1 \quad \text{and} \quad x_j = 0, \quad j \neq p,$$

when $\|\mathbf{x}\|_1 = 1$ and

$$\|\mathbf{Ax}\|_1 = \sum_{i=1}^{n} |a_{ip}|.$$

Equality is obtained in (9.11) for this choice of $\mathbf{x}$ and, therefore,

$$\|\mathbf{A}\|_1 = \max_{1 \leqslant j \leqslant n} \sum_{i=1}^{n} |a_{ij}|.$$

Thus $\|\mathbf{A}\|_1$ is the maximum column sum of the absolute values of elements. Similarly, from (9.5),

$$\|\mathbf{A}\|_\infty = \max_{\|\mathbf{x}\|_\infty=1} \|\mathbf{Ax}\|_\infty$$

$$\leqslant \max_{\|\mathbf{x}\|_\infty=1} \left[\max_i \left| \sum_{j=1}^{n} a_{ij} x_j \right| \right]$$

$$\leqslant \max_i \sum_{j=1}^{n} |a_{ij}|, \qquad (9.12)$$

as $|x_j| \leqslant 1$ for $\|\mathbf{x}\|_\infty = 1$. Suppose the maximum in (9.12) occurs for $i = p$. Assuming $\mathbf{A}$ is real, we now choose $\mathbf{x}$ such that

$$x_j = \text{sign } a_{pj}.$$

We obtain equality in (9.12) and, therefore,

$$\|\mathbf{A}\|_\infty = \max_i \sum_{j=1}^n |a_{ij}|.$$

(If $\mathbf{A}$ is complex we make a different choice of $\mathbf{x}$. See Problem 9.5.) Thus $\|\mathbf{A}\|_\infty$ is the maximum row sum of the absolute values of elements.

The evaluation of $\|\mathbf{A}\|_2$ is much more difficult and involves eigenvalues. A full account is given by Noble.

The following lemma will be useful in subsequent error analyses.

Lemma 9.3 If $\|\mathbf{A}\| < 1$ then $\mathbf{I} + \mathbf{A}$ and $\mathbf{I} - \mathbf{A}$ are non-singular and

$$\frac{1}{1 + \|\mathbf{A}\|} \leqslant \|(\mathbf{I} \pm \mathbf{A})^{-1}\| \leqslant \frac{1}{1 - \|\mathbf{A}\|}. \tag{9.13}$$

Proof We restrict ourselves to $\mathbf{I} + \mathbf{A}$ since we can always replace $\mathbf{A}$ by $-\mathbf{A}$. We first note from (9.10) that

$$\|\mathbf{I}\| = \max_{\|\mathbf{x}\|=1} \|\mathbf{I}.\mathbf{x}\| = \max_{\|\mathbf{x}\|=1} \|\mathbf{x}\| = 1.$$

If $\mathbf{I} + \mathbf{A}$ is singular, there is a vector $\mathbf{x} \neq \mathbf{0}$ such that

$$(\mathbf{I} + \mathbf{A})\mathbf{x} = \mathbf{0}.$$

We will assume that $\mathbf{x}$ is scaled so that $\|\mathbf{x}\| = 1$, when

$$\mathbf{A}\mathbf{x} = -\mathbf{I}\mathbf{x} = -\mathbf{x}$$

and

$$\|\mathbf{A}\mathbf{x}\| = |-1|.\|\mathbf{x}\| = 1.$$

From (9.10), it follows that $\|\mathbf{A}\| \geqslant 1$. Thus $\mathbf{I} + \mathbf{A}$ is non-singular for $\|\mathbf{A}\| < 1$. From

$$\mathbf{I} = (\mathbf{I} + \mathbf{A})^{-1}(\mathbf{I} + \mathbf{A}) \tag{9.14}$$

and Lemma 9.2, we have

$$1 = \|\mathbf{I}\| \leqslant \|(\mathbf{I} + \mathbf{A})^{-1}\| \, \|\mathbf{I} + \mathbf{A}\| \leqslant \|(\mathbf{I} + \mathbf{A})^{-1}\|(\|\mathbf{I}\| + \|\mathbf{A}\|).$$

Dividing by the last factor gives the left side of (9.13).

We rearrange (9.14) as

$$(\mathbf{I} + \mathbf{A})^{-1} = \mathbf{I} - (\mathbf{I} + \mathbf{A})^{-1}\mathbf{A}$$

and, on using Lemma 9.2,

$$\|(I + A)^{-1}\| \leqslant \|I\| + \|(I + A)^{-1}\| \cdot \|A\|.$$

Thus

$$(1 - \|A\|)\|(I + A)^{-1}\| \leqslant 1$$

and, as $\|A\| < 1$, the right side of (9.13) follows. $\square$

The convergence of a sequence of matrices $\{A_m\}_{m=0}^{\infty}$ is defined in an identical manner to that for vectors and we write

$$\lim_{m \to \infty} A_m = B$$

if, for all i and j, the sequence of (i, j)th elements of the A_m converges to the (i, j)th element of B. Corresponding to Lemma 9.1 we have the following.

Lemma 9.4 For any type of matrix norm

$$\lim_{m \to \infty} A_m = B$$

if, and only if,

$$\lim_{m \to \infty} \|A_m - B\| = 0.$$

Proof The lemma is clearly true for the norms $\|\cdot\|_1$ and $\|\cdot\|_\infty$ because of the continuous dependence of the norms on each element and (i) of Lemma 9.2. The proof for general norms is again more difficult and is given by Isaacson and Keller. $\square$

One important sequence is

$$I, A, A^2, A^3, A^4, \ldots. \tag{9.15}$$

It follows from Lemma 9.2(v) that

$$\|A^m\| \leqslant \|A\| \|A^{m-1}\|$$

and, by induction,

$$\|A^m\| \leqslant \|A\|^m.$$

Hence, if for some norm

$$\|A\| < 1,$$

then the sequence (9.15) converges to the zero matrix.
The *infinite series*

$$I + A + A^2 + A^3 + \cdots \tag{9.16}$$

is said to be convergent if the sequence of partial sums is convergent. The partial sums are

$$S_m = I + A + A^2 + \cdots + A^m$$

and, if

$$\lim_{m \to \infty} S_m = S,$$

we say that (9.16) is convergent with sum S. Now,

$$
\begin{aligned}
(I - A)S_m &= (I - A)(I + A + A^2 + \cdots + A^m) \\
&= (I + A + \cdots + A^m) - (A + A^2 + \cdots + A^{m+1}) \\
&= I - A^{m+1}
\end{aligned}
$$

and, if $\|A\| < 1$, from Lemma 9.3 $I - A$ is non-singular, so that

$$S_m = (I - A)^{-1}(I - A^{m+1}).$$

Thus

$$S_m - (I - A)^{-1} = -(I - A)^{-1}A^{m+1}$$

and

$$\|S_m - (I - A)^{-1}\| \leqslant \|(I - A)^{-1}\| . \|A\|^{m+1}.$$

For $\|A\| < 1$, the right side tends to zero as m increases and by Lemma 9.4

$$\lim_{m \to \infty} S_m = (I - A)^{-1}.$$

Thus, for $\|A\| < 1$, the series (9.16) converges with sum $(I - A)^{-1}$.

Example 9.2   If

$$A = \begin{bmatrix} 0.6 & 0.5 \\ 0.1 & 0.3 \end{bmatrix}$$

then

$$\|A\|_1 = 0.8, \qquad \|A\|_\infty = 1.1.$$

Using Lemma 9.3 with $\| \cdot \|_1$, we deduce that $I + A$ is non-singular and

$$\frac{1}{1.8} \leqslant \|(I + A)^{-1}\|_1 \leqslant 5.$$

The sequence $\{A^m\}_{m=0}^{\infty}$ converges to the zero matrix and the series $\sum A^m$ is convergent with sum $(I - A)^{-1}$. Notice that we cannot make these deductions using $\|A\|_\infty$. $\square$

9.3 Rounding errors in solving linear equations

We have seen in Chapter 8 that, if we are not careful, rounding errors may seriously affect the accuracy of the calculation of a solution of a system of linear equations. Examples 8.6 and 8.7 illustrate the importance of pivoting. We should also consider the effect of rounding errors even in apparently well behaved cases. A large number of additions and multiplications are used in the elimination process and we need to investigate whether the error caused by repeated rounding will build up to serious proportions, particularly for large systems of equations.

We shall use *backward error analysis* in our investigation. This approach was pioneered by J. H. Wilkinson and is suitable for various different algebraic problems.

We will assume that, instead of calculating $\mathbf{x}$, the solution of

$$\mathbf{Ax} = \mathbf{b} \qquad (9.17)$$

where $\mathbf{A}$ is $n \times n$ and non-singular, we actually calculate $\mathbf{x} + \boldsymbol{\delta}\mathbf{x}$, because of rounding errors. We will show that $\mathbf{x} + \boldsymbol{\delta}\mathbf{x}$ is the (exact) solution of the perturbed set of equations

$$(\mathbf{A} + \boldsymbol{\delta}\mathbf{A})(\mathbf{x} + \boldsymbol{\delta}\mathbf{x}) = \mathbf{b} + \boldsymbol{\delta}\mathbf{b}. \qquad (9.18)$$

In this section we will derive *a posteriori* bounds on the elements of the $n \times n$ perturbation matrix $\boldsymbol{\delta}\mathbf{A}$ and the perturbation column vector $\boldsymbol{\delta}\mathbf{b}$ and thus also derive bounds on $\|\boldsymbol{\delta}\mathbf{A}\|$ and $\|\boldsymbol{\delta}\mathbf{b}\|$. Backward error analysis of this type does not immediately provide bounds on the final error $\boldsymbol{\delta}\mathbf{x}$ but in the next section we will obtain a relation between $\|\boldsymbol{\delta}\mathbf{x}\|$ and the perturbations $\|\boldsymbol{\delta}\mathbf{A}\|$ and $\|\boldsymbol{\delta}\mathbf{b}\|$.

Much of the calculation in the elimination method consists of determining inner products of the form

$$u_1v_1 + u_2v_2 + \cdots + u_rv_r. \qquad (9.19)$$

Many computers are equipped with a double-length accumulator in which it is possible to calculate inner products to double-length accuracy. Single-length quantities u_i and v_i are multiplied to produce a double-length product without rounding. This product is then added to the double-length accumulator with possibly some rounding at the least significant end of the double-length result. By successively adding u_iv_i into the accumulator for $i = 1, 2, \ldots, r$, we produce a double-length inner product. The final result is usually rounded to single-length and placed in a normal single-length word in the store. We will assume that the rounding errors introduced in producing the double-length inner product are negligible when compared with the error introduced in the final rounding to single-length. Thus, for floating point arithmetic, we produce

$$(u_1 v_1 + u_2 v_2 + \cdots + u_r v_r)(1 + \varepsilon),$$

where bounds on ε are determined by the type of rounding. (See Appendix.)

In the compact elimination method, we seek triangular factors $\mathbf{L}$ and $\mathbf{U}$ ($= \mathbf{DV}$) such that

$$\mathbf{A} = \mathbf{LU}, \tag{9.20}$$

where $\mathbf{L}$ has units on the diagonal. Let us suppose that because of rounding errors we actually compute matrices $\mathbf{L}'$ and $\mathbf{U}'$. We now seek $\boldsymbol{\delta}\mathbf{A}$ such that

$$\mathbf{A} + \boldsymbol{\delta}\mathbf{A} = \mathbf{L}'\mathbf{U}'. \tag{9.21}$$

First consider the attempt to compute $\mathbf{U}$. No errors will be introduced in determining the first row of $\mathbf{U}$, as this row is identical to that of $\mathbf{A}$. To compute u_{ij}, for $1 < i \leqslant j$, we try to use

$$l_{i1} u_{1j} + l_{i2} u_{2j} + \cdots + l_{i,i-1} u_{i-1,j} + 1 . u_{ij} = a_{ij}$$

in the form

$$u_{ij} = a_{ij} - l_{i1} u_{1j} - \cdots - l_{i,i-1} u_{i-1,j}.$$

If the inner product is calculated with only one rounding error, we obtain

$$u'_{ij} = (a_{ij} - l'_{i1} u'_{1j} - \cdots - l'_{i,i-1} u'_{i-1,j})(1 + \varepsilon_{ij}).$$

Thus

$$u'_{ij}\left(1 - \frac{\varepsilon_{ij}}{1 + \varepsilon_{ij}}\right) = a_{ij} - l'_{i1} u'_{1j} - \cdots - l'_{i,i-1} u'_{i-1,j}$$

and

$$l'_{i1} u'_{1j} + \cdots + l'_{i,i-1} u'_{i-1,j} + u'_{ij} = a_{ij} + \varepsilon'_{ij} u'_{ij}, \tag{9.22}$$

where

$$\varepsilon'_{ij} = \frac{\varepsilon_{ij}}{1 + \varepsilon_{ij}}. \tag{9.23}$$

The term $\varepsilon'_{ij} u'_{ij}$ in (9.22) is therefore δa_{ij}, the amount that is added to a_{ij} as a result of rounding. If the rounding errors are bounded by ε (< 1), that is,

$$|\varepsilon_{ij}| \leqslant \varepsilon < 1$$

then

$$|\varepsilon'_{ij}| \leqslant \frac{\varepsilon}{1 - \varepsilon}. \tag{9.24}$$

The first column of $\mathbf{L}$ is obtained by dividing the first column of $\mathbf{A}$ by u'_{11}, so with rounding

$$l'_{11} = \frac{a_{11}}{u'_{11}} (1 + \varepsilon_{11})$$

and

$$l'_{11} u'_{11} = a_{11} + \varepsilon'_{11} l'_{11} u'_{11}.$$

As before, ε'_{ij} is given by (9.23) and hence is bounded as in (9.24).

To compute l_{ij} for $i > j > 1$ we try to use

$$l_{i1} u_{1j} + l_{i2} u_{2j} + \cdots + l_{ij} u_{jj} = a_{ij},$$

in the form

$$l_{ij} = \frac{a_{ij} - l_{i1} u_{1j} - \cdots - l_{i,j-1} u_{j-1,j}}{u_{jj}}.$$

There will be two rounding errors: one due to the formation of the numerator (an inner product) and the other due to dividing by u_{jj}. (If there is no pivoting it is possible to avoid the second rounding error. See Problem 9.12.) We therefore obtain

$$l'_{ij} = \left(\frac{a_{ij} - l'_{i1} u'_{1j} - \cdots - l'_{i,j-1} u'_{j-1,j}}{u'_{jj}} \right) (1 + \varepsilon_{ij})(1 + \delta_{ij}),$$

where ε_{ij} and δ_{ij} are due to rounding. Rearrangement yields

$$l'_{i1} u'_{1j} + \cdots + l'_{ij} u'_{jj} = a_{ij} + \varepsilon'_{ij} l'_{ij} u'_{jj}, \tag{9.25}$$

where

$$\varepsilon'_{ij} = \frac{\varepsilon_{ij} + \delta_{ij} + \varepsilon_{ij} \delta_{ij}}{(1 + \varepsilon_{ij})(1 + \delta_{ij})}.$$

The last term of (9.25) is therefore δa_{ij} for $i > j > 1$. If $|\varepsilon_{ij}|$ and $|\delta_{ij}|$ are bounded by ε, then

$$|\varepsilon'_{ij}| < \frac{\varepsilon(2 + \varepsilon)}{(1 - \varepsilon)^2}. \tag{9.26}$$

Hence

$$\delta \mathbf{A} = \begin{bmatrix} 0 & 0 & 0 & 0 & \cdots \\ \varepsilon'_{21} l'_{21} u'_{11} & \varepsilon'_{22} u'_{22} & \varepsilon'_{23} u'_{23} & \varepsilon'_{24} u'_{24} & \cdots \\ \varepsilon'_{31} l'_{31} u'_{31} & \varepsilon'_{32} l'_{32} u'_{32} & \varepsilon'_{33} u'_{33} & \varepsilon'_{34} u'_{34} & \cdots \\ \vdots & \vdots & \vdots & \vdots & \end{bmatrix}, \tag{9.27}$$

where ε'_{ij} is bounded by (9.26) for $i > j > 1$, and by (9.24) for $1 < i \leqslant j$ and $j = 1, i > 1$.

We can make a similar analysis of the solution of

$$\mathbf{Ly} = \mathbf{b} \tag{9.28}$$

and

$$\mathbf{Ux} = \mathbf{y}. \tag{9.29}$$

In place of (9.28) we obtain

$$\mathbf{L'y'} = \mathbf{b} + \delta\mathbf{b}_1$$

where $\mathbf{y'}$ is the vector calculated instead of $\mathbf{y}$. The effects of rounding errors are similar to those in the calculation of $\mathbf{U}$ and

$$\delta\mathbf{b}_1 = \begin{bmatrix} 0 \\ \varepsilon'_2 y'_2 \\ \vdots \\ \varepsilon'_n y'_n \end{bmatrix}.$$

The first element is zero as $y'_1 = y_1 = b_1$ and the ε'_i, $i = 2, \ldots, n$, are bounded as in (9.24).

The analysis of (9.29) is similar to that for the calculation of $\mathbf{L}$. If $\mathbf{x'} = \mathbf{x} + \delta\mathbf{x}$ is computed instead of $\mathbf{x}$, then

$$\mathbf{U'x'} = \mathbf{y'} + \delta\mathbf{y}, \tag{9.30}$$

where $\delta\mathbf{y}$ has components $\delta'_i x'_i u'_{ii}$, $i = 1, 2, \ldots, n$ with

$$|\delta'_i| < \frac{\varepsilon(2 + \varepsilon)}{(1 - \varepsilon)^2} \quad \text{for } i = 1, 2, \ldots, n - 1 \quad \text{and} \quad |\delta'_n| \leqslant \frac{\varepsilon}{1 - \varepsilon}. \tag{9.31}$$

The former are due to two rounding errors and the latter is due to one. In order to relate $\delta\mathbf{y}$ to the original equations, we premultiply (9.30) by $\mathbf{L'}$.

$$\mathbf{L'U'x'} = \mathbf{L'(y' + \delta y)} = \mathbf{b} + \delta\mathbf{b}_1 + \delta\mathbf{b}_2 \tag{9.32}$$

where

$$\delta\mathbf{b}_2 = \mathbf{L'\delta y}.$$

From (9.21) and (9.32), it follows that we have a perturbation of the form (9.18). To obtain bounds on $\delta\mathbf{A}$ and $\delta\mathbf{b}$ we will assume that partial pivoting is used and consequently that the off-diagonal elements of $\mathbf{L'}$ do not exceed unity in modulus.

Theorem 9.1 Suppose that the compact elimination method is used and that

(1) arithmetic is floating point with relative rounding errors bounded by ε;

(2) inner products are computed with only one rounding error;

(3) partial pivoting is employed;

(4) the elements of the final upper triangular matrix $\mathbf{U}'$, the reduced right side $\mathbf{y}'$ and the calculated approximate solution $\mathbf{x}'$ satisfy

$$|u'_{ii}| \leqslant \bar{u}, \qquad |u'_{ij}| \leqslant u, \qquad |y'_i| \leqslant y, \qquad |x'_i| \leqslant x,$$
$$\text{for } i, j = 1, 2, \ldots, n.$$

Then the computed approximation $\mathbf{x}' = \mathbf{x} + \delta\mathbf{x}$ satisfies

$$(\mathbf{A} + \delta\mathbf{A})(\mathbf{x} + \delta\mathbf{x}) = \mathbf{b} + \delta\mathbf{b}_1 + \delta\mathbf{b}_2 \tag{9.33}$$

where

$$|\delta\mathbf{A}| \leqslant u \begin{bmatrix} 0 & 0 & 0 & 0 & \cdots & 0 \\ \varepsilon' & \varepsilon' & \varepsilon' & \varepsilon' & \cdots & \varepsilon' \\ \varepsilon' & \varepsilon'' & \varepsilon' & \varepsilon' & \cdots & \varepsilon' \\ \varepsilon' & \varepsilon'' & \varepsilon'' & \varepsilon' & \cdots & \varepsilon' \\ \cdot & \cdot & & \cdot & & \cdot \\ \cdot & \cdot & & & \cdot & \cdot \\ \cdot & \cdot & & & \cdot & \cdot \\ \varepsilon' & \varepsilon'' & \varepsilon'' & \cdots\cdots & \varepsilon'' & \varepsilon' \end{bmatrix},$$

$$|\delta\mathbf{b}_1| \leqslant y \begin{bmatrix} 0 \\ c' \\ \vdots \\ \varepsilon' \end{bmatrix} \quad \text{and} \quad |\delta\mathbf{b}_2| \leqslant x\bar{u} \begin{bmatrix} \varepsilon'' \\ 2\varepsilon'' \\ \vdots \\ (n-1)\varepsilon'' \\ (n-1)\varepsilon'' + \varepsilon' \end{bmatrix}$$

with

$$\varepsilon' = \frac{\varepsilon}{1 - \varepsilon} \quad \text{and} \quad \varepsilon'' = \frac{\varepsilon(2 + \varepsilon)}{(1 - \varepsilon)^2}. \tag{9.34}$$

(We use $|\mathbf{A}|$ to denote the matrix with elements $|a_{ij}|$, that is, the matrix of absolute values of elements of $\mathbf{A}$. By $\mathbf{A} \leqslant \mathbf{B}$, we mean that $a_{ij} \leqslant b_{ij}$ for all i and j.) Note that, for small ε,

$$\varepsilon' \simeq \varepsilon \quad \text{and} \quad \varepsilon'' \simeq 2\varepsilon.$$

Proof The bounds for $\delta\mathbf{A}$ and $\delta\mathbf{b}_1$ may be obtained from the earlier analysis with $|l_{ij}| \leqslant 1$. Also

$$\delta\mathbf{b}_2 = \mathbf{L}'\delta\mathbf{y},$$

whose ith element has modulus

$$|l'_{i1}\,\delta y_1 + l'_{i2}\,\delta y_2 + \cdots + l'_{i,i-1}\,\delta y_{i-1} + 1.\delta y_i| \tag{9.35}$$
$$\leqslant |\delta y_1| + |\delta y_2| + \cdots + |\delta y_i|$$
$$= |\delta'_1 x'_1 u'_{11}| + |\delta'_2 x'_2 u'_{22}| + \cdots + |\delta'_i x'_i u'_{ii}|$$
$$\leqslant \begin{cases} i\varepsilon'' x\bar{u}, & \text{for } i = 1, 2, \ldots, (n-1); \\ ((n-1)\varepsilon'' + \varepsilon')x\bar{u}, & \text{for } i = n. \end{cases}$$

The last step is a consequence of (9.31). $\square$

The largest perturbation is likely to be δb_2, which is due to errors in the back substitution process. From (9.35), it may be seen that the magnitudes of elements of δb_2 are dependent on the magnitudes of elements of L'. This provides a very good reason for using partial pivoting with row interchanges instead of column interchanges. Partial pivoting with row interchanges yields a matrix L' whose off-diagonal elements are less than unity in modulus. Partial pivoting with column interchanges yields a matrix U' in which the moduli of off-diagonal elements of any row do not exceed that of the diagonal element (pivot). There is no limit on the magnitudes of elements of L' (multipliers). Whichever method of pivoting is employed, δb_2 is likely to be larger than the other perturbations, δA and δb_1, and, therefore, it may be desirable to use more accurate arithmetic in the back substitution, especially as it forms only a small part of the whole calculation.

Strictly speaking, the bounds of Theorem 9.1 are *a posteriori* bounds, as they depend on the magnitude of elements determined in the calculation. However, the bounds may be used to decide, *a priori*, to what accuracy the calculations should be made, assuming the equations are "well-behaved". If the elements of U' and A are of similar magnitudes, so that

$$a = \max_{i,j} |a_{ij}| \simeq u,$$

the perturbation δA, in (9.33), has elements whose moduli cannot be much greater than that of a double rounding error determined by the floating point representation of a. Similarly, elements of δb_1 may not be much greater than a single rounding error. For a well-behaved set of equations there is therefore no point in computing with more than one extra "guarding" decimal digit (or two binary digits) during the main part of the reduction process. Only the back substitution may need more accuracy. Even without guarding digits, the error introduced in the main part of the process will be only a little more than that due to representing the coefficients by floating point numbers.

The calculation for the simple elimination process of §8.3, in which many more intermediate coefficients are determined, is identical to that of the compact elimination method without the full use of a double-length accumulator. Extra rounding errors are introduced in computing intermediate coefficients. There are similar perturbations to those in Theorem 9.1, but the bounds on their elements are much larger. If a double-length accumulator is not used, the errors in both the simple and the compact methods are identical as the methods are algebraically equivalent. Rounding errors are introduced in forming each term in an inner product. If all initial, intermediate and final coefficients are bounded by a, we find (see Forsythe and Moler) that, provided pivoting is used,

$$|\delta \mathbf{A}| \leqslant a\varepsilon' \begin{bmatrix} 0 & 0 & 0 & 0 & \cdots & 0 & 0 \\ 1 & 2 & 2 & 2 & \cdots & 2 & 2 \\ 1 & 3 & 4 & 4 & \cdots & 4 & 4 \\ 1 & 3 & 5 & 6 & \cdots & 6 & 6 \\ \vdots & & & & & & \vdots \\ 1 & 3 & 5 & 7 & \cdots & (2n-4) & (2n-4) \\ 1 & 3 & 5 & 7 & \cdots & (2n-3) & (2n-2) \end{bmatrix}$$

where ε' is given by (9.34) and ε is the usual error for single-length floating point arithmetic. The perturbations $\delta \mathbf{b}_1$ and $\delta \mathbf{b}_2$ are similarly bounded.

A comparison of the perturbation bounds for the simple elimination and compact elimination methods provides a strong reason for preferring the latter when calculating on a computer with a double-length accumulator. It must be stressed that, in spite of the apparent extra length of simple elimination, the methods are identical in terms of computing effort, which consists primarily of approximately $n^3/3$ multiplications. One advantage of simple elimination is that complete pivoting is possible, which may be desirable for troublesome sets of equations.

So far we have not discussed the effects of perturbations, due to rounding errors, on the accuracy of the computed solution, that is, we have not discussed $\delta \mathbf{x}$ in (9.18). We have already seen in §8.5 that these effects may be very serious but that pivoting does help. In the next section we consider the "sensitivity" of equations to small changes in the coefficients by determining bounds on $\|\delta \mathbf{x}\|$ in terms of $\|\delta \mathbf{A}\|$ and $\|\delta \mathbf{b}\|$.

9.4 Conditioning

Definition 9.3 We define the *condition number* of an $n \times n$ non-singular matrix $\mathbf{A}$ for the norm $\|\cdot\|_p$ to be

$$k_p(\mathbf{A}) = \|\mathbf{A}\|_p \|\mathbf{A}^{-1}\|_p.$$

If the particular choice of norm is immaterial, we will omit the subscript p. $\square$

The condition number of a matrix **A** gives a measure of how sensitive systems of equations, with coefficient matrix **A**, are to small perturbations, such as those caused by rounding. We shall see that, for large $k(\mathbf{A})$, perturbations may have a large effect on the solution. For reasons of clarity, we will consider the effects of perturbing the coefficient matrix **A** and the right side vector **b** separately.

Suppose first that

$$\mathbf{Ax} = \mathbf{b} \tag{9.36}$$

is perturbed so that **A** (non-singular) is kept fixed and **δb** is added to **b**. Thus

$$\mathbf{A}(\mathbf{x} + \mathbf{\delta x}) = \mathbf{b} + \mathbf{\delta b},$$

$$\mathbf{\delta x} = \mathbf{A}^{-1}\mathbf{\delta b}$$

and

$$\|\mathbf{\delta x}\| \leqslant \|\mathbf{A}^{-1}\|\,\|\mathbf{\delta b}\|. \tag{9.37}$$

Also, from (9.36),

$$\|\mathbf{b}\| \leqslant \|\mathbf{A}\|\,\|\mathbf{x}\|$$

and, provided $\mathbf{b} \neq \mathbf{0}$ (and, therefore, $\mathbf{x} \neq \mathbf{0}$),

$$\frac{1}{\|\mathbf{x}\|} \leqslant \frac{\|\mathbf{A}\|}{\|\mathbf{b}\|}. \tag{9.38}$$

Multiplying (9.37) and (9.38) provides

$$\frac{\|\mathbf{\delta x}\|}{\|\mathbf{x}\|} \leqslant \|\mathbf{A}^{-1}\|\,\|\mathbf{A}\|\,\frac{\|\mathbf{\delta b}\|}{\|\mathbf{b}\|} = k(\mathbf{A}).\frac{\|\mathbf{\delta b}\|}{\|\mathbf{b}\|}. \tag{9.39}$$

The quantity on the left of (9.39) may be considered a measure of the relative disturbance of **x**. The inequality provides a bound on this relative disturbance in terms of the relative disturbance of **b**. You will notice that the bound increases as $k(\mathbf{A})$ increases.

Secondly, we keep **b** fixed in (9.36) and perturb **A** so that

$$(\mathbf{A} + \mathbf{\delta A})(\mathbf{x} + \mathbf{\delta x}) = \mathbf{b},$$

whence, assuming that $\mathbf{A} + \mathbf{\delta A}$ is non-singular,

$$\mathbf{x} + \mathbf{\delta x} = (\mathbf{A} + \mathbf{\delta A})^{-1}\mathbf{b}$$

and, from (9.36),

$$\mathbf{\delta x} = [(\mathbf{A} + \mathbf{\delta A})^{-1} - \mathbf{A}^{-1}]\mathbf{b}. \tag{9.40}$$

Thus **δx** depends on the effect of **δA** on the inverse of the coefficient matrix.

To simplify (9.40), we use the identity

$$(\mathbf{A} + \mathbf{\delta A}) - \mathbf{A} = \mathbf{\delta A},$$

which, when premultiplied by $(A + \delta A)^{-1}$ and postmultiplied by A^{-1}, becomes

$$
\begin{aligned}
A^{-1} - (A + \delta A)^{-1} &= (A + \delta A)^{-1}.\delta A.A^{-1} \\
&= [A(I + A^{-1} \delta A)]^{-1}.\delta A A^{-1} \\
&= (I + A^{-1} \delta A)^{-1}.A^{-1}.\delta A.A^{-1}. \quad (9.41)
\end{aligned}
$$

From Lemma 9.3,

$$
\|(I + A^{-1} \delta A)^{-1}\| \leq \frac{1}{1 - \|A^{-1} \delta A\|} \leq \frac{1}{1 - \|A^{-1}\| \|\delta A\|} \quad (9.42)
$$

provided

$$
\|A^{-1}.\delta A\| \leq \|A^{-1}\| \|\delta A\| < 1, \quad (9.43)
$$

which will be satisfied if $\|\delta A\|$ is sufficiently small. (This condition and Lemma 9.3 also ensure that $(I + A^{-1} \delta A)^{-1}$ exists and thus that $(A + \delta A)^{-1}$ exists.)

We combine (9.40) and (9.41) to obtain

$$
\begin{aligned}
\delta x &= -(I + A^{-1} \delta A)^{-1}.A^{-1}.\delta A.A^{-1}.b \\
&= -(I + A^{-1} \delta A)^{-1}.A^{-1}.\delta A.x
\end{aligned}
$$

and thus, from (9.42),

$$
\begin{aligned}
\|\delta x\| &\leq \frac{1}{1 - \|A^{-1}\| \|\delta A\|} \|A^{-1}\| \|\delta A\| \|x\| \\
&= \frac{1}{1 - \dfrac{\|\delta A\|}{\|A\|}.\|A^{-1}\| \|A\|} \|A^{-1}\| \|A\|.\frac{\|\delta A\|}{\|A\|}.\|x\|. \quad (9.44)
\end{aligned}
$$

Let

$$
\frac{\|\delta A\|}{\|A\|} = e, \quad (9.45)
$$

so that e is a measure of the relative disturbance of A. From (9.44) the relative disturbance of the solution satisfies

$$
\frac{\|\delta x\|}{\|x\|} \leq \frac{e.k(A)}{1 - e.k(A)}, \quad (9.46)
$$

provided (9.43) holds, that is

$$
e.k(A) < 1.
$$

It can be seen from (9.46) that, if $k(A)$ is very large, the bound in (9.46) will be much larger than e. Finally we note that similar bounds are obtainable when A and b are perturbed simultaneously. (See Problem 9.15.)

We have thus shown that, if the condition number of a matrix is large, the effects of rounding errors in the solution process may be serious. Even if a matrix or its inverse has large elements, the condition number is not necessarily large. It is easily seen that for any non-zero scalar λ,

$$k(\lambda \mathbf{A}) = k(\mathbf{A}).$$

Increasing (or decreasing) λ will increase the elements of $\lambda \mathbf{A}$ (or $(\lambda \mathbf{A})^{-1}$) but the condition number will not change. If $k(\mathbf{A}) \gg 1$ we say that $\mathbf{A}$ is *ill-conditioned*.

Example 9.3   The Hilbert matrices (see also §5.4)

$$\mathbf{H}_n = \begin{bmatrix} 1 & \frac{1}{2} & \frac{1}{3} & \frac{1}{4} & \cdots & \frac{1}{n} \\ \frac{1}{2} & \frac{1}{3} & \frac{1}{4} & \frac{1}{5} & \cdots & \frac{1}{n+1} \\ \frac{1}{3} & \frac{1}{4} & \frac{1}{5} & \frac{1}{6} & \cdots & \frac{1}{n+2} \\ \vdots & & & & & \vdots \\ \frac{1}{n} & \frac{1}{n+1} & \frac{1}{n+2} & \frac{1}{n+3} & \cdots & \frac{1}{2n-1} \end{bmatrix},$$

$n = 1, 2, 3, \ldots$, are notoriously ill-conditioned and $k(\mathbf{H}_n) \to \infty$ very rapidly as $n \to \infty$. For example,

$$\mathbf{H}_3 = \begin{bmatrix} 1 & \frac{1}{2} & \frac{1}{3} \\ \frac{1}{2} & \frac{1}{3} & \frac{1}{4} \\ \frac{1}{3} & \frac{1}{4} & \frac{1}{5} \end{bmatrix}, \qquad \mathbf{H}_3^{-1} = \begin{bmatrix} 9 & -36 & 30 \\ -36 & 192 & -180 \\ 30 & -180 & 180 \end{bmatrix},$$

$$\|\mathbf{H}_3\|_1 = \|\mathbf{H}_3\|_\infty = 11/6, \qquad \|\mathbf{H}_3^{-1}\|_1 = \|\mathbf{H}_3^{-1}\|_\infty = 408$$

and $k_1(\mathbf{H}_3) = k_\infty(\mathbf{H}_3) = 748$.

For n as large as 6, the ill-conditioning is extremely bad, with

$$k_1(\mathbf{H}_6) = k_\infty(\mathbf{H}_6) \simeq 29 \times 10^6.$$

Even for $n = 3$, the effects of rounding the coefficients are serious. For example the solution of

$$\mathbf{H}_3 \mathbf{x} = \begin{bmatrix} \frac{11}{6} \\ \frac{13}{12} \\ \frac{47}{60} \end{bmatrix} \quad \text{is} \quad \mathbf{x} = \begin{bmatrix} 1 \\ 1 \\ 1 \end{bmatrix}.$$

If we round the coefficients in the equations to three correct significant decimal digits, we obtain

$$\begin{bmatrix} 1.00 & 0.500 & 0.333 \\ 0.500 & 0.333 & 0.250 \\ 0.333 & 0.250 & 0.200 \end{bmatrix} \mathbf{x} = \begin{bmatrix} 1.83 \\ 1.08 \\ 0.783 \end{bmatrix} \tag{9.47}$$

and these have as solution (correct to four significant figures)

$$\mathbf{x} = \begin{bmatrix} 1.090 \\ 0.4880 \\ 1.491 \end{bmatrix}. \tag{9.48}$$

The relative disturbance of the coefficients never exceeds 0.3% but the solution is changed by over 50%.

The main symptom of ill-conditioning is that the magnitudes of the pivots become very small even if pivoting is used. Consider, for example, the equations (9.47) in which the last two rows are interchanged if partial pivoting is employed. If we use the compact elimination method and work to three significant decimal digits with a double-length accumulator, we obtain the triangular matrices

$$\begin{bmatrix} 1 & 0 & 0 \\ 0.333 & 1 & 0 \\ 0.500 & 0.994 & 1 \end{bmatrix} \begin{bmatrix} 1.00 & 0.500 & 0.333 \\ 0 & 0.0835 & 0.0891 \\ 0 & 0 & -0.00507 \end{bmatrix}.$$

The last pivot, -0.00507, is very small in magnitude compared with other elements. ☐

Scaling equations (or unknowns) has an effect on the condition number of a coefficient matrix. It is often desirable to scale so as to reduce any disparity in the magnitude of coefficients. Such scaling does not always improve the accuracy of the elimination method but may be important, especially if only partial pivoting is employed, as the next example demonstrates.

Example 9.4 Consider the equations

$$\begin{bmatrix} 1 & 10^4 \\ 1 & 1 \end{bmatrix} \begin{bmatrix} x_1 \\ x_2 \end{bmatrix} = \begin{bmatrix} 10^4 \\ 2 \end{bmatrix}$$

for which

$$\mathbf{A}^{-1} = \left(\frac{1}{10^4 - 1} \right) \begin{bmatrix} -1 & 10^4 \\ 1 & -1 \end{bmatrix}$$

and

$$k_\infty(\mathbf{A}) = \frac{(10^4 + 1)^2}{10^4 - 1} \simeq 10^4.$$

The elimination method with partial pivoting does not involve interchanges, so that, working to three decimal digits, we obtain

$$x_1 + 10^4 x_2 = 10^4$$
$$- 10^4 x_2 = - 10^4.$$

On back substituting, we obtain the very poor result

$$x_2 = 1, \qquad x_1 = 0.$$

If the first equation is scaled by 10^{-4}, the coefficient matrix becomes

$$\mathbf{B} = \begin{bmatrix} 10^{-4} & 1 \\ 1 & 1 \end{bmatrix}, \quad \text{with } \mathbf{B}^{-1} = \left(\frac{1}{1 - 10^{-4}}\right) \begin{bmatrix} -1 & 1 \\ 1 & -10^{-4} \end{bmatrix}$$

and

$$k_\infty(\mathbf{B}) = \frac{4}{1 - 10^{-4}} \simeq 4.$$

This time partial pivoting interchanges the rows, so that the equations reduce to

$$x_1 + x_2 = 2$$
$$x_2 = 1.$$

These yield $x_1 = x_1 = 1$, a good approximation to the solution. $\square$

It must be stressed that the inequalities (9.39) and (9.46) can rarely be used to provide a precise bound on $\|\delta\mathbf{x}\|$ as only rarely is the condition number $k(\mathbf{A})$ known. When solving linear equations, it is usually impracticable to determine $k(\mathbf{A})$ as this requires a knowledge of $\mathbf{A}^{-1}$, the computation of which would require more calculation than the original problem. However, the two inequalities (9.39) and (9.46) when combined with the results of §9.3 do provide qualitative information regarding $\delta\mathbf{x}$, the error in the computed solution due to the effect of rounding error.

9.5 Iterative correction from residual vectors

Suppose that $\mathbf{x}_0$ is an approximation to the solution of non-singular equations

$$\mathbf{A}\mathbf{x} = \mathbf{b}.$$

Corresponding to $\mathbf{x}_0$ there is a *residual vector* $\mathbf{r}_0$ given by

$$\mathbf{r}_0 = \mathbf{A}\mathbf{x}_0 - \mathbf{b}. \tag{9.49}$$

We sometimes use r_0 to assess the accuracy of x_0 as an approximation to x, but this may be misleading, especially if the matrix A is ill-conditioned, as the following example illustrates.

Example 9.5 The residual vector for the equations (9.47), when

$$x_0 = \begin{bmatrix} 1 \\ 1 \\ 1 \end{bmatrix}, \quad \text{is} \quad r_0 = \begin{bmatrix} 0.003 \\ 0.003 \\ 0 \end{bmatrix}.$$

Notice that for these equations

$$\frac{\|x - x_0\|_\infty}{\|x\|_\infty} = \frac{0.512}{1.491} > 0.3, \tag{9.50}$$

whereas

$$\frac{\|r_0\|_\infty}{\|b\|_\infty} = \frac{0.003}{1.83} < 0.002. \tag{9.51}$$

The left of (9.50) is a measure of the relative accuracy of x_0 and the left of (9.51) is a measure of the relative magnitude of r_0. □

If a bound for $k(A)$ is known, then the residual vector may be used as a guide to the accuracy of the solution. (See Problem 9.19.) The residual vector is also useful in enabling us to make a correction to the approximation to the solution. If

$$\delta x = x - x_0, \tag{9.52}$$

then

$$\begin{aligned} r_0 = Ax_0 - b &= A(x - \delta x) - b \\ &= -A.\delta x, \end{aligned}$$

so that δx is the solution of the system

$$A.\delta x = -r_0. \tag{9.53}$$

We attempt to solve (9.53) and so determine $x = x_0 + \delta x$. In practice, rounding errors are introduced again and, therefore, only an approximation to δx is computed. Hopefully, however, we obtain a better approximation to x. We may iterate by repeating this process and, to investigate convergence, we need the following theorem.

Theorem 9.2 Suppose that for arbitrary x_0 the sequence of n-dimensional vectors $\{x_m\}_{m=0}^\infty$ satisfies

$$x_{m+1} = Mx_m + d, \quad m = 0, 1, 2, \ldots. \tag{9.54}$$

If, for some choice of norm,

$$\|\mathbf{M}\| < 1$$

then the sequence $\{\mathbf{x}_m\}$ converges to $\mathbf{x}$, the unique solution of

$$(\mathbf{I} - \mathbf{M})\mathbf{x} = \mathbf{d}. \tag{9.55}$$

Proof The theorem is a special case of the contraction mapping theorem 10.1, but because the proof is much simpler we give it here.

First we note from Lemma 9.3 that, for $\|\mathbf{M}\| < 1$, $\mathbf{I} - \mathbf{M}$ is non-singular and therefore (9.55) has a unique solution $\mathbf{x}$.

Now suppose that

$$\mathbf{x}_m = \mathbf{x} + \mathbf{e}_m$$

when, from (9.54),

$$\mathbf{x} + \mathbf{e}_{m+1} = \mathbf{M}(\mathbf{x} + \mathbf{e}_m) + \mathbf{d}$$

and, using (9.55),

$$\mathbf{e}_{m+1} = \mathbf{M}\mathbf{e}_m.$$

Hence

$$\|\mathbf{e}_{m+1}\| \leqslant \|\mathbf{M}\|\, \|\mathbf{e}_m\|. \tag{9.56}$$

Using induction, we have

$$\|\mathbf{e}_m\| \leqslant \|\mathbf{M}\|^m \|\mathbf{e}_0\|$$

and, as $\|\mathbf{M}\| < 1$,

$$\|\mathbf{x}_m - \mathbf{x}\| = \|\mathbf{e}_m\| \to 0 \quad \text{as} \quad m \to \infty,$$

that is,

$$\mathbf{x}_m \to \mathbf{x} \quad \text{as} \quad m \to \infty. \ \square$$

We note from (9.56) that for $\mathbf{e}_0 \neq \mathbf{0}$ (and hence $\mathbf{e}_m \neq \mathbf{0}$ for all m),

$$\frac{\|\mathbf{e}_{m+1}\|}{\|\mathbf{e}_m\|} \leqslant \|\mathbf{M}\| < 1$$

and therefore we say that the convergence is of at least first order (cf. §7.6).

We return to the problem of correcting an approximation $\mathbf{x}_0$ to a solution vector $\mathbf{x}$. We assume that $\mathbf{x}_0$ has been calculated by a factorization method with rounding errors, so that

$$(\mathbf{A} + \delta\mathbf{A}) = \mathbf{L}'\mathbf{U}' \tag{9.57}$$

and

$$(A + \delta A)x_0 = b + \delta b.$$

We compute the residual vector r_0 from

$$r_0 = Ax_0 - b \qquad (9.58)$$

using high accuracy arithmetic so that rounding errors in this step are negligible. This is not too difficult as each element of r_0 consists of an inner product which may be formed in a double-length accumulator. We now attempt to solve (9.53) and so find δx. In practice, we use the known approximate factorization of A, (9.57), but we do use more accurate arithmetic in the forward and back substitutions involving the right side r_0, so that the perturbation of r_0 due to rounding errors is negligible. Thus we calculate δx_0, say, the solution of

$$(A + \delta A)\, \delta x_0 = -r_0. \qquad (9.59)$$

Finally we compute x_1 from

$$x_1 = x_0 + \delta x_0, \qquad (9.60)$$

which we again assume is achieved with negligible rounding error. In general $x_1 \neq x$, due to the perturbation δA in (9.59) and, therefore, we repeat the above process, starting with x_1 instead of x_0.

From (9.60), (9.59) and (9.58),

$$\begin{aligned}
x_1 &= x_0 - (A + \delta A)^{-1}r_0 \\
&= x_0 - (A + \delta A)^{-1}(Ax_0 - b) \\
&= (A + \delta A)^{-1}(A + \delta A - A)x_0 + (A + \delta A)^{-1}b \\
&= (A + \delta A)^{-1}\, \delta A . x_0 + (A + \delta A)^{-1}b.
\end{aligned}$$

If the process is repeated, we obtain a sequence of vectors $\{x_m\}_{m=0}^{\infty}$ related by

$$x_{m+1} = (A + \delta A)^{-1}\, \delta A . x_m + (A + \delta A)^{-1}b, \qquad m = 0, 1, 2, \ldots.$$

This is of the form (9.54) and, by Theorem 9.2, the sequence $\{x_m\}$ will converge if

$$\|(A + \delta A)^{-1} . \delta A\| < 1. \qquad (9.61)$$

This will be satisfied if the elements of δA are sufficiently small, as

$$\|(A + \delta A)^{-1} . \delta A\| \leqslant \|(A + \delta A)^{-1}\|\, \|\delta A\|.$$

The sequence $\{x_m\}$ converges to the solution of

$$[I - (A + \delta A)^{-1}\, \delta A]x = (A + \delta A)^{-1}b,$$

that is, on premultiplying by $(A + \delta A)$,

$$Ax = b.$$

As was observed after Theorem 9.2, the convergence is of at least first order.

We normally apply this correction process only once or twice after an approximate solution has been calculated by one of the elimination methods. The intention is to reduce the effects of rounding errors.

Example 9.6 Consider the equations

$$\begin{bmatrix} 33 & 25 & 20 \\ 20 & 17 & 14 \\ 25 & 20 & 17 \end{bmatrix} \mathbf{x} = \begin{bmatrix} 78 \\ 51 \\ 62 \end{bmatrix},$$

which have solution

$$\mathbf{x} = [1 \quad 1 \quad 1]^T.$$

If we use the simple elimination method with arithmetic accurate to only two significant decimal digits except in a double-length accumulator, we obtain

$$\begin{bmatrix} 33 & 25 & 20 & : & 78 \\ 0 & 1.8 & 1.8 & : & 3.4 \\ 0 & 0 & 0.79 & : & 0.80 \end{bmatrix}.$$

The first three columns form $\mathbf{U}'$ and the last column is the reduced right side $(\mathbf{L}')^{-1}\mathbf{b}$, where the multipliers are elements of

$$\mathbf{L}' = \begin{bmatrix} 1 & 0 & 0 \\ 0.61 & 1 & 0 \\ 0.76 & 0.56 & 1 \end{bmatrix}.$$

The computed approximation to $\mathbf{x}$ is

$$\mathbf{x}_0 = [1.1 \quad 0.89 \quad 1.0]^T.$$

We now find, working to four significant digits, that

$$\mathbf{r}_0 = [0.5500 \quad 0.1300 \quad 0.3000]^T$$

which, when treated by the multipliers as above, becomes

$$(\mathbf{L}')^{-1}\mathbf{r}_0 = [0.5500 \quad -0.2055 \quad -0.002920]^T.$$

On solving

$$\mathbf{U}'\boldsymbol{\delta}\mathbf{x}_0 = -(\mathbf{L}')^{-1}\mathbf{r}_0,$$

we obtain

$$\boldsymbol{\delta}\mathbf{x}_0 = [-0.1026 \quad 0.1105 \quad 0.003696]^T,$$

so that

$$\mathbf{x}_1 = \mathbf{x}_0 + \delta\mathbf{x}_0 = [0.9974 \quad 1.000 \quad 1.004]^T.$$

Notice that

$$\|\mathbf{x} - \mathbf{x}_0\|_\infty = 0.11,$$

whereas

$$\|\mathbf{x} - \mathbf{x}_1\|_\infty = 0.004,$$

showing that there has been a considerable improvement in the approximation.

A second application of the process yields

$$\delta\mathbf{x}_1 = [0.002737 \quad -0.0001538 \quad -0.004034]^T$$

so that if $\mathbf{x}_2 = \mathbf{x}_1 + \delta\mathbf{x}_1$,

$$\|\mathbf{x} - \mathbf{x}_2\|_\infty < 0.0002. \quad \square$$

9.6 Iterative correction for an inverse

In Chapter 8 we saw how the inverse of a matrix may be computed. This process also will introduce rounding errors and it may be necessary to use a correction procedure. We could use the process of §9.5 by applying it to each column of the inverse. However, there is a better procedure if the complete matrix inverse is required. Suppose that $\mathbf{W}_0$ is a known approximation to $\mathbf{A}^{-1}$. We calculate a sequence of matrices $\mathbf{W}_0, \mathbf{W}_1, \mathbf{W}_2, \ldots$ using

$$\mathbf{W}_{m+1} = \mathbf{W}_m(2\mathbf{I} - \mathbf{A}\mathbf{W}_m), \qquad m = 0, 1, 2, \ldots. \tag{9.62}$$

We will assume that the rounding errors introduced in using (9.62) are negligible. We show that the sequence converges to $\mathbf{A}^{-1}$ provided $\mathbf{W}_0$ is a sufficiently good approximation to $\mathbf{A}^{-1}$.

Let $\mathbf{E}_m$ denote the error matrix

$$\mathbf{E}_m = \mathbf{W}_m - \mathbf{A}^{-1}$$

so that, in (9.62),

$$\begin{aligned}
\mathbf{A}^{-1} + \mathbf{E}_{m+1} &= (\mathbf{A}^{-1} + \mathbf{E}_m)[2\mathbf{I} - \mathbf{A}(\mathbf{A}^{-1} + \mathbf{E}_m)] \\
&= (\mathbf{A}^{-1} + \mathbf{E}_m)(\mathbf{I} - \mathbf{A}\mathbf{E}_m) \\
&= \mathbf{A}^{-1} + \mathbf{E}_m - \mathbf{A}^{-1}\mathbf{A}\mathbf{E}_m - \mathbf{E}_m\mathbf{A}\mathbf{E}_m.
\end{aligned}$$

Thus

$$\mathbf{E}_{m+1} = -\mathbf{E}_m\mathbf{A}\mathbf{E}_m. \tag{9.63}$$

It follows that

$$\|\mathbf{E}_{m+1}\| \leqslant \|\mathbf{E}_m\| \|\mathbf{AE}_m\| \tag{9.64}$$

and that

$$\mathbf{AE}_{m+1} = -\mathbf{AE}_m\mathbf{AE}_m,$$

whence

$$\|\mathbf{AE}_{m+1}\| \leqslant \|\mathbf{AE}_m\|^2. \tag{9.65}$$

Suppose that $\|\mathbf{AE}_0\| = a$, when from (9.65)

$$\|\mathbf{AE}_1\| \leqslant a^2,$$
$$\|\mathbf{AE}_2\| \leqslant \|\mathbf{AE}_1\|^2 \leqslant a^4$$

and

$$\|\mathbf{AE}_m\| \leqslant a^{2^m}. \tag{9.66}$$

Also, from (9.64),

$$\|\mathbf{E}_1\| \leqslant a\|\mathbf{E}_0\|,$$
$$\|\mathbf{E}_2\| \leqslant \|\mathbf{E}_1\| \|\mathbf{AE}_1\| \leqslant a^3\|\mathbf{E}_0\|,$$
$$\|\mathbf{E}_3\| \leqslant \|\mathbf{E}_2\| \|\mathbf{AE}_2\| \leqslant a^7\|\mathbf{E}_0\|$$

and

$$\|\mathbf{E}_m\| \leqslant a^{2^m-1}\|\mathbf{E}_0\|. \tag{9.67}$$

The reader should verify by an induction argument that (9.66) and (9.67) follow from (9.64) and (9.65). We conclude from (9.67) that, if

$$\|\mathbf{AE}_0\| = a < 1, \tag{9.68}$$

then

$$\|\mathbf{E}_m\| \to 0 \quad \text{as} \quad m \to \infty$$

and, therefore,

$$\mathbf{W}_m \to \mathbf{A}^{-1} \quad \text{as} \quad m \to \infty.$$

The condition (9.68) is not quite the same as that for convergence of the correction procedure for linear equations. If

$$\mathbf{W}_0 = (\mathbf{A} + \delta\mathbf{A})^{-1},$$
$$\begin{aligned}
\mathbf{AE}_0 &= \mathbf{A}(\mathbf{W}_0 - \mathbf{A}^{-1}) \\
&= \mathbf{A}(\mathbf{A} + \delta\mathbf{A})^{-1} - \mathbf{I} = [\mathbf{A} - (\mathbf{A} + \delta\mathbf{A})](\mathbf{A} + \delta\mathbf{A})^{-1} \\
&= -\delta\mathbf{A}(\mathbf{A} + \delta\mathbf{A})^{-1}.
\end{aligned}$$

Thus we require

$$\|\delta\mathbf{A}(\mathbf{A} + \delta\mathbf{A})^{-1}\| < 1.$$

This is not the same as (9.61) but both are satisfied if

$$\|\delta\mathbf{A}\| \cdot \|(\mathbf{A} + \delta\mathbf{A})^{-1}\| < 1.$$

From (9.63) we observe that

$$\|\mathbf{E}_{m+1}\| \leqslant \|\mathbf{A}\| \, \|\mathbf{E}_m\|^2$$

so that if, $\mathbf{E}_m \neq \mathbf{0}$,

$$\frac{\|\mathbf{E}_{m+1}\|}{\|\mathbf{E}_m\|^2} \leqslant \|\mathbf{A}\|,$$

which indicates that the method has at least second order (or quadratic) rate of convergence. It is interesting to notice that for 1×1 matrices we have in (9.62) an iterative method of finding an inverse without division. (See Problem 7.20.)

9.7 Iterative methods

We again consider the problem of solving n non-singular equations in n unknowns

$$\mathbf{Ax} = \mathbf{b}. \tag{9.69}$$

If $\mathbf{E}$ and $\mathbf{F}$ are $n \times n$ matrices such that

$$\mathbf{A} = \mathbf{E} - \mathbf{F}, \tag{9.70}$$

we call (9.70) a *splitting* of $\mathbf{A}$. For such a splitting, (9.69) may be written as

$$\mathbf{Ex} = \mathbf{Fx} + \mathbf{b}.$$

This form of the equations suggest an iterative procedure

$$\mathbf{Ex}_{m+1} = \mathbf{Fx}_m + \mathbf{b}, \qquad m = 0, 1, 2, \ldots, \tag{9.71}$$

for arbitrary $\mathbf{x}_0$. If the sequence is to be uniquely defined for a given $\mathbf{x}_0$, we require $\mathbf{E}$ to be non-singular, when

$$\mathbf{x}_{m+1} = \mathbf{E}^{-1}\mathbf{Fx}_m + \mathbf{E}^{-1}\mathbf{b}.$$

This is of the form (9.54) and Theorem 9.2 states that the sequence $\{\mathbf{x}_m\}_{m=0}^{\infty}$ converges if

$$\|\mathbf{E}^{-1}\mathbf{F}\| < 1. \tag{9.72}$$

It can also be seen that, in this case, $\{\mathbf{x}_m\}$ converges to the solution vector $\mathbf{x}$.

The important question is: what is a suitable choice of $\mathbf{E}$ and $\mathbf{F}$? If $\mathbf{E} = \mathbf{A}$ and $\mathbf{F} = \mathbf{0}$ we achieve nothing. We repeatedly solve (9.71) in order to determine $\mathbf{x}_{m+1}$ and we would need to use one of the elimination methods unless

E is of a particularly simple form. If **E** is not of a simpler form than **A**, the work involved for each iteration will be identical to that required in solving the original equations (9.69) by an elimination method from the start.

The simplest choice of **E** is a diagonal matrix, usually the diagonal of **A** provided all the diagonal elements are non-zero. We obtain the splitting

$$\mathbf{A} = \mathbf{D} - \mathbf{B},$$

where **D** is diagonal with non-zero diagonal elements and **B** has zeros on its diagonal. The relation (9.71) becomes

$$\mathbf{x}_{m+1} = \mathbf{D}^{-1}\mathbf{B}\mathbf{x}_m + \mathbf{D}^{-1}\mathbf{b}, \qquad m = 0, 1, 2, \ldots. \tag{9.73}$$

This is known as the *Jacobi* iterative method. $\mathbf{D}^{-1}$ is simply the diagonal matrix whose diagonal elements are the inverses of those of **D**. The method is convergent if

$$\|\mathbf{D}^{-1}\mathbf{B}\| < 1.$$

This condition is certainly satisfied if the matrix **A** is a *strictly diagonally dominant* matrix, that is, a matrix such that

$$|a_{ii}| > \sum_{\substack{j=1 \\ j \neq i}}^{n} |a_{ij}| \quad \text{for all } i = 1, 2, \ldots, n.$$

$\mathbf{D}^{-1}\mathbf{B}$ has off-diagonal elements a_{ij}/a_{ii} and zeros on the diagonal, so that for a strictly diagonally dominant matrix **A**,

$$\|\mathbf{D}^{-1}\mathbf{B}\|_{\infty} = \max_{1 \leq i \leq n} \sum_{\substack{j=1 \\ j \neq i}}^{n} \left| \frac{a_{ij}}{a_{ii}} \right| = \max_{1 \leq i \leq n} \frac{1}{|a_{ii}|} \sum_{\substack{j=1 \\ j \neq i}}^{n} |a_{ij}| < 1.$$

Another suitable choice of **E** is a triangular matrix. We split **A** into

$$\mathbf{A} = (\mathbf{D} - \mathbf{L}) - \mathbf{U},$$

where **D** is diagonal with non-zero diagonal elements, **L** is lower triangular with zeros on the diagonal and **U** is upper triangular with zeros on the diagonal.† We obtain in place of (9.71),

$$(\mathbf{D} - \mathbf{L})\mathbf{x}_{m+1} = \mathbf{U}\mathbf{x}_m + \mathbf{b}, \qquad m = 0, 1, 2, \ldots. \tag{9.74}$$

This is known as the *Gauss–Seidel* iterative method. The coefficient matrix $(\mathbf{D} - \mathbf{L})$ is lower triangular and $\mathbf{x}_{m+1}$ is easily found by forward substitution. There is no need to compute $(\mathbf{D} - \mathbf{L})^{-1}\mathbf{U}$ explicitly.

The Gauss–Seidel method converges if

$$\|(\mathbf{D} - \mathbf{L})^{-1}\mathbf{U}\| < 1.$$

You will notice that both the Jacobi and the Gauss–Seidel methods require the elements on the diagonal of **A** to be non-zero.

---

† The **L** and **U** described here should not be confused with the triangular factors of **A**.

Example 9.7 Solve the equations

$$
\begin{bmatrix}
5 & -1 & -1 & -1 \\
-1 & 10 & -1 & -1 \\
-1 & -1 & 5 & -1 \\
-1 & -1 & -1 & 10
\end{bmatrix} \mathbf{x} =
\begin{bmatrix}
-4 \\ 12 \\ 8 \\ 34
\end{bmatrix}
$$

by an iterative process. (The solution is $\mathbf{x} = [1 \quad 2 \quad 3 \quad 4]^T$.) The coefficient matrix is diagonally dominant and, therefore, the Jacobi method is convergent. The equations (9.73) are:

$$
\mathbf{x}_{m+1} =
\begin{bmatrix}
0 & 0.2 & 0.2 & 0.2 \\
0.1 & 0 & 0.1 & 0.1 \\
0.2 & 0.2 & 0 & 0.2 \\
0.1 & 0.1 & 0.1 & 0
\end{bmatrix} \mathbf{x}_m +
\begin{bmatrix}
-0.8 \\ 1.2 \\ 1.6 \\ 3.4
\end{bmatrix}
$$

and $\|\mathbf{D}^{-1}\mathbf{B}\|_\infty = 0.6$. Table 9.1 shows six successive iterates starting with $\mathbf{x}_0 = \mathbf{0}$ and working to three decimal places.

Table 9.1   The iterative solution of the equations in Example 9.7 by (a) the Jacobi method, (b) the Gauss–Seidel method.

(a)

m	0	1	2	3	4	5
	0	−0.800	0.440	0.716	0.883	0.948
$\mathbf{x}_m$	0	1.200	1.620	1.840	1.929	1.969
	0	1.600	2.360	2.732	2.880	2.948
	0	3.400	3.600	3.842	3.929	3.969

(b)

m	0	1	2	3	4	5
	0	−0.800	0.476	0.889	0.977	0.995
$\mathbf{x}_m$	0	1.120	1.774	1.956	1.990	1.998
	0	1.664	2.770	2.949	2.989	2.998
	0	3.598	3.902	3.979	3.996	3.999

The Gauss–Seidel method (9.74) is

$$
\begin{bmatrix}
5 & 0 & 0 & 0 \\
-1 & 10 & 0 & 0 \\
-1 & -1 & 5 & 0 \\
-1 & -1 & -1 & 10
\end{bmatrix} \mathbf{x}_{m+1} =
\begin{bmatrix}
0 & 1 & 1 & 1 \\
0 & 0 & 1 & 1 \\
0 & 0 & 0 & 1 \\
0 & 0 & 0 & 0
\end{bmatrix} \mathbf{x}_m +
\begin{bmatrix}
-4 \\ 12 \\ 8 \\ 34
\end{bmatrix}
$$

and

$$(\mathbf{D} - \mathbf{L})^{-1}\mathbf{U} = \begin{bmatrix} 0 & 0.2 & 0.2 & 0.2 \\ 0 & 0.02 & 0.12 & 0.12 \\ 0 & 0.044 & 0.064 & 0.264 \\ 0 & 0.0264 & 0.0384 & 0.0584 \end{bmatrix}.$$

Thus $\|(\mathbf{D} - \mathbf{L})^{-1}\mathbf{U}\|_\infty = 0.6 < 1$ and this method is convergent also. Table 9.1 shows six successive iterates starting with $\mathbf{x}_0 = \mathbf{0}$ and working to three decimal places. □

You will notice that in Example 9.7 the iterates for the Gauss–Seidel method appear to converge faster than those for the Jacobi method. This is often the case and might be expected in that the Gauss–Seidel method makes use of the most recently calculated approximations to elements of $\mathbf{x}$ at all times. If $(x_i)_m$ denotes the ith component of $\mathbf{x}_m$, the Jacobi method is

$$(x_i)_{m+1} = \frac{1}{a_{ii}} [b_i - a_{i1}(x_1)_m - a_{i2}(x_2)_m - \cdots - a_{i,i-1}(x_{i-1})_m$$

$$- a_{i,i+1}(x_{i+1})_m - \cdots - a_{i,n}(x_n)_m], \qquad i = 1, 2, \ldots, n,$$

whereas the Gauss–Seidel method is

$$(x_i)_{m+1} = \frac{1}{a_{ii}} [b_i - a_{i1}(x_1)_{m+1} - a_{i2}(x_2)_{m+1} - \cdots - a_{i,i-1}(x_{i-1})_{m+1}$$

$$- a_{i,i+1}(x_{i+1})_m - \cdots - a_{i,n}(x_n)_m], \qquad i = 1, 2, \ldots, n.$$

The Jacobi method does not make use of the known components $(x_1)_{m+1}, \ldots, (x_{i-1})_{m+1}$ when calculating $(x_i)_{m+1}$. It must be stressed that the Gauss–Seidel method is not always better than the Jacobi method. Indeed the former may diverge whilst the latter converges.

We now consider the equation

$$\mathbf{D}\tilde{\mathbf{x}}_{m+1} = \mathbf{L}\mathbf{x}_{m+1} + \mathbf{U}\mathbf{x}_m + \mathbf{b}. \qquad (9.75)$$

The Gauss–Seidel method (9.74) is obtained by putting $\mathbf{x}_{m+1} = \tilde{\mathbf{x}}_{m+1}$. Instead we let

$$\mathbf{x}_{m+1} = \omega\tilde{\mathbf{x}}_{m+1} + (1 - \omega)\mathbf{x}_m, \qquad (9.76)$$

where ω is a parameter to be fixed. On combining (9.75) and (9.76), we now have

$$\mathbf{D}\mathbf{x}_{m+1} = \omega\mathbf{L}\mathbf{x}_{m+1} + \omega\mathbf{U}\mathbf{x}_m + \omega\mathbf{b} + (1 - \omega)\mathbf{D}\mathbf{x}_m$$

so that

$$(\mathbf{D} - \omega\mathbf{L})\mathbf{x}_{m+1} = ((1 - \omega)\mathbf{D} + \omega\mathbf{U})\mathbf{x}_m + \omega\mathbf{b}. \qquad (9.77)$$

The iterative method (9.77) is called the *successive over relaxation* method. It can be shown that for convergence we must choose ω in the range $0 < \omega < 2$. We try to choose ω so that the sequence $\{\mathbf{x}_m\}$ converges as rapidly as possible. For a wide range of problems arising from differential equations we find that a careful choice of ω can yield a drastic reduction in the number of iterations required for a given accuracy. We usually need to choose $\omega \in (1, 2)$ but the precise value depends on the problem. Much of the work on this subject is due to D. M. Young and R. S. Varga and a simplified account may be found in Isaacson and Keller. Of course for $\omega = 1$ we obtain the Gauss–Seidel method. Isaacson and Keller also discuss methods in which ω is varied from iteration to iteration. We can show that for $\omega \neq 0$ the method (9.77) can be derived from a splitting of $\mathbf{A}$. (See Problem 9.25.)

Rounding errors do not seriously affect the accuracy of a convergent iterative process unless $\|\mathbf{M}\| \simeq 1$ (see Problem 9.26). One advantage of an iterative method is that the effects of an isolated error will rapidly decay as the iterations advance if $\|\mathbf{M}\|$ is small. As the number of completed iterations increases and the accuracy of the iterates improves, it is desirable to increase the accuracy of the calculation.

Problems

Section 9.1

9.1 Prove that (9.2) is valid for all vectors $\mathbf{x}$ and $\mathbf{y}$ and any norm.

9.2 Show that if $\mathbf{x}$ and $\mathbf{y}$ are n-dimensional vectors with real components then, for any real number λ,

$$\sum_{i=1}^{n} x_i^2 + 2\lambda \sum_{i=1}^{n} x_i y_i + \lambda^2 \sum_{i=1}^{n} y_i^2 = \sum_{i=1}^{n} (x_i + \lambda y_i)^2 \geqslant 0.$$

If $\mathbf{y} \neq \mathbf{0}$ put $\lambda = -\sum_{i=1}^{n} x_i y_i / \sum_{i=1}^{n} y_i^2$ and thus prove the Cauchy–Schwarz inequality

$$(x_1 y_1 + \cdots + x_n y_n)^2 \leqslant (x_1^2 + \cdots + x_n^2)(y_1^2 + \cdots + y_n^2).$$

Hence prove that

$$\|\mathbf{x} + \mathbf{y}\|_2^2 \leqslant \|\mathbf{x}\|_2^2 + 2\|\mathbf{x}\|_2\|\mathbf{y}\|_2 + \|\mathbf{y}\|_2^2$$

and thus verify (9.1) for the Euclidean norm of real vectors.

9.3 Prove that (9.4) and (9.5) do define norms, that is they satisfy the conditions of Definition 9.1.

9.4 Show that for $p = \frac{1}{2}$ we do not obtain a norm in (9.6) by finding an example for which condition (iii) fails.

Section 9.2

9.5 (Assumes knowledge of complex numbers.) Show that, for a complex matrix, we may obtain equality in (9.12) by choosing

$$x_j = e^{-i\theta_j},$$

where θ_j is the argument of a_{pj}, that is, $a_{pj} = |a_{pj}|e^{i\theta_j}$ and the maximum occurs in (9.12) for $i = p$.

9.6 Show that, for all non-singular matrices **A** and **B**,

$$\|\mathbf{A}^{-1}\| \geqslant \frac{1}{\|\mathbf{A}\|},$$

$$\|\mathbf{A}^{-1} - \mathbf{B}^{-1}\| \leqslant \|\mathbf{A}^{-1}\| \cdot \|\mathbf{B}^{-1}\| \cdot \|\mathbf{A} - \mathbf{B}\|.$$

9.7 Show that

$$\mathbf{B} = \begin{bmatrix} 1 & 0.4 & 0.7 \\ 0.1 & 1 & 0.2 \\ 0.3 & 0.2 & 1 \end{bmatrix}$$

is non-singular and find an upper bound for $\|\mathbf{B}^{-1}\|_1$ without calculating the inverse.

9.8 Show that if **A** and $(\mathbf{A} - \mathbf{B})$ are non-singular and $\|\mathbf{A}^{-1}\| \cdot \|\mathbf{B}\| < 1$ then

$$\|(\mathbf{A} - \mathbf{B})^{-1}\| \leqslant \frac{1}{\|\mathbf{A}^{-1}\|^{-1} - \|\mathbf{B}\|}.$$

9.9 If $p(x)$ is the polynomial

$$p(x) = a_0 + a_1 x + a_2 x^2 + \cdots + a_r x^r,$$

with non-negative coefficients ($a_i \geqslant 0$, $i = 1, 2, \ldots, r$) and

$$p(\mathbf{A}) = a_0 \mathbf{I} + a_1 \mathbf{A} + a_2 \mathbf{A}^2 + \cdots + a_r \mathbf{A}^r,$$

show that

$$\|p(\mathbf{A})\| \leqslant p(\|\mathbf{A}\|).$$

9.10 If

$$\mathbf{A} = \begin{bmatrix} 0 & 0 \\ 2 & 0 \end{bmatrix},$$

show that the series $\mathbf{I} + \mathbf{A} + \mathbf{A}^2 + \mathbf{A}^3 + \cdots$ is convergent even though

$$\|\mathbf{A}\|_1 = \|\mathbf{A}\|_\infty > 1.$$

9.11   Show that the sequence

$$I, A, \frac{1}{2!} A^2, \frac{1}{3!} A^3, \frac{1}{4!} A^4, \cdots$$

is convergent to **0** for any choice of **A**.

Section 9.3

9.12   On some computers it is possible to divide a double-length number in an accumulator by a single-length number and produce a single-length result. This makes it possible to compute quantities of the form

$$(u_1 v_1 + u_2 v_2 + \cdots + u_r v_r)/t$$

with only one rounding error. Show that, if such a device is used in determining **L′** in the factorization process, the **δA** of (9.27) is such that ε'_{ij} is bounded by (9.24) for all i and j.

Partial pivoting will introduce extra errors unless the intermediate values w_j of §8.9 are stored as double-length numbers. Show that if such pivoting is made and the division of double-length numbers is used to determine **x** in the back substitution then the bounds of the perturbations in Theorem 9.1 are the same, except that

$$\varepsilon'' = \varepsilon' = \frac{\varepsilon}{1 - \varepsilon} \simeq \varepsilon.$$

Section 9.4

9.13   Find the condition numbers of

$$A = \begin{bmatrix} 1 & 2 \\ 1.001 & 2.001 \end{bmatrix}$$

for the 1 and ∞ norms. This matrix is ill-conditioned because the second row is almost a multiple of the first row.

9.14   Use the identity

$$(A + \delta A)^{-1} - A^{-1} = -A^{-1} \delta A (A + \delta A)^{-1}$$

to show that the **δx** of (9.40) satisfies

$$\frac{\|\delta x\|}{\|x + \delta x\|} \leqslant k(A) \cdot \frac{\|\delta A\|}{\|A\|} = e.k(A),$$

where $e = \|\delta A\|/\|A\|$. This provides a bound on the relative size of **δx** with respect to the computed **x** + **δx**.

9.15   Show that, if

$$(\mathbf{A} + \mathbf{\delta A})(\mathbf{x} + \mathbf{\delta x}) = \mathbf{b} + \mathbf{\delta b}$$

with $e = \|\mathbf{\delta A}\|/\|\mathbf{A}\|$, then

$$\frac{\|\mathbf{\delta x}\|}{\|\mathbf{x}\|} \leqslant \frac{k(\mathbf{A})}{1 - e.k(\mathbf{A})} \left(e + \frac{\|\mathbf{\delta b}\|}{\|\mathbf{b}\|} \right),$$

provided $e.k(\mathbf{A}) < 1$.

9.16   The linear equations

$$\begin{bmatrix} 2.01 & 1.01 \\ 1 & 0.5 \end{bmatrix} \mathbf{x} = \begin{bmatrix} -0.01 \\ 0 \end{bmatrix}$$

have solution $\mathbf{x} = [1 \quad -2]^T$. Show that, if the elimination method is used with multipliers and coefficients computed to only three significant digits, the equations reduce to

$$\begin{bmatrix} 2.01 & 1.01 \\ 0 & -0.00298 \end{bmatrix} \mathbf{x} = \begin{bmatrix} -0.01 \\ 0.00498 \end{bmatrix}$$

and hence the computed solution is

$$[0.834 \quad -1.67]^T.$$

9.17   The first row of a matrix is scaled so that

$$\mathbf{A} = \begin{bmatrix} 2\lambda & \lambda \\ 1 & 1 \end{bmatrix}.$$

Show that $k_\infty(\mathbf{A})$ is a minimum for $\lambda = \pm\frac{2}{3}$.

9.18   Show that if $(\mathbf{A} + \mathbf{\delta A})^{-1}$ is computed as an approximation to $\mathbf{A}^{-1}$ and

$$e = \|\mathbf{\delta A}\|/\|\mathbf{A}\|,$$

then

$$\frac{\|(\mathbf{A} + \mathbf{\delta A})^{-1} - \mathbf{A}^{-1}\|}{\|\mathbf{A}^{-1}\|} \leqslant \frac{k(\mathbf{A}).e}{1 - k(\mathbf{A}).e},$$

provided that $k(\mathbf{A}).e < 1$. This is a bound on the relative error of $(\mathbf{A} + \mathbf{\delta A})^{-1}$.

Section 9.5

9.19   Show that with $\mathbf{r}_0$ and $\mathbf{\delta x}$ as in (9.49) and (9.52)

$$\frac{\|\mathbf{\delta x}\|}{\|\mathbf{x}\|} \leqslant k(\mathbf{A}) \frac{\|\mathbf{r}_0\|}{\|\mathbf{b}\|}.$$

9.20   Use the correction process described in §9.5 to improve the computed approximate solution in Problem 9.16.

9.21   Draw a flow diagram for the compact elimination method followed by one correction of the solution using the process of §9.5.

Section 9.6

9.22   Show that if

$$\|I - \alpha A\| < 1$$

where α is a scalar, the iterative procedure of §9.6 for determining A^{-1} will converge for $W_0 = \alpha I$.
  Use this result to find an approximation to the inverse of

$$\begin{bmatrix} 3 & 1 \\ 2 & 4 \end{bmatrix}.$$

Section 9.7

9.23   Carry out three iterations of
  (a) the Jacobi method,
  (b) the Gauss–Seidel method
of solving

$$\begin{bmatrix} -8 & 1 & 1 \\ 1 & -5 & 1 \\ 1 & 1 & -4 \end{bmatrix} x = \begin{bmatrix} 1 \\ 16 \\ 7 \end{bmatrix}.$$

(Solution $x = [-1 \quad -4 \quad -3]^T$.)
  In each case check that the methods converge and start with $x_0 = [0 \ 0 \ 0]^T$.

9.24   Draw flow diagrams for the Jacobi and Gauss–Seidel iterative methods.

9.25   Show that for $\omega \neq 0$ the successive over relaxation method (9.77) may be obtained from the splitting (see (9.70))

$$A = \left(\frac{1}{\omega} D - L \right) - \left(U - \left(1 - \frac{1}{\omega} \right) D \right).$$

9.26   Suppose that an iterative process

$$x_{m+1} = M x_m + d, \qquad m = 0, 1, 2, \ldots, \tag{9.78}$$

with $\|\mathbf{M}\| < 1$, is affected by rounding errors so that a sequence $\{\tilde{\mathbf{x}}_m\}_{m=0}^{\infty}$ is computed with

$$\tilde{\mathbf{x}}_{m+1} = \mathbf{M}\tilde{\mathbf{x}}_m + \mathbf{d} + \mathbf{r}_m, \qquad m = 0, 1, 2, \ldots, \tag{9.79}$$

where the error vectors $\mathbf{r}_m$ satisfy $\|\mathbf{r}_m\| \leqslant \varepsilon$ and $\tilde{\mathbf{x}}_0 = \mathbf{x}_0$.

Show that

$$\|\tilde{\mathbf{x}}_m - \mathbf{x}_m\| \leqslant \frac{\varepsilon}{1 - \|\mathbf{M}\|}, \qquad m = 0, 1, 2, \ldots.$$

(*Hint:* If $\boldsymbol{\rho}_m = \tilde{\mathbf{x}}_m - \mathbf{x}_m$, show from (9.78) and (9.79) that

$$\|\boldsymbol{\rho}_{m+1}\| \leqslant \|\mathbf{M}\| \, \|\boldsymbol{\rho}_m\| + \varepsilon$$

and thus, by an induction argument, that

$$\|\boldsymbol{\rho}_m\| \leqslant \frac{\varepsilon}{1 - \|\mathbf{M}\|}.$$

See also Lemma 11.1.)

Chapter 10

Systems of Non-Linear Equations

10.1 Contraction mapping theorem

We now consider a system of algebraic equations of the form

$$\mathbf{f(x)} = \mathbf{0}, \tag{10.1}$$

where $\mathbf{x}$ is an n-dimensional vector and $\mathbf{f}$, a function of $\mathbf{x}$, is an m-dimensional vector. The vector $\mathbf{0}$ is also m-dimensional. We assume that all elements are real. The equations (10.1) may be written in full as

$$
\begin{aligned}
f_1(x_1, \ldots, x_n) &= 0 \\
f_2(x_1, \ldots, x_n) &= 0 \\
&\vdots \\
f_m(x_1, \ldots, x_n) &= 0.
\end{aligned}
\tag{10.2}
$$

A special case of (10.1) is the system of *linear* equations

$$\mathbf{Ax} = \mathbf{b}, \tag{10.3}$$

where $\mathbf{A}$ is an $m \times n$ matrix. (See Chapters 8 and 9.) As an example of a system of *non-linear* equations, we have

$$x_1 + 2x_2 - 3 = 0 \tag{10.4a}$$

$$2x_1{}^2 + x_2{}^2 - 5 = 0 \tag{10.4b}$$

which is of the form (10.2) with $m = n = 2$. Geometrically, the solutions of (10.4) are the points in the x_1x_2-plane where the straight line with equation $x_1 + 2x_2 - 3 = 0$ cuts the ellipse with equation $2x_1{}^2 + x_2{}^2 - 5 = 0$. We see from a graph that the system (10.4) has two solutions, corresponding to the points A and B of Fig. 10.1. However, there is no general rule which tells us how many solutions to expect. A general system of equations (10.2) may have no solution, an infinite number of solutions or any finite number of solutions. For example, the system of two equations

$$
\begin{aligned}
x_1 - 1 &= 0 \\
(x_2 - 1)(x_2 - 2) \cdots (x_2 - N) &= 0
\end{aligned}
$$

has exactly N solutions.

We restrict our attention to the case where $m = n$ in (10.2), so that the number of equations is the same as the number of unknowns, and write (10.1) in the form

$$\mathbf{x} = \mathbf{g}(\mathbf{x}). \tag{10.5}$$

This is an extension of the one variable form $x = g(x)$, which we used (see §7.4) in discussing iterative methods for a single equation. Notice that (10.5) is also a generalization of the equation

$$\mathbf{x} = \mathbf{Mx} + \mathbf{d}$$

which we used (see §9.5 and §9.7) in solving *linear* equations by iterative methods.

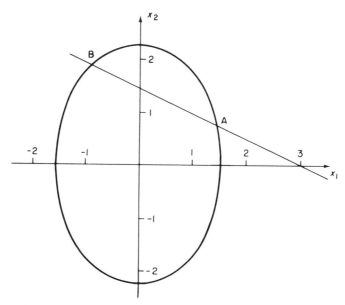

Fig. 10.1. The straight line and ellipse have equations (10.4a)
and (10.4b) respectively.

Given the non-linear equations (10.5), we construct a sequence of vectors $\{\mathbf{x}_m\}_{m=0}^{\infty}$ from

$$\mathbf{x}_{r+1} = \mathbf{g}(\mathbf{x}_r),$$

$r = 0, 1, \ldots$, beginning with some initial vector $\mathbf{x}_0$. We are interested in whether the sequence $\{\mathbf{x}_m\}_{m=0}^{\infty}$ converges to a vector $\boldsymbol{\alpha}$ which is a solution of the equations (10.5). We recall from §9.1 that a sequence of vectors $\{\mathbf{x}_m\}_{m=0}^{\infty}$ is said to converge to $\boldsymbol{\alpha}$ if

$$\lim_{m \to \infty} \mathbf{x}_m = \boldsymbol{\alpha}.$$

In Lemma 9.1 we saw that this is equivalent to the condition

$$\lim_{m \to \infty} \|\mathbf{x}_m - \boldsymbol{\alpha}\| = 0$$

for *any* vector norm, including the norms $\|\cdot\|_p$, $1 \leqslant p \leqslant \infty$.

Definition 10.1 A region R in n-dimensional space is said to be *closed* if every convergent sequence $\{\mathbf{x}_m\}$, with each $\mathbf{x}_m \in R$, is such that its limit $\boldsymbol{\alpha} \in R$. $\square$

For example, in two-dimensional space with coordinates x_1, x_2, the region

$$a \leqslant x_1 \leqslant b, \qquad c \leqslant x_2 \leqslant d$$

is closed. (See Problem 10.3.) The region

$$a \leqslant x_1 \leqslant b, \qquad -\infty < x_2 < \infty$$

is also closed. But

$$a \leqslant x_1 < b, \qquad c \leqslant x_2 \leqslant d$$

is not closed as we can have sequences whose first coordinates converge to b.

Definition 10.2 A function $\mathbf{g}$, which maps real n-dimensional vectors into real n-dimensional vectors, is said to be a *contraction mapping* with respect to a norm $\|\cdot\|$ on a closed region R if

(i) $\mathbf{x} \in R \Rightarrow \mathbf{g}(\mathbf{x}) \in R$, $\qquad\qquad\qquad\qquad\qquad\qquad$ (10.6)

(ii) $\|\mathbf{g}(\mathbf{x}) - \mathbf{g}(\mathbf{x}')\| \leqslant L\|\mathbf{x} - \mathbf{x}'\|$, with $0 \leqslant L < 1$, $\qquad$ (10.7)

for all $\mathbf{x}$, $\mathbf{x}' \in R$. $\square$

We have extended the closure and Lipschitz conditions to n variables and thus extended the notion of contraction mapping from Definition 7.1. We can now generalize Theorem 7.3.

Theorem 10.1 If there is a closed region R on which $\mathbf{g}$ is a contraction mapping with respect to some norm $\|\cdot\|$, then

(i) the equation $\mathbf{x} = \mathbf{g}(\mathbf{x})$ has a unique solution (say $\boldsymbol{\alpha}$) belonging to R,

(ii) for any $\mathbf{x}_0 \in R$, the sequence $\{\mathbf{x}_r\}$ defined by

$$\mathbf{x}_{r+1} = \mathbf{g}(\mathbf{x}_r), \qquad r = 0, 1, \ldots,$$

converges to $\boldsymbol{\alpha}$.

Proof. The only essential difference between this proof and that of Theorem 7.3 is that it is slightly more difficult, when $n > 1$, to show the *existence* of a solution. Otherwise, the proof is the same except that we write $\|\cdot\|$ in place of $|\cdot|$ throughout.

If $\mathbf{x}_0 \in R$, it follows from the closure condition that $\mathbf{x}_r \in R$, $r = 0, 1, \ldots$. Then, as in Problems 7.24 and 7.25, we obtain the inequality

$$\|\mathbf{x}_{m+k} - \mathbf{x}_m\| \leqslant \frac{L^m}{1 - L} \|\mathbf{x}_1 - \mathbf{x}_0\|, \qquad (10.8)$$

for any fixed $m \geqslant 0$ and all $k = 0, 1, \ldots$. Thus, for any k, we have

$$\lim_{m \to \infty} \|\mathbf{x}_{m+k} - \mathbf{x}_m\| = 0$$

and, from Lemma 9.1, this implies that the sequence of vectors $\{\mathbf{x}_{m+k} - \mathbf{x}_m\}_{m=0}^{\infty}$ converges to the zero vector. If $x_{m,i}$ denotes the ith element of the vector $\mathbf{x}_m$, it follows that, for each $i = 1, 2, \ldots, n$, the numbers $|x_{m+k,i} - x_{m,i}|$ may be made arbitrarily small *for all $k \geqslant 0$* by choosing m sufficiently large. Thus, for any i, it follows from Cauchy's principle for convergence (see Theorem 2.7) that the sequence $x_{0,i}, x_{1,i}, x_{2,i}, \ldots$ converges to some number, say α_i. Therefore $\{\mathbf{x}_m\}$ converges to a vector $\boldsymbol{\alpha}$ and, since R is closed, $\boldsymbol{\alpha} \in R$. To show that $\boldsymbol{\alpha}$ is a solution of $\mathbf{x} = \mathbf{g}(\mathbf{x})$, we write

$$\boldsymbol{\alpha} - \mathbf{g}(\boldsymbol{\alpha}) = \boldsymbol{\alpha} - \mathbf{x}_{m+1} + \mathbf{g}(\mathbf{x}_m) - \mathbf{g}(\boldsymbol{\alpha}),$$

for any $m \geqslant 0$. Thus

$$\|\boldsymbol{\alpha} - \mathbf{g}(\boldsymbol{\alpha})\| \leqslant \|\boldsymbol{\alpha} - \mathbf{x}_{m+1}\| + \|\mathbf{g}(\mathbf{x}_m) - \mathbf{g}(\boldsymbol{\alpha})\|$$
$$\leqslant \|\boldsymbol{\alpha} - \mathbf{x}_{m+1}\| + L\|\mathbf{x}_m - \boldsymbol{\alpha}\|.$$

Since both terms on the right may be made arbitrarily small by choosing m sufficiently large, it follows that $\|\boldsymbol{\alpha} - \mathbf{g}(\boldsymbol{\alpha})\| = 0$. Thus $\boldsymbol{\alpha} - \mathbf{g}(\boldsymbol{\alpha}) = \mathbf{0}$ and $\boldsymbol{\alpha}$ is a solution of $\mathbf{x} = \mathbf{g}(\mathbf{x})$. The uniqueness is shown as in Theorem 7.3. $\square$

We now suppose that on some region R, $\partial g_i / \partial x_j$, $1 \leqslant i, j \leqslant n$, are continuous, and write $g_i(\mathbf{x})$ to denote $g_i(x_1, \ldots, x_n)$. Then by Taylor's theorem in several variables (Theorem 3.2) we have

$$g_i(\mathbf{x}) - g_i(\mathbf{x}') = (x_1 - x_1') . \frac{\partial g_i(\boldsymbol{\xi})}{\partial x_1} + \cdots + (x_n - x_n') . \frac{\partial g_i(\boldsymbol{\xi})}{\partial x_n}, \qquad (10.9)$$

where $\mathbf{x}, \mathbf{x}' \in R$ and $\boldsymbol{\xi}$ is "between" $\mathbf{x}$ and $\mathbf{x}'$, that is $\boldsymbol{\xi} = (1 - \theta)\mathbf{x} + \theta\mathbf{x}'$ for some $\theta \in (0, 1)$. We shall assume that $\boldsymbol{\xi} \in R$, which will be so if the line segment joining any two points of R is also in R. (A region satisfying this property is said to be *convex*.) If $\mathbf{G}$ denotes the $n \times n$ matrix whose (i, j)th element is

$$g_{ij} = \sup_{\mathbf{x} \in R} \left| \frac{\partial g_i(\mathbf{x})}{\partial x_j} \right|, \qquad (10.10)$$

it follows that

$$|\mathbf{g}(\mathbf{x}) - \mathbf{g}(\mathbf{x}')| \leqslant \mathbf{G}|\mathbf{x} - \mathbf{x}'| \qquad (10.11)$$

where, as in §9.3, $|\mathbf{u}|$ denotes a vector whose elements are the absolute values of the elements of the vector $\mathbf{u}$. Thus, for any of the norms $\|\cdot\|_p$,

$$\|\mathbf{g}(\mathbf{x}) - \mathbf{g}(\mathbf{x}')\|_p \leqslant \|\mathbf{G}\|_p \cdot \|\mathbf{x} - \mathbf{x}'\|_p.$$

With the choice $p = 1$ or $p = \infty$, we can take the Lipschitz constant L as $\|\mathbf{G}\|_1$ or $\|\mathbf{G}\|_\infty$ respectively. As we saw in §9.2, both of these matrix norms are easily evaluated. In practice, the only other norm we might wish to use is $\|\cdot\|_2$ and, as remarked in Chapter 9, the corresponding matrix norm is not easily evaluated. However, we can write (see Problem 10.4)

$$\|\mathbf{G}(\mathbf{x} - \mathbf{x}')\|_2 \leqslant \left[\sum_{i,j} g_{ij}^2\right]^{1/2} \cdot \|\mathbf{x} - \mathbf{x}'\|_2$$

so that in working with $\|\cdot\|_2$ we can take

$$L = \left[\sum_{i,j} g_{ij}^2\right]^{1/2}. \tag{10.12}$$

Example 10.1 We apply Theorem 10.1 to equations (10.4), although these are easily solved by eliminating x_2 between the two equations and solving the resulting quadratic equation for x_1. We rewrite equations (10.4) as

$$x_1 = \left(\frac{5 - x_2^2}{2}\right)^{1/2}$$
$$x_2 = \tfrac{1}{2}(3 - x_1) \tag{10.13}$$

which are in the form $\mathbf{x} = \mathbf{g}(\mathbf{x})$. If R denotes the closed rectangular region $1 \leqslant x_1 \leqslant 2$, $0.5 \leqslant x_2 \leqslant 1.5$, we see from (10.13) that if $\mathbf{x} \in R$, $\mathbf{g}(\mathbf{x}) \in R$. We have

$$\frac{\partial g_1}{\partial x_2} = -x_2/(10 - 2x_2^2)^{1/2},$$

whose maximum modulus on R occurs for $x_2 = 1.5$, with value $3/\sqrt{22}$. Thus the matrix $\mathbf{G}$, defined by (10.10), is

$$\mathbf{G} = \begin{pmatrix} 0 & 3/\sqrt{22} \\ \tfrac{1}{2} & 0 \end{pmatrix}.$$

We have $\|\mathbf{G}\|_1 = \|\mathbf{G}\|_\infty = 3/\sqrt{22}$ and (10.12) gives $L = (29/44)^{1/2}$. Thus we have a contraction mapping on R with any of the three common norms and Theorem 10.1 is applicable. Taking $x_1 = 1.5$, $x_2 = 1.0$, the centre point of R, as the initial iterate, we obtain the next six iterates as shown in Table 10.1. Because of the simple form of the equations, it is possible to make the iterations computing values of only x_1 for even iterations and x_2 for odd iterations

Table 10.1   Results for Example 10.1.

Iteration number	0	1	2	3	4	5	6
x_1	1.5	1.414	1.490	1.478	1.488	1.487	1.488
x_2	1.0	0.750	0.793	0.755	0.761	0.756	0.756

or vice versa. For the other solution of equations (10.4) (which is, of course, *outside R*) see Problem 10.7. □

Example 10.2   Consider the equations

$$x_1 = \tfrac{1}{12}(-1 + \sin x_2 + \sin x_3)$$
$$x_2 = \tfrac{1}{3}(x_1 - \sin x_2 + \sin x_3) \qquad (10.14)$$
$$x_3 = \tfrac{1}{12}(1 - \sin x_1 + x_2).$$

The closure condition is satisfied with R as the cube $-1 \leqslant x_1, x_2, x_3 \leqslant 1$. The matrix **G**, defined by (10.10), is

$$\mathbf{G} = \tfrac{1}{12}\begin{bmatrix} 0 & 1 & 1 \\ 4 & 4 & 4 \\ 1 & 1 & 0 \end{bmatrix}$$

In this case, $\|\mathbf{G}\|_1 = \tfrac{1}{2}$, $\|\mathbf{G}\|_\infty = 1$ and, for $\|.\|_2$, (10.12) gives $L = \sqrt{13}/6$. Thus we have a contraction mapping for norms $\|.\|_1$ and $\|.\|_2$. □

The concept of a contraction mapping, which has been applied in this section to a system of *algebraic* equations, applies also to other types of equations. For example, in Chapter 11, we apply it to a first order ordinary differential equation.

10.2 Newton's method

Suppose that $\boldsymbol{\alpha}$ is a solution of the system of equations $\mathbf{x} = \mathbf{g}(\mathbf{x})$ and that $\partial g_i(\boldsymbol{\alpha})/\partial x_j = 0$ for $1 \leqslant i, j \leqslant n$. We assume also that, on a closed region R containing $\boldsymbol{\alpha}$, the second derivatives $\partial^2 g_i/\partial x_j\,\partial x_k\,(1 \leqslant i, j, k \leqslant n)$ are bounded and continuous. Then from Taylor's theorem we have

$$g_i(\mathbf{x}) - g_i(\boldsymbol{\alpha}) = \frac{1}{2}\sum_{j=1}^{n}\sum_{k=1}^{n}(x_j - \alpha_j)(x_k - \alpha_k)\frac{\partial^2 g(\boldsymbol{\xi})}{\partial x_j\,\partial x_k}. \qquad (10.15)$$

As for (10.9), we assume that $\boldsymbol{\xi} \in R$ whenever $\mathbf{x} \in R$. If

$$\sup_{\mathbf{x}\in R}\left|\frac{\partial^2 g_i(\mathbf{x})}{\partial x_j\,\partial x_k}\right| \leqslant M, \qquad 1 \leqslant i, j, k \leqslant n,$$

then we deduce from (10.15) that

$$|g_i(\mathbf{x}) - g_i(\boldsymbol{\alpha})| \leqslant \tfrac{1}{2}Mn^2\|\mathbf{x} - \boldsymbol{\alpha}\|_\infty^2,$$

since $|x_i - \alpha_i| \leqslant \|\mathbf{x} - \boldsymbol{\alpha}\|_\infty$, $1 \leqslant i \leqslant n$. It follows that

$$\|\mathbf{g}(\mathbf{x}) - \mathbf{g}(\boldsymbol{\alpha})\|_\infty \leqslant \tfrac{1}{2}Mn^2\|\mathbf{x} - \boldsymbol{\alpha}\|_\infty^2. \tag{10.16}$$

Thus, if $\mathbf{x}_{r+1} = \mathbf{g}(\mathbf{x}_r)$,

$$\|\mathbf{x}_{r+1} - \boldsymbol{\alpha}\|_\infty \leqslant \tfrac{1}{2}Mn^2\|\mathbf{x}_r - \boldsymbol{\alpha}\|_\infty^2$$

and we have at least second order convergence of the iterative method.

We now construct an iterative method which has second order convergence. Consider the system of equations

$$\mathbf{f}(\mathbf{x}) = \mathbf{0} \tag{10.17}$$

and write

$$\mathbf{x}_m = \boldsymbol{\alpha} + \mathbf{e}_m, \tag{10.18}$$

where $\boldsymbol{\alpha}$ is a solution of (10.17). We have

$$f_i(\mathbf{x}_m) = e_{m,1}\frac{\partial f_i(\boldsymbol{\xi})}{\partial x_1} + \cdots + e_{m,n}\frac{\partial f_i(\boldsymbol{\xi})}{\partial x_n}, \qquad 1 \leqslant i \leqslant n, \tag{10.19}$$

where $e_{m,j}$ denotes the jth element of the vector $\mathbf{e}_m$. If we knew the values of $\boldsymbol{\xi}$ (which depend on i), we could solve the system of linear equations (10.19) to find $\mathbf{e}_m$ and thus find $\boldsymbol{\alpha}$ from (10.18). If instead we evaluate the partial derivatives $\partial f_i/\partial x_j$ at $\mathbf{x}_m$ and replace $\mathbf{e}_m$ in (10.19) by $\mathbf{x}_m - \mathbf{x}_{m+1}$, rather than $\mathbf{x}_m - \boldsymbol{\alpha}$, we obtain

$$f_i(\mathbf{x}_m) = (x_{m,1} - x_{m+1,1})\frac{\partial f_i(\mathbf{x}_m)}{\partial x_1} + \cdots + (x_{m,n} - x_{m+1,n})\frac{\partial f_i(\mathbf{x}_m)}{\partial x_n},$$
$$1 \leqslant i \leqslant n.$$

We write this system of linear equations, with unknown vector $\mathbf{x}_{m+1}$, as

$$\mathbf{f}(\mathbf{x}_m) = \mathbf{J}(\mathbf{x}_m)(\mathbf{x}_m - \mathbf{x}_{m+1}),$$

where $\mathbf{J}(\mathbf{x})$ is the *Jacobian* matrix whose (i,j)th element is $\partial f_i(\mathbf{x})/\partial x_j$. If $\mathbf{J}(\mathbf{x}_m)$ is non-singular, we have

$$\mathbf{x}_{m+1} = \mathbf{x}_m - [\mathbf{J}(\mathbf{x}_m)]^{-1}\mathbf{f}(\mathbf{x}_m), \tag{10.20}$$

which is called Newton's method. Note that each step requires the solution of a system of linear equations with coefficient matrix $\mathbf{J}(\mathbf{x}_m)$. When $n = 1$, (10.20) coincides with the familiar Newton method for a single equation. To verify that we have second order convergence, we write (10.20) as

$$\mathbf{x}_{m+1} = \mathbf{g}(\mathbf{x}_m),$$

where

$$g(x) = x - [J(x)]^{-1}f(x). \tag{10.21}$$

We denote by $\partial g/\partial x_j$, the vector whose ith element is $\partial g_i/\partial x_j$. Then from (10.21)

$$\frac{\partial g(x)}{\partial x_j} = \frac{\partial x}{\partial x_j} - [J(x)]^{-1}\frac{\partial f(x)}{\partial x_j} - \left(\frac{\partial}{\partial x_j}[J(x)]^{-1}\right)f(x). \tag{10.22}$$

The partial derivatives of J^{-1} (see Problem 10.9) satisfy the equation

$$\frac{\partial}{\partial x_j}(J^{-1}) = -J^{-1}\frac{\partial J}{\partial x_j}J^{-1}.$$

Thus these exist if J is non-singular and if the partial derivatives of J exist. The second condition is equivalent to asking for the existence of the second derivatives of f. Premultiplying (10.22) by $J(x)$, and noting that $f(\alpha) = 0$, we have

$$\left(J(x)\cdot\frac{\partial g(x)}{\partial x_j}\right)_{x=\alpha} = \left(J(x)\cdot\frac{\partial x}{\partial x_j} - \frac{\partial f(x)}{\partial x_j}\right)_{x=\alpha}. \tag{10.23}$$

Since

$$\frac{\partial x_i}{\partial x_j} = \begin{cases} 0, & i \neq j \\ 1, & i = j \end{cases}$$

we see that each element of the vector on the right of (10.23) is zero. Thus, from (10.23), if $J(\alpha)$ is non-singular,

$$\left(\frac{\partial g}{\partial x_j}\right)_{x=\alpha} = 0$$

and Newton's method is at least second order.

If for any x_m, $J(x_m)$ is singular, then x_{m+1} is not defined. There is a theorem, proved by L. V. Kantorovich in 1937, which states conditions under which the sequence $\{x_m\}$ is defined and converges to the solution α. (See Henrici [1962].)

In §7.6 we saw how Newton's method for a single equation and unknown can be described in terms of tangents to the graph of the function involved. In the case of two equations in two unknowns, we can again visualize the process using graphs. This time we have two surfaces $z = f_1(x_1, x_2)$ and $z = f_2(x_1, x_2)$ and we are trying to find a point where both of these are zero, that is we seek a point at which both surfaces cut the plane $z = 0$. Given an approximate solution $x_{m,1}$, $x_{m,2}$, we construct the tangent planes to f_1 and f_2 at the point $(x_{m,1}, x_{m,2})$. We now calculate the point at which these two planes and the plane $z = 0$ intersect. This point is taken as the next iterate.

Example 10.3 For the equations (10.4), we have

$$J(x) = \begin{bmatrix} 1 & 2 \\ 4x_1 & 2x_2 \end{bmatrix}.$$

With $x_1 = 1.5$, $x_2 = 1.0$ initially, the next few iterates obtained by Newton's method (10.20) are as shown in Table 10.2. For the last iterate in the table, the residuals for equations (10.4) are less than 10^{-6}. □

Table 10.2 Results for Example 10.3.

Iteration number	0	1	2	3
x_1	1.5	1.5	1.488095	1.488034
x_2	1.0	0.75	0.755952	0.755983

Problems

Section 10.1

10.1 Show that the closure condition and contraction mapping property apply to the following equations on the region $0 \leqslant x_1, x_2, x_3 \leqslant 1$.

$$x_1 = \tfrac{1}{12}(x_2^2 + 2e^{-x_3})$$
$$x_2 = \tfrac{1}{6}(1 - x_1 + \sin x_3)$$
$$x_3 = \tfrac{1}{6}(x_1^2 + x_2^2 + x_3^2).$$

10.2 Show that the equations

$$x = \tfrac{1}{2}\cos y$$
$$y = \tfrac{1}{2}\sin x$$

have a unique solution. Find the solution to two decimal places.

10.3 If **a**, **b** and **x** denote vectors with n real elements and R denotes the region $a \leqslant x \leqslant b$, show that R is closed. (*Hint:* consider the sequences composed from corresponding components of vectors.)

10.4 If **A** is an $m \times n$ (real) matrix and **x** is an n-dimensional real vector show that

$$\|Ax\|_2 = \left(\sum_{i=1}^{m} \left(\sum_{j=1}^{n} a_{ij} x_j \right)^2 \right)^{1/2}$$
$$\leqslant \left(\sum_{i=1}^{m} \left(\sum_{j=1}^{n} a_{ij}^2 \sum_{j=1}^{n} x_j^2 \right) \right)^{1/2}$$
$$= \left(\sum_{i=1}^{m} \sum_{j=1}^{n} a_{ij}^2 \right)^{1/2} \left(\sum_{j=1}^{n} x_j^2 \right)^{1/2}$$

using the Cauchy–Schwarz inequality (see Problem 9.2.) Thus

$$\|\mathbf{Ax}\|_2 \leqslant N(\mathbf{A}).\|\mathbf{x}\|_2$$

where

$$N(\mathbf{A}) = \left(\sum_{i,j} a_{ij}^2\right)^{1/2}.$$

(Because $N(\mathbf{A})$ satisfies all the properties mentioned in Lemma 9.2, it is sometimes called the *Schur norm* of the matrix $\mathbf{A}$. Note however that $N(\mathbf{A})$ is not a *subordinate* norm, as in Definition 9.2.)

10.5   Show that, working with the norm $\|.\|_2$, the contraction mapping property holds for the following system of equations on the region $-1 \leqslant x_1, x_2, x_3 \leqslant 1$.

$$x_1 = \tfrac{1}{3}(x_1 + x_2 - x_3)$$
$$x_2 = \tfrac{1}{6}(2x_1 + \sin x_2 + x_3)$$
$$x_3 = \tfrac{1}{6}(2x_1 - x_2 + \sin x_3).$$

10.6   In Example 10.1, eliminate x_2 between the two equations and show that the sequence $\{x_{2m,1}\}$ satisfies

$$x_{2m+2,1} = \tfrac{1}{4}[40 - 2(3 - x_{2m,1})^2]^{1/2}, \qquad m = 0, 1, \ldots.$$

If this relation is denoted by

$$y_{m+1} = g(y_m)$$

where $y_m = x_{2m,1}$, show that $|g'(y)| \leqslant \sqrt{2}/8$ for $1 \leqslant y \leqslant 2$. (Note that for the three common norms 1, 2 and ∞ we obtained in Example 10.1 Lipschitz constants of $3/\sqrt{22}$, $(29/44)^{1/2}$ and $3/\sqrt{22}$ respectively. The value $L = \sqrt{2}/8$ is in fairly close agreement with the apparent rate of convergence of alternate iterates in Table 10.1.)

Section 10.2

10.7   The equations (10.4) have a solution which is near to $x_1 = -1$, $x_2 = 2$. Use Newton's method to find this solution to two decimal places.

10.8   An $m \times n$ matrix $\mathbf{A}$ has elements $a_{ij}(x)$ which depend on x and an n-dimensional vector $\mathbf{y}$ has elements $y_j(x)$ dependent on x. If all elements of $\mathbf{A}$ and $\mathbf{y}$ are differentiable with respect to x, show that

$$\frac{d}{dx}(\mathbf{Ay}) = \frac{d\mathbf{A}}{dx}\cdot\mathbf{y} + \mathbf{A}\frac{d\mathbf{y}}{dx},$$

where $d\mathbf{A}/dx$ is an $m \times n$ matrix with elements da_{ij}/dx and $d\mathbf{y}/dx$ is an n-dimensional vector with elements dy_j/dx.

10.9   Assuming that $\mathbf{J}^{-1}(\mathbf{x})$ exists, by differentiating the product $\mathbf{J}\mathbf{J}^{-1} = \mathbf{I}$ partially with respect to x_j, show that

$$\frac{\partial}{\partial x_j}(\mathbf{J}^{-1}) = -\mathbf{J}^{-1}\frac{\partial \mathbf{J}}{\partial x_j}\mathbf{J}^{-1}.$$

10.10   Use Newton's method to obtain the solution of the system of equations in Problem 10.2, beginning with $\mathbf{x} = \mathbf{y} = \mathbf{0}$. Show that $\mathbf{J}^{-1}$ exists for all $\mathbf{x}$ and $\mathbf{y}$.

10.11   What happens if Newton's method is applied to the system of equations

$$\mathbf{A}\mathbf{x} = \mathbf{b},$$

where $\mathbf{A}$ is an $n \times n$ non-singular matrix?

Chapter 11

Ordinary Differential Equations

11.1 Introduction

Let $f(x, y)$ be a real valued function of two variables defined for $a \leqslant x \leqslant b$ and all real y. Suppose now that y is a real valued function defined on $[a, b]$. The equation

$$y' = f(x, y) \tag{11.1}$$

is called an ordinary differential equation of the first order. Any real valued function y, which is differentiable and satisfies (11.1) for $a \leqslant x \leqslant b$, is said to be a *solution* of this differential equation. In general such a solution, if it exists, will not be unique because an arbitrary constant is introduced on integrating (11.1). To make the solution unique we need to impose an extra condition on the solution. This condition usually takes the form of requiring $y(x)$ to have a specified value for a particular value of x, that is,

$$y(x_0) = s \tag{11.2}$$

where $x_0 \in [a, b]$ is the particular value of x and s is the specified value of $y(x_0)$. Equations (11.1) and (11.2) are collectively called an *initial value problem*. Equation (11.2) is called the initial condition and s the initial value of y.

Example 11.1 The general solution of

$$y' = ky, \tag{11.3}$$

where k is a given constant, is

$$y(x) = ce^{kx} \tag{11.4}$$

where c is an arbitrary constant. If, for example, we are given the extra condition

$$y(0) = s \tag{11.5}$$

then, on putting $x = 0$ in (11.4), we see that we require $c = s$. The unique solution satisfying (11.3) and (11.5) is

$$y(x) = se^{kx}.$$

The differential equation (11.3) occurs in the description of many physical phenomena, for example, uninhibited bacteriological growth, a condenser discharging, a gas escaping under pressure from a container and radioactive decay. In all these examples, time is represented by x. The condition (11.5) gives the initial value of the physical quantity being measured, for example, the volume of bacteria at time $x = 0$. Many more complex phenomena may be described by equations of the form (11.1). $\square$

During most of this chapter we shall discuss numerical methods of solving (11.1) subject to (11.2). By this, we mean that we seek approximations to the numerical values of $y(x)$ for particular values of x or we seek a function, for example a polynomial, which approximates to y over some range of values of x. Only for rare special cases can we obtain an explicit form of the solution, as in Example 11.1. In most problems, $f(x, y)$ will be far too complicated for us to obtain an explicit solution and we must therefore resort to numerical methods. We shall also discuss *systems* of ordinary differential equations and higher order differential equations, in which derivatives of higher order than the first occur. The methods for both of these are natural extensions of those for a single first order equation.

The first question we ask of (11.1) and (11.2) is whether a solution y necessarily exists and, if not, what additional restrictions must be imposed on $f(x, y)$ to ensure this. The following example illustrates that a solution does not necessarily exist.

Example 11.2 For $0 \leqslant x \leqslant 2$, solve

$$y' = y^2 \quad \text{with } y(0) = 1.$$

The general solution of the differential equation is

$$y(x) = -\frac{1}{(x + c)},$$

where c is arbitrary. The initial condition implies $c = -1$, when

$$y(x) = \frac{1}{(1 - x)}.$$

This solution is not defined for $x = 1$ and therefore the given initial value problem has *no solution* on $[0, 2]$. $\square$

You will notice that in Example 11.2 the function $f(x, y)$ is the well-behaved function y^2. Obviously continuity of $f(x, y)$ is not sufficient to ensure the existence of a unique solution of (11.1) and (11.2). We find that a Lipschitz condition on $f(x, y)$ with respect to y is sufficient and we have the following theorem.

Theorem 11.1   The first order differential equation

$$y' = f(x, y)$$

with initial condition

$$y(x_0) = s,$$

where $x_0 \in [a, b]$, has a unique solution y on $[a, b]$ if $f(x, y)$ is continuous in x and if there exists L, independent of x, such that

$$|f(x, y) - f(x, z)| \leqslant L|y - z| \qquad (11.6)$$

for all $x \in [a, b]$ and all real y and z.

Proof   We shall consider an outline of the proof, which is based on the contraction mapping theorem. We may write the problem in the integral form

$$y(x) = s + \int_{x_0}^{x} f(x, y(x)) \, dx. \qquad (11.7)$$

We consider the sequence of functions $\{y_m(x)\}_{m=0}^{\infty}$ defined recursively by

$$y_0(x) = s, \qquad (11.8)$$

$$y_{m+1}(x) = s + \int_{x_0}^{x} f(x, y_m(x)) \, dx, \qquad m = 0, 1, \dots. \qquad (11.9)$$

We find that, for $m \geqslant 1$,

$$|y_{m+1}(x) - y_m(x)| \leqslant \int_{x_0}^{x} |f(x, y_m(x)) - f(x, y_{m-1}(x))| \, dx$$

$$\leqslant \int_{x_0}^{x} L|y_m(x) - y_{m-1}(x)| \, dx.$$

Now

$$|y_1(x) - y_0(x)| = |y_1(x) - s| = \left| \int_{x_0}^{x} f(x, s) \, dx \right| \leqslant M|x - x_0|,$$

where $M \geqslant |f(x, s)|$ for $x \in [a, b]$. M, an upper bound of $|f(x, s)|$, exists as f is continuous in x. Thus

$$|y(x_2) - y(x_1)| \leqslant \int_{x_0}^{x} L|y_1(x) - y_0(x)| \, dx$$

$$\leqslant \int_{x_0}^{x} LM|x - x_0| \, dx = \frac{LM}{2!} |x - x_0|^2$$

and

$$|y_3(x) - y_2(x)| \leqslant \int_{x_0}^{x} L|y_2(x) - y_1(x)| \, dx \leqslant \int_{x_0}^{x} \frac{L^2 M}{2!} |x - x_0|^2 \, dx$$

$$= \frac{L^2 M}{3!} |x - x_0|^3.$$

We may prove by induction that

$$|y_{m+1}(x) - y_m(x)| \leqslant \frac{ML^m}{(m+1)!} |x - x_0|^{m+1}$$

for $m \geqslant 0$. The series

$$y_0(x) + (y_1(x) - y_0(x)) + (y_2(x) - y_1(x)) + \cdots$$

is thus uniformly convergent on $[a, b]$, as the absolute value of each term is less than the corresponding term of the uniformly convergent series

$$|s| + \frac{M}{L} \sum_{m=1}^{\infty} \frac{L^m |x - x_0|^m}{m!}.$$

The partial sums of the first series are $y_0(x), y_1(x), y_2(x), \ldots$ and this sequence has a limit, say $y(x)$. This limit $y(x)$ can be shown to be a differentiable function which is the solution of the initial value problem. By assuming the existence of two distinct solutions and obtaining a contradiction, the uniqueness of y can be shown. $\square$

The sequence $\{y_m(x)\}_{m=0}^{\infty}$ defined in the proof of the last theorem may be used to find approximations to the solution. The resulting process is called *Picard* iteration, but is only practicable for simple functions $f(x, y)$ for which the integration in (11.9) can be performed explicitly.

Example 11.3 Use Picard iteration to find an approximate solution of

$$y' = y \quad \text{with } y(0) = 1$$

for $0 \leqslant x \leqslant 2$.
In this case, $f(x, y) = y$ and thus

$$|f(x, y) - f(x, z)| = |y - z|,$$

so we choose $L = 1$. We start with $y_0(x) \equiv 1$ when, from (11.9),

$$y_1(x) = 1 + \int_0^x y_0(x) \, dx = 1 + x,$$

$$y_2(x) = 1 + \int_0^x (1 + x) \, dx = 1 + x + \frac{x^2}{2!},$$

$$y_3(x) = 1 + \int_0^x \left(1 + x + \frac{x^2}{2!}\right) dx = 1 + x + \frac{x^2}{2!} + \frac{x^3}{3!},$$

and, in general,

$$y_r(x) = 1 + x + \frac{x^2}{2!} + \cdots + \frac{x^r}{r!}.$$

The exact solution in this case is

$$y(x) = e^x = 1 + x + \frac{x^2}{2!} + \cdots$$

so that the approximations obtained by the Picard process are the partial sums of this infinite series. □

Example 11.4 We show that

$$y' = \frac{1}{1 + y^2} \quad \text{with } y(x_0) = s$$

has a unique solution for $x \in [a, b]$. We have

$$f(x, y) = \frac{1}{1 + y^2}$$

and from the mean value theorem

$$f(x, y) - f(x, z) = (y - z)\frac{\partial f(x, \xi)}{\partial y},$$

where ξ lies between y and z. Thus to satisfy (11.6) we choose

$$L \geqslant \left| \frac{\partial f(x, y)}{\partial y} \right|$$

for all $x \in [a, b]$ and $y \in (-\infty, \infty)$, provided such an upper bound L exists. In this case,

$$\left| \frac{\partial f}{\partial y} \right| = \left| \frac{-2y}{(1 + y^2)^2} \right|$$

and, as may be seen by sketching a graph, this is a maximum for $y = \pm 1/\sqrt{3}$. We therefore choose

$$L = \left| \frac{\partial f(x, 1/\sqrt{3})}{\partial y} \right| = \frac{3\sqrt{3}}{8}. \quad \square$$

The basis of most numerical methods of solving (11.1) and (11.2) is a *step-by-step* or *marching* process. Approximations $y_0, y_1, y_2, \ldots$ are sought† for the values $y(x_0), y(x_1), y(x_2), \ldots$ of the solution y at distinct points $x = x_0, x_1, x_2, \ldots$. We calculate the y_n by means of a recurrence relation derived from (11.1), which expresses y_n in terms of $y_{n-1}, y_{n-2}, \ldots, y_0$. We

† We shall use the notation y_r to denote an approximation to $y(x_r)$, the exact solution at $x = x_r$, throughout the rest of this chapter.

take $y_0 = y(x_0) = s$, since this value is given. We usually take the points $x_0, x_1, \ldots$ to be equally spaced so that $x_r = x_0 + rh$, where $h > 0$ is called the *step size* or *mesh length*. We also take $a = x_0$ as we are only concerned with points $x_n \geqslant a$. (We could equally well take $h < 0$ but this may be avoided by the transformation $x' = -x$.)

The simplest suitable recurrence relation is derived by replacing the derivative in (11.1) by a forward difference. We obtain

$$\frac{y_{n+1} - y_n}{h} = f(x_n, y_n),$$

which on rearrangement yields *Euler's* method

$$y_0 = s, \tag{11.10}$$

$$y_{n+1} = y_n + hf(x_n, y_n), \quad n = 0, 1, 2, \ldots. \tag{11.11}$$

The latter is a particular type of *difference* equation.

11.2 Difference equations

In order to discuss the accuracy of approximations obtained by methods such as (11.11) in solving initial value problems, we must use some properties of difference equations. If $\{F_n(w_0, w_1)\}$ is a sequence of functions defined for $n \in N$, some set of consecutive integers, and for all w_0 and w_1 in some interval I of the real numbers, then the system of equations

$$F_n(y_n, y_{n+1}) = 0, \tag{11.12}$$

for all $n \in N$, is called a *difference equation of order 1*. A sequence $\{y_n\}$, with all $y_n \in I$ and satisfying (11.12) for all $n \in N$, is called a solution of this difference equation.

Example 11.5   If

$$F_n(w_0, w_1) - w_0 - w_1 + 2n$$

for all integers n, we obtain the difference equation

$$y_n - y_{n+1} + 2n = 0.$$

The general solution of this equation is

$$y_n = n(n - 1) + c,$$

where c is an arbitrary constant. Such an arbitrary constant may be determined if the value of one element of the solution sequence is given. For example, the "initial" value y_0 may be given, when clearly we must choose

$c = y_0$. The arbitrary constant is analogous to that occurring in the integration of first order differential equations. □

The difference equation of order m is obtained from a sequence

$$\{F_n(w_0, w_1, \ldots, w_m)\}$$

of functions of $m + 1$ variables defined for all $n \in N$, a set of consecutive integers, and all $w_r \in I$, $r = 0, \ldots, m$, for some interval I. We seek a sequence $\{y_n\}$ with all $y_n \in I$ and such that

$$F_n(y_n, y_{n+1}, \ldots, y_{n+m}) = 0,$$

for all $n \in N$. The simplest difference equation of order m is of the form

$$a_0 y_n + a_1 y_{n+1} + \cdots + a_m y_{n+m} = 0, \tag{11.13}$$

where $a_0, a_1, \ldots, a_m$ are independent of n. Equation (11.13) is called an mth *order homogeneous linear* difference equation with constant coefficients. It is homogeneous because if $\{y_n\}$ is a solution, then so is $\{\alpha y_n\}$, where α is any scalar. Equation (11.13) is called linear because it consists of a linear combination of elements of $\{y_n\}$. We illustrate the method of solution by first considering the second order equation,

$$a_0 y_n + a_1 y_{n+1} + a_2 y_{n+2} = 0. \tag{11.14}$$

We seek a solution of the form

$$y_n = z^n,$$

where z is independent of n. On substituting z^n in (11.14), we have

$$a_0 z^n + a_1 z^{n+1} + a_2 z^{n+2} = 0,$$

that is,

$$z^n(a_0 + a_1 z + a_2 z^2) = 0.$$

We conclude that for such a solution we require either

$$z^n = 0$$

or

$$a_0 + a_1 z + a_2 z^2 = 0. \tag{11.15}$$

The former only yields $z = 0$ and hence the trivial solution $y_n = 0$. The latter, a quadratic equation, is called the *characteristic* equation of the difference equation (11.14) and, in general, yields two possible values of z. We suppose for the moment that (11.15) has two distinct roots, z_1 and z_2. We then have two independent solutions

$$y_n = z_1{}^n \quad \text{and} \quad y_n = z_2{}^n.$$

It is easily verified that

$$y_n = c_1 z_1{}^n + c_2 z_2{}^n, \tag{11.16}$$

where c_1 and c_2 are *arbitrary* constants, is also a solution of (11.14). This is the general solution of (11.14) for the case when the characteristic equation has distinct roots. You will notice that this general solution contains two arbitrary constants and those familiar with differential equations will know that two such constants occur in the solution of second order differential equations.

Now suppose that in (11.15)

$$a_1{}^2 = 4a_0 a_2 \quad \text{and} \quad a_2 \neq 0,$$

so that the characteristic equation has two equal roots. We only have one solution

$$y_n = z_1{}^n.$$

We try, as a possible second solution,

$$y_n = n z_1{}^n \tag{11.17}$$

in (11.14). The left side becomes

$$a_0 n z_1{}^n + a_1(n + 1)z_1^{n+1} + a_2(n + 2)z_1^{n+2}$$
$$= z_1{}^n[n(a_0 + a_1 z_1 + a_2 z_1{}^2) + z_1(a_1 + 2a_2 z_1)]$$
$$= 0$$

as $z_1 = -\tfrac{1}{2}a_1/a_2$ is the repeated root of (11.15). Clearly (11.17) is a solution and the general solution of (11.14) is

$$y_n = c_1 z_1{}^n + c_2 n z_1{}^n. \tag{11.18}$$

The only remaining case is $a_2 = 0$, when the difference equation is of only first order and has general solution

$$y_n = c\left(-\frac{a_0}{a_1}\right)^n,$$

where c is arbitrary.

If $a_2 \neq 0$ we may write (11.14) as the recurrence relation

$$y_{n+2} = -(a_0/a_2)y_n - (a_1/a_2)y_{n+1}.$$

If we are to use this to calculate members of a sequence $\{y_n\}$ we need two starting (or initial) values, for example, y_0 and y_1. Such starting values fix the arbitrary constants in the general solution. It can also be seen from this that (11.16) or (11.18) is the general solution in that any particular solution may be expressed in this form. Given any particular solution we can always use

y_0 and y_1 to determine c_1 and c_2. The remaining elements of the solution are then uniquely determined by the recurrence relation above.

Example 11.6 The difference equation

$$y_n - 2 \cos \theta . y_{n+1} + y_{n+2} = 0,$$

where θ is a constant, has characteristic equation

$$1 - 2 \cos \theta . z + z^2 = 0$$

which has roots $z = \cos \theta \pm i \sin \theta = e^{\pm i\theta}$. The general solution is:

$$\theta \neq m\pi, \quad \text{is } y_n = c_1 e^{ni\theta} + c_2 e^{-ni\theta}$$
$$= k_1 \cos n\theta + k_2 \sin n\theta;$$

$$\theta = 2m\pi, \quad \text{is } y_n = c_1 + c_2 n;$$

$$\theta = (2m + 1)\pi, \quad \text{is } y_n = (c_1 + c_2 n)(-1)^n;$$

m is any integer and c_1, c_2 and, hence, k_1, k_2 are arbitrary. □

To determine the general solution of the mth order equation

$$a_0 y_n + a_1 y_{n+1} + \cdots + a_m y_{n+m} = 0, \tag{11.19}$$

we again try $y_n = z^n$. We require

$$a_0 z^n + a_1 z^{n+1} + \cdots + a_m z^{n+m} = 0$$

and thus either $z = 0$ or

$$p_m(z) \equiv a_0 + a_1 z + a_2 z^2 + \cdots + a_m z^m = 0. \tag{11.20}$$

Equation (11.20) is called the characteristic equation and p_m the characteristic polynomial of (11.19). In general, (11.20) has m roots $z_1, z_2, \ldots, z_m$ and, if these are distinct, the general solution of (11.19) is

$$y_n = c_1 z_1{}^n + c_2 z_2{}^n + \cdots + c_m z_m{}^n,$$

where the m constants $c_1, c_2, \ldots, c_m$ are arbitrary. Again we may expect m arbitrary constants in the general solution of (11.19), because we need m starting values when it is expressed as a recurrence relation. For example, $y_0, y_1, \ldots, y_{m-1}$ are needed to calculate $y_m, y_{m+1}, \ldots$.

If p_m has a repeated zero, for example if $z_2 = z_1$, then we note that $p'_m(z_1) = 0$, where the prime denotes differentiation. (At a repeated zero of a polynomial, the first derivative must also be zero. See Problem 2.5.) We try a solution of the form $y_n = nz_1{}^n$ in (11.19). The left side becomes

$$a_0 n z_1{}^n + a_1(n + 1)z_1^{n+1} + \cdots + a_m(n + m)z_1^{n+m}$$
$$= z_1{}^n[n(a_0 + a_1 z_1 + a_2 z_1{}^2 + \cdots + a_m z_1{}^m)$$
$$\qquad + z_1(a_1 + 2a_2 z_1 + 3a_3 z_1{}^2 + \cdots + m a_m z_1^{m-1})]$$
$$= z_1{}^n[n p_m(z_1) + z_1 p'_m(z_1)] = 0.$$

If the other zeros of p_m are distinct, we obtain the general solution

$$y_n = c_1 z_1{}^n + c_2 n z_1{}^n + c_3 z_3{}^n + \cdots + c_m z_m{}^n.$$

If a zero of p_m is repeated r times, for example if

$$z_1 = z_2 = \cdots = z_r,$$

the general solution is

$$y_n = c_1 z_1{}^n + c_2 n z_1{}^n + c_3 n^2 z_1{}^n + \cdots + c_r n^{r-1} z_1{}^n + c_{r+1} z_{r+1}^n + \cdots + c_m z_m{}^n,$$

again assuming that $z_{r+1}, \ldots, z_m$ are distinct.

Example 11.7 Find the solution of the third order equation

$$y_n - y_{n+1} - y_{n+2} + y_{n+3} = 0, \tag{11.21}$$

with $y_{-1} = 0$, $y_0 = 1$ and $y_1 = 2$.

The characteristic equation is

$$1 - z - z^2 + z^3 = 0$$

with roots $z = -1, 1, 1$. The general solution of (11.21) is

$$y_n = c_1(-1)^n + c_2 + nc_3$$

and thus we require

$$
\begin{aligned}
y_{-1} &= -c_1 + c_2 - c_3 = 0 \\
y_0 &= c_1 + c_2 \phantom{{}- c_3} = 1 \\
y_1 &= -c_1 + c_2 + c_3 = 2.
\end{aligned}
$$

We obtain

$$c_1 = 0, c_2 = c_3 = 1,$$

and the solution required is

$$y_n = 1 + n. \ \square$$

The non-homogeneous difference equation

$$a_0 y_n + a_1 y_{n+1} + \cdots + a_m y_{n+m} = b_n \tag{11.22}$$

is solved by first finding the general solution of the homogeneous equation

$$a_0 y_n + a_1 y_{n+1} + \cdots + a_m y_{n+m} = 0.$$

We denote this general solution by g_n. Secondly we find any particular solution $\tilde{y}_n$ of the complete equation (11.22) by some means (usually by inspection). The general solution of (11.22) is then

$$y_n = \tilde{y}_n + g_n.$$

There are m arbitrary constants in g_n and hence m in y_n. (This method is entirely analogous to the "particular integral plus complementary function" solution of a linear differential equation.)

Example 11.8   Find the general solution of

$$2y_n + y_{n+1} - y_{n+2} = 1.$$

The homogeneous equation is

$$2y_n + y_{n+1} - y_{n+2} = 0,$$

which has general solution

$$y_n = c_1(-1)^n + c_2.2^n.$$

A particular solution is

$$\tilde{y}_n = \tfrac{1}{2} \quad \text{for all } n,$$

so that the general solution of the complete equation is

$$y_n = c_1(-1)^n + c_2.2^n + \tfrac{1}{2}. \quad \square$$

In the error analysis of this chapter we shall meet *recurrent inequalities*. We consider first:

Lemma 11.1   If the sequence $\{u_n\}_{n=0}^{\infty}$ of positive reals satisfies the recurrent inequality

$$u_{n+1} \leqslant au_n + b \tag{11.23}$$

for $n = 0, 1, 2, \ldots$, where $a, b \geqslant 0$, then

$$u_n \leqslant a^n u_0 + \begin{cases} \left(\dfrac{a^n - 1}{a - 1}\right)b, & \text{if } a \neq 1 \\[2mm] nb, & \text{if } a = 1. \end{cases} \tag{11.24}$$

If $a = 1 + \alpha$, where $\alpha > 0$, then

$$u_n \leqslant e^{n\alpha}u_0 + \left(\dfrac{e^{n\alpha} - 1}{\alpha}\right).b. \tag{11.25}$$

Proof   We can prove (11.24) holds by mathematical induction directly from (11.23) but we will consider a proof based on difference equations. Let $\{y_n\}_{n=0}^{\infty}$ be the solution of the difference equation

$$y_{n+1} = ay_n + b, \quad \text{with } y_0 = u_0.$$

We see by induction that $u_n \leqslant y_n$ for all n.

The solution of the homogeneous equation

$$y_{n+1} = ay_n$$

is

$$y_n = ca^n,$$

where c is arbitrary and a particular solution of the complete equation is

$$\tilde{y}_n = -\left(\frac{b}{a-1}\right) \quad \text{for } a \neq 1.$$

Thus the general solution of the complete equation is

$$y_n = ca^n - \frac{b}{a-1} \quad \text{for } a \neq 1.$$

For $y_0 = u_0$, we require

$$c = u_0 + \frac{b}{a-1}$$

so that

$$u_n \leqslant y_n = u_0 a^n + \left(\frac{a^n - 1}{a - 1}\right)b$$

for $a \neq 1$. When $a = 1$ we use the particular solution $\tilde{y}_n = nb$.
For $a = 1 + \alpha$, with $\alpha > 0$, we have

$$a = 1 + \alpha < e^\alpha$$

and (11.25) follows from

$$a^n = (1 + \alpha)^n < (e^\alpha)^n = e^{n\alpha}. \quad \square$$

The inequality (11.25) is a special case of the next lemma.

Lemma 11.2 If the sequence $\{u_n\}_{n=0}^\infty$ of positive reals satisfies the recurrent inequality

$$u_{n+1} \leqslant u_n + \alpha_0 u_n + \alpha_1 u_{n-1} + \cdots + \alpha_m u_{n-m} + b \tag{11.26}$$

for $n = m, m+1, m+2, \ldots$, where $\alpha_r \geqslant 0, r = 0, 1, \ldots, m$ and $b \geqslant 0$, then

$$u_n \leqslant \left(\delta + \frac{b}{A}\right)e^{nA} - \frac{b}{A} \tag{11.27}$$

for $n = 0, 1, 2, \ldots$, where δ is any real number satisfying

$$\delta \geqslant u_r, \qquad r = 0, 1, \ldots, m$$

and

$$A = \alpha_0 + \alpha_1 + \cdots + \alpha_m \neq 0.$$

(If $A = 0$, (11.26) becomes (11.23) with $a = 1$, which is dealt with in Lemma 11.1.)

Proof Let $\{y_n\}_{n=0}^{\infty}$ be the solution of the difference equation

$$y_{n+1} = (1 + A)y_n + b, \quad \text{with } y_0 = \delta.$$

Thus

$$y_{n+1} \geqslant y_n \tag{11.28}$$

for all $n \geqslant 0$ and

$$y_r \geqslant y_0 = \delta \geqslant u_r \quad \text{for } r = 0, 1, \ldots, m. \tag{11.29}$$

We show by induction that $y_n \geqslant u_n$ for all $n \geqslant 0$. Assume that for some $N \geqslant m$ we have $y_n \geqslant u_n$ for $n = 0, 1, \ldots, N$. Now

$$
\begin{aligned}
u_{N+1} &\leqslant u_N + (\alpha_0 u_N + \cdots + \alpha_m u_{N-m}) + b \\
&\leqslant y_N + (\alpha_0 y_N + \cdots + \alpha_m y_{N-m}) + b \\
&\leqslant y_N + (\alpha_0 y_N + \cdots + \alpha_m y_N) + b,
\end{aligned}
$$

from (11.28). Thus

$$u_{N+1} \leqslant y_N + Ay_N + b = y_{N+1} \tag{11.30}$$

and, therefore, by induction for all $n \geqslant 0$ we have

$$u_n \leqslant y_n \leqslant \left(\delta + \frac{b}{A}\right)e^{nA} - \frac{b}{A},$$

where the bound on y_n is obtained from the last part of Lemma 11.1. $\square$

11.3 One-step methods

We return to the numerical solution of the initial value problem

$$y' = f(x, y)$$

with

$$y(x_0) = s,$$

where $x \in [x_0, b]$. We shall assume that f satisfies the conditions of Theorem 11.1, so that the existence of a unique solution is ensured. We shall also assume that f is "sufficiently" differentiable, that is, all the derivatives used in the analysis exist. We look for a sequence $y_0, y_1, y_2, \ldots$, whose elements are approximations to $y(x_0), y(x_1), y(x_2), \ldots$, where $x_n = x_0 + nh$. We seek a first order difference equation to determine $\{y_n\}$. A first order difference equation yields a *one-step* process, in that we evaluate y_{n+1} from only y_n and do not use $y_{n-1}, y_{n-2}, \ldots$ explicitly.

One approach is to consider the Taylor series for y. We have

$$y(x_{n+1}) = y(x_n) + hy'(x_n) + \frac{h^2}{2!}y''(x_n) + \cdots + \frac{h^p}{p!}y^{(p)}(x_n) + h^{p+1}R_{p+1}(x_n),$$

$$(11.31)$$

where

$$R_{p+1}(x_n) = \frac{1}{(p+1)!}y^{(p+1)}(\xi) \quad \text{and} \quad x_n < \xi < x_{n+1}.$$

We find we can determine the derivatives of y from f and its partial derivatives. For example,

$$y'(x) = f(x, y(x))$$

and, on using the chain rule for partial differentiation,

$$\begin{aligned} y''(x) &= \frac{\partial f}{\partial x} + \frac{\partial f}{\partial y} \cdot \frac{dy}{dx} \\ &= f_x + f_y y' \\ &= f_x + f_y f. \end{aligned} \quad (11.32)$$

We have written f_x and f_y to denote the partial derivatives of f with respect to x and y respectively.

If we define the functions $f^{(r)}(x, y)$ recursively by

$$f^{(r)}(x, y) = f_x^{(r-1)}(x, y) + f_y^{(r-1)}(x, y) \cdot f(x, y) \qquad (11.33)$$

for $r \geq 1$, with $f^{(0)}(x, y) = f(x, y)$, then

$$y^{(r+1)}(x) = f^{(r)}(x, y(x)) \qquad (11.34)$$

for $r \geq 0$. The proof of this is easily obtained by induction. The result is true for $r = 0$ and on differentiating (11.34) we have

$$\begin{aligned} y^{(r+2)}(x) &= f_x^{(r)}(x, y(x)) + f_y^{(r)}(x, y(x)) \cdot y'(x) \\ &= f_x^{(r)}(x, y(x)) + f_y^{(r)}(x, y(x)) \cdot f(x, y(x)) \\ &= f^{(r+1)}(x, y(x)), \end{aligned}$$

showing that we can extend the result to the $(r+2)$th derivative of y. Note that the $f^{(r)}(x, y)$ are *not* derivatives of f in the *usual* sense.

If we substitute (11.34) in (11.31), we obtain

$$y(x_{n+1}) = y(x_n) + hf(x_n, y(x_n)) + \cdots + \frac{h^p}{p!} f^{(p-1)}(x_n, y(x_n))$$

$$+ h^{p+1} R_{p+1}(x_n). \quad (11.35)$$

The approximation obtained by neglecting the remainder term is

$$y_{n+1} = y_n + hf(x_n, y_n) + \cdots + \frac{h^p}{p!} f^{(p-1)}(x_n, y_n) \quad (11.36)$$

and the Taylor series method consists of using this recurrence relation for $n = 0, 1, 2, \ldots$ together with $y_0 = s$. The disadvantage of this method is that the $f^{(r)}(x, y)$ are difficult to determine. For example,

$$f^{(1)}(x, y) = f_x + f_y f, \quad (11.37)$$
$$f^{(2)}(x, y) = f_x^{(1)} + f_y^{(1)} f$$

$$= f_{xx} + 2f_{xy}f + f_{yy}f^2 + (f_x + f_y f)f_y. \quad (11.38)$$

The method might, however, be useful for $p \leqslant 2$. For $p = 1$ we obtain Euler's method. For most problems the accuracy of the method increases with p since the remainder term decreases as p increases if the derivatives of $y(x)$ are well behaved.

Example 11.9   For the problem

$$y' = y \quad \text{with } y(0) = 1,$$
$$f(x, y) = y, \quad f^{(1)}(x, y) = f_x + f_y f = 0 + 1 . y = y,$$
$$f^{(2)}(x, y) = f_x^{(1)} + f_y^{(1)} f = y \quad \text{and} \quad f^{(r)}(x, y) = y.$$

From (11.36), the Taylor series method is

$$y_{n+1} = y_n + hy_n + \frac{h^2}{2!} y_n + \cdots + \frac{h^p}{p!} y_n$$

$$= \left(1 + h + \frac{h^2}{2!} + \cdots + \frac{h^p}{p!} \right) y_n$$

with $y_0 = 1$. Thus we have

$$y_n = \left(1 + h + \cdots + \frac{h^p}{p!} \right)^n. \quad \Box$$

Example 11.10   For the problem

$$y' = \frac{1}{1 + y^2}, \quad \text{with } y(0) = 1,$$

we obtain

$$f(x, y) = \frac{1}{1 + y^2},$$

$$f^{(1)}(x, y) = f_x + f_y f = \frac{-2y}{(1 + y^2)^3},$$

$$f^{(2)}(x, y) = \frac{2(5y^2 - 1)}{(1 + y^2)^5}.$$

Thus, for $p = 3$, the Taylor series method (11.36) is

$$y_{n+1} = y_n + \frac{h}{1 + y_n^2} + \frac{h^2}{2!} \frac{(-2y_n)}{(1 + y_n^2)^3} + \frac{h^3}{3!} \frac{2(5y_n^2 - 1)}{(1 + y_n^2)^5},$$

with $y_0 = 1$. $\square$

A much more useful class of methods is derived by considering approximations to the first $p + 1$ terms on the right of (11.35). We seek an expression of the form

$$y_{n+1} = y_n + h \sum_{i=1}^{r} \alpha_i k_i, \tag{11.39}$$

where $k_1 = f(x_n, y_n)$ and

$$k_i = f(x_n + h\lambda_i, y_n + h\mu_i k_{i-1}), \qquad i > 1.$$

We try to determine the coefficients α_i, λ_i and μ_i so that the first $p + 1$ terms in the Taylor series expansion of (11.39) agree with the first $p + 1$ terms in (11.35). Taylor's theorem for two variables gives

$$f(x_n + h\lambda, y_n + h\mu k)$$

$$= f(x_n, y_n) + h\left(\lambda \frac{\partial}{\partial x} + \mu k \frac{\partial}{\partial y}\right) f(x_n, y_n) + \frac{h^2}{2!} \left(\lambda \frac{\partial}{\partial x} + \mu k \frac{\partial}{\partial y}\right)^2 f(x_n, y_n)$$

$$+ \cdots + \frac{h^{p-1}}{(p-1)!} \left(\lambda \frac{\partial}{\partial x} + \mu k \frac{\partial}{\partial y}\right)^{p-1} f(x_n, y_n) + h^p S_p(x_n, y_n, \lambda, \mu k), \tag{11.40}$$

where $h^p S_p$ is the remainder term. It is clear that equating terms of (11.39), using the expansions (11.40), with the terms of the expansion of (11.35), using (11.33), is very tedious. We shall illustrate only the case $p = 2$, when (11.35) becomes

$$y(x_{n+1}) = y(x_n) + hf + \frac{h^2}{2}(f_x + f_y f) + h^3 R_3(x_n), \tag{11.41}$$

where f, f_x and f_y have argument $(x_n, y(x_n))$. Clearly there are an infinity of ways of choosing r, α_i, λ_i, and μ_i.

If we choose $r = 2$, (11.39) when expanded becomes

$$y_{n+1} = y_n + h(\alpha_1 + \alpha_2)f + h^2\alpha_2(\lambda_2 f_x + \mu_2 f f_y)$$
$$+ h^3\alpha_2 S_2(x_n, y_n, \lambda_2, \mu_2 f), \quad (11.42)$$

where f, f_x and f_y have arguments (x_n, y_n). On comparing (11.41) and (11.42) we see that we require

$$\alpha_1 + \alpha_2 = 1, \qquad \alpha_2\lambda_2 = \tfrac{1}{2}, \qquad \alpha_2\mu_2 = \tfrac{1}{2}. \qquad (11.43)$$

One solution is

$$\alpha_1 = 0, \qquad \alpha_2 = 1, \qquad \lambda_2 = \mu_2 = \tfrac{1}{2},$$

when (11.39) becomes

$$y_{n+1} = y_n + hf(x_n + \tfrac{1}{2}h, y_n + \tfrac{1}{2}hf(x_n, y_n)). \qquad (11.44)$$

This is known as the *modified Euler method*.

If we take $\alpha_1 = \alpha_2 = \tfrac{1}{2}$, then $\lambda_2 = \mu_2 = 1$ and we obtain

$$y_{n+1} = y_n + \tfrac{1}{2}h[f(x_n, y_n) + f(x_{n+1}, y_n + hf(x_n, y_n))], \qquad (11.45)$$

which is known as the *simple Runge–Kutta method*.

For $p = 4$ and $r = 4$, we obtain the *classical Runge–Kutta* method

$$y_{n+1} = y_n + \frac{h}{6}[k_1 + 2k_2 + 2k_3 + k_4], \qquad (11.46)$$

where

$$k_1 = f(x_n, y_n),$$
$$k_2 = f(x_n + \tfrac{1}{2}h, y_n + \tfrac{1}{2}hk_1),$$
$$k_3 = f(x_n + \tfrac{1}{2}h, y_n + \tfrac{1}{2}hk_2),$$
$$k_4 = f(x_{n+1}, y_n + hk_3).$$

This is not the only choice of method for $p = r = 4$. Ralston gives a complete account of the derivation of this and other similar methods.

11.4 Truncation errors of one-step methods

The general one-step method of solving (11.1) may be expressed as

$$y_{n+1} = y_n + h\phi(x_n, y_n; h) \qquad (11.47)$$

where $\phi(x, y; h)$ is called the *increment* function. For $h \neq 0$, we rewrite (11.47) as

$$\frac{y_{n+1} - y_n}{h} = \phi(x_n, y_n; h). \qquad (11.48)$$

We are interested in the accuracy of (11.48) when considered as an approximation to (11.1). If $y(x)$ is the solution of (11.1) and (11.2), we define the truncation error, $t(x; h)$, of (11.47) to be

$$t(x; h) = \frac{y(x + h) - y(x)}{h} - \phi(x, y(x); h), \qquad (11.49)$$

where $x \in [x_0, b]$ and $h > 0$. We say that (11.48) is a *consistent* approximation to (11.1) if

$$t(x; h) \to 0 \quad \text{as } h \to 0,$$

uniformly for $x \in [x_0, b]$. From (11.49) we see that, for a consistent method,

$$y'(x) = \phi(x, y(x); 0) = f(x, y(x)). \qquad (11.50)$$

More precisely, we say that the method (11.47) is *consistent of order p*, if there exists $N \geqslant 0$, $h_0 > 0$ and a positive integer p such that

$$\sup_{x_0 \leqslant x \leqslant b} |t(x; h)| \leqslant Nh^p, \quad \text{for all } h \in (0, h_0]. \qquad (11.51)$$

The Taylor series method (11.36) is consistent of order p as, on comparing (11.35) and (11.36), we see from (11.49) that

$$t(x; h) = h^p R_{p+1}(x) = \frac{h^p}{(p + 1)!} y^{(p+1)}(\xi), \qquad (11.52)$$

where $x < \xi < x + h$. We choose

$$N = \frac{1}{(p + 1)!} \max_{x_0 \leqslant x \leqslant b} |y^{(p+1)}(x)|. \qquad (11.53)$$

The Runge–Kutta method (11.39) with $p + 1$ terms of its expansion agreeing with the Taylor series will also be consistent of order p. It is difficult to obtain an explicit form of N. We just observe from (11.40) and (11.35) that the truncation error takes the form

$$t(x; h) = h^p(R_{p+1}(x) + Q_p(x, h)), \qquad (11.54)$$

where Q_p depends on p, r and the coefficients $\alpha_i, \lambda_i, \mu_i$. Provided sufficient derivatives of y and f are bounded, we can choose a suitable N. Full details are given by Henrici [1962].

11.5 Convergence of one-step methods

So far we have considered truncation errors. These give a measure of the accuracy of the difference equation used to approximate to a differential equation. We have not considered how accurately the solution sequence

$\{y_n\}$, of the difference equation, approximates to the solution $y(x)$ of the differential equation. We shall also be interested in knowing whether the solution of the difference equation converges to the solution of the differential equation as the step size h is decreased.

We must define convergence carefully. If we look at the behaviour of y_n as $h \to 0$ and keep n fixed, we will not obtain a useful concept. Clearly

$$x_n = x_0 + nh \to x_0, \quad \text{as } h \to 0, \quad \text{with } n \text{ fixed,}$$

yet we are interested in obtaining solutions for values of x other than $x = x_0$. We must therefore consider the behaviour of y_n as $h \to 0$ with $x_n = x_0 + nh$ kept fixed. In order to obtain a solution at a fixed value of $x \neq x_0$, we must increase the number of steps required to reach x from x_0 if the step size h is decreased. If

$$\lim_{\substack{h \to 0 \\ x_n = x \text{ fixed}}} y_n = y(x) \tag{11.55}$$

for all $x \in [x_0, b]$, we say that the method is convergent. This definition is due to G. Dahlquist.

Example 11.11   The Taylor series method (11.36) applied to $y' = y$ with $y(0) = 1$, $x \in [0, b]$, is convergent. In Example 11.9 we obtained

$$y_n = \left(1 + h + \cdots + \frac{h^p}{p!}\right)^n.$$

Now by Taylor's theorem

$$e^h = 1 + h + \cdots + \frac{h^p}{p!} + \frac{h^{p+1}}{(p+1)!} e^{h'}, \quad \text{where } 0 < h' < h,$$

and thus

$$y_n = \left(e^h - \frac{h^{p+1}}{(p+1)!} e^{h'}\right)^n$$

$$= e^{nh}\left(1 - \frac{h^{p+1}}{(p+1)!} e^{h'-h}\right)^n.$$

As $x_n = nh$ and $y(x_n) = e^{nh}$, we have

$$|y(x_n) - y_n| = e^{nh}\left|\left(1 - \frac{h^{p+1}}{(p+1)!} e^{h'-h}\right)^n - 1\right|$$

$$\leqslant e^{x_n} \frac{nh^{p+1}}{(p+1)!} e^{h'-h}$$

for h sufficiently small. (See Example 2.13.) Thus, as $e^{h'-h} < 1$,

$$|y(x_n) - y_n| \leqslant e^{x_n} \frac{x_n h^p}{(p+1)!}. \tag{11.56}$$

Hence $|y(x_n) - y_n| \to 0$ as $h \to 0$ with x_n fixed.

We also remark that there exists M such that

$$|y(x_n) - y_n| \leqslant M h^p. \tag{11.57}$$

We choose any

$$M \geqslant \max_{0 \leqslant x \leqslant b} \frac{xe^x}{(p+1)!}. \quad \square$$

The following theorem provides an error bound for a consistent one-step method, provided the increment function ϕ satisfies a Lipschitz condition with respect to y. From this error bound it follows that such a method is convergent.

Theorem 11.2   The initial value problem

$$y' = f(x, y), \qquad y(x_0) = s, \qquad x \in [x_0, b],$$

is replaced by a one-step method:

$$y_0 = s$$
$$\left. \begin{array}{l} x_n = x_0 + nh \\ y_{n+1} = y_n + h\phi(x_n, y_n; h) \end{array} \right\} \quad n = 0, 1, 2, \ldots$$

where ϕ satisfies the Lipschitz condition

$$|\phi(x, y; h) - \phi(x, z; h)| \leqslant L_\phi |y - z| \tag{11.58}$$

for all $x \in [x_0, b]$, $\quad -\infty < y, z < \infty$, $\quad h \in [0, h_0]$, for some L_ϕ and $h_0 > 0$.

If the method is consistent of order p, so that the truncation error defined by (11.49) satisfies (11.51), then the *global* error $y(x_n) - y_n$ satisfies

$$|y(x_n) - y_n| \leqslant \begin{cases} \left(\dfrac{e^{(x_n - x_0)L_\phi} - 1}{L_\phi} \right) Nh^p, & \text{for } L_\phi \neq 0, \\[4mm] (x_n - x_0)Nh^p, & \text{for } L_\phi = 0, \end{cases} \tag{11.59}$$

for all $x_n \in [x_0, b]$ and all $h \in (0, h_0]$, and thus the method is convergent, as (11.55) is satisfied.

Proof We let

$$e_n = |y(x_n) - y_n|.$$

From (11.49),

$$y(x_{n+1}) = y(x_n) + h\phi(x_n, y(x_n); h) + ht(x_n; h) \qquad (11.60)$$

and

$$y(x_{n+1}) - y_{n+1} = y(x_n) - y_n + h\phi(x_n, y(x_n); h) - h\phi(x_n, y_n; h) + ht(x_n; h).$$

Thus

$$|y(x_{n+1}) - y_{n+1}| \leqslant |y(x_n) - y_n| + h|\phi(x_n, y(x_n); h) - \phi(x_n, y_n; h)|$$
$$+ h|t(x_n; h)| \quad (11.61)$$
$$\leqslant |y(x_n) - y_n| + hL_\phi |y(x_n) - y_n| + Nh^{p+1},$$

where we have used (11.51) and the Lipschitz condition (11.58). Hence

$$e_{n+1} \leqslant (1 + hL_\phi)e_n + Nh^{p+1}.$$

This is a recurrence inequality of the type considered in Lemma 11.1. In this case

$$e_0 = |y(x_0) - y_0| = 0$$

and from (11.25), for $L_\phi \neq 0$,

$$e_n \leqslant \frac{e^{nhL_\phi} - 1}{hL_\phi} Nh^{p+1}$$

$$= \left(\frac{e^{(x_n - x_0)L_\phi} - 1}{L_\phi} \right) Nh^p.$$

For $L_\phi = 0$, Lemma 11.1 yields

$$e_n \leqslant n.Nh^{p+1} = (x_n - x_0)Nh^p. \quad \square$$

You will notice from (11.59) that, for all $h \in (0, h_0]$ and all $x_n \in [x_0, b]$, there is a constant $M \geqslant 0$ such that

$$|y(x_n) - y_n| \leqslant Mh^p. \qquad (11.62)$$

A method for which the global error satisfies (11.62), is said to be *convergent of order p*. We shall normally expect that a method which is consistent of order p is also convergent of order p and we say simply that such a method is *of order p*.

It follows from (11.50) that, for a consistent method, (11.58) reduces to

$$|f(x, y) - f(x, z)| \leqslant L_\phi |y - z|$$

when $h = 0$. Thus (11.58) implies the existence of a Lipschitz condition on $f(x, y)$, of the type required to ensure the existence of a unique solution of the differential equation.

In order to obtain expressions for L_ϕ for the Taylor series and Runge–Kutta methods, we assume that bounds

$$L_k \geqslant \sup_{\substack{x_0 \leqslant x \leqslant b \\ -\infty < y < \infty}} \left| \frac{\partial f^{(k)}(x, y)}{\partial y} \right|$$

exist for the values of k required in the following analysis.

As in Example 11.4 we observe, using the mean value theorem, that we may choose $L = L_0$. For the Taylor series method (11.36),

$$\phi(x, y; h) = f(x, y) + \frac{h}{2!} f^{(1)}(x, y) + \cdots + \frac{h^{p-1}}{p!} f^{(p-1)}(x, y)$$

and, on using the mean value theorem for each of the $f^{(k)}$, we obtain

$$|\phi(x, y; h) - \phi(x, z; h)| \leqslant \left(L_0 + \frac{h}{2!} L_1 + \frac{h^2}{3!} L_2 + \cdots + \frac{h^{p-1}}{p!} L_{p-1} \right) |y - z|.$$

We therefore choose

$$L_\phi = L_0 + \frac{h_0}{2!} L_1 + \cdots + \frac{h_0^{p-1}}{p!} L_{p-1}, \quad \text{for } h \in [0, h_0]. \qquad (11.63)$$

For the simple Runge–Kutta method (11.45),

$$\phi(x, y; h) = \tfrac{1}{2}[f(x, y) + f(x + h, y + hf(x, y))].$$

Now with $L = L_0$ we have the Lipschitz condition on f,

$$|f(x, y) - f(x, z)| \leqslant L_0 |y - z|$$

and, for $h \in [0, h_0]$,

$$
\begin{aligned}
|f(x + h, y + hf(x, y)) &- f(x + h, z + hf(x, z))| \\
&\leqslant L_0 |y + hf(x, y) - z - hf(x, z)| \\
&\leqslant L_0[|y - z| + h|f(x, y) - f(x, z)|] \\
&\leqslant L_0[|y - z| + hL_0 |y - z|] \\
&\leqslant L_0(1 + h_0 L_0) |y - z|.
\end{aligned}
$$

Thus

$$
\begin{aligned}
|\phi(x, y; h) - \phi(x, z; h)| &\leqslant \tfrac{1}{2}[L_0 + L_0(1 + h_0 L_0)]|y - z| \\
&= L_0(1 + \tfrac{1}{2}h_0 L_0)|y - z|,
\end{aligned}
$$

so we may choose

$$L_\phi = L_0(1 + \tfrac{1}{2}h_0 L_0). \qquad (11.64)$$

For the classical Runge–Kutta method (11.46), we may choose

$$L_\phi = L_0\left(1 + \frac{h_0 L_0}{2} + \frac{h_0{}^2 L_0{}^2}{6} + \frac{h_0{}^3 L_0{}^3}{24}\right). \tag{11.65}$$

Full details are given by Henrici [1962].

Example 11.12   We reconsider Example 11.11, only this time we will obtain a global error bound, using Theorem 11.2.
  We know that $y(x) = e^x$ and thus from (11.53)

$$N = \frac{1}{(p + 1)!} \max_{0 \le x \le b} |y^{(p+1)}(x)| = \frac{e^b}{(p + 1)!}.$$

From Example 11.9, $f^{(k)} = y$ for all k and thus $L_k = 1$ for all k. We therefore choose, from (11.63),

$$L_\phi = 1 + \frac{h_0}{2!} + \cdots + \frac{h_0^{p-1}}{p!}$$

and (11.59) becomes

$$|y(x_n) - y_n| \le \left(\frac{e^{x_n L_\phi} - 1}{L_\phi}\right) \frac{e^b}{(p + 1)!} h^p.$$

For $x \in [0, b]$ this bound is larger and hence not as good as that of (11.56). $\square$

Example 11.13   Find a bound of the global error for Euler's method applied to the problem

$$y' = \frac{1}{1 + y^2} \quad \text{with } y(0) = 1 \quad \text{and} \quad 0 \le x \le 1.$$

From (11.52) with $p = 1$ the truncation error is

$$t(x; h) = \frac{h}{2!} y''(\xi) = \frac{h}{2} f^{(1)}(\xi, y(\xi)).$$

By sketching a graph, we find that

$$f^{(1)} = f_x + f f_y = \frac{-2y}{(1 + y^2)^3}$$

has maximum modulus for $y = 1/\sqrt{5}$ and thus

$$|f^{(1)}(x, y(x))| \le \frac{25\sqrt{5}}{108}.$$

We therefore obtain

$$|t(x; h)| \le h \cdot \frac{25\sqrt{5}}{216}.$$

Since $\phi(x, y; h) = f(x, y)$ for Euler's method, we choose $L_\phi = L = 3\sqrt{3}/8$. (See Example 11.4.) The bound (11.59) becomes

$$|y_n - y(x_n)| \leq (e^{x_n L_\phi} - 1) \frac{25\sqrt{5}}{81\sqrt{3}} h.$$

For $x_n = 1$,

$$|y_n - y(1.0)| < 0.365\, h.$$

This is a very pessimistic bound for the error. With $h = 0.1$ the absolute value of the actual error is 0.00687. $\square$

From the last example it can be seen that Theorem 11.2 may not give a useful bound for the error. In practice this is often true and in any case it is usually impossible to determine (11.59) because of the difficulty of obtaining suitable values of N and L_ϕ. The importance of Theorem 11.2 is that it describes the general behaviour of the global errors as $h \to 0$.

11.6 Effect of rounding errors on one-step methods

One difficulty when computing with a recurrence relation such as

$$y_{n+1} = y_n + h\phi(x_n, y_n; h)$$

is that an error will be introduced at each step due to rounding. If we reduce h so as to reduce the truncation error, then for a given x_n we increase the number of steps and the number of rounding errors. For a given arithmetical accuracy, there is clearly a *minimum* step size h below which rounding errors will produce inaccuracies larger than those due to truncation errors. Suppose that instead of $\{y_n\}$ we actually compute a sequence $\{z_n\}$ with

$$z_{n+1} = z_n + h\psi(x_n, z_n; h) + \varepsilon_n, \tag{11.66}$$

where ε_n is due to rounding. From (11.60) and (11.66) we obtain, analogous to (11.61),

$$|y(x_{n+1}) - z_{n+1}| \leq |y(x_n) - z_n| + h|\phi(x_n, y(x_n); h) - \phi(x_n, z_n; h)| + h|t(x_n; h)| + |\varepsilon_n|.$$

Thus, if $|\varepsilon_n| \leq \varepsilon$ for all n, a quantity ε/h must be added to Nh^p in (11.59) to bound $|y(x_n) - z_n|$. The choice of method and arithmetical accuracy to which we work should therefore depend on whether ε/h is appreciable in magnitude when compared with Nh^p.

11.7 Methods based on numerical integration; explicit methods

On integrating the differential equation

$$y' = f(x, y)$$

between limits x_n and x_{n+1}, we obtain

$$y(x_{n+1}) = y(x_n) + \int_{x_n}^{x_{n+1}} f(x, y(x)) \, dx. \tag{11.67}$$

The basis of many numerical methods of solving the differential equation is to replace the integral in (11.67) by a suitable approximation. For example, we may replace $f(x, y(x))$ by its Taylor polynomial constructed at $x = x_n$ and integrate this polynomial. This leads to the Taylor series method described earlier. (See Problem 11.16.) A more successful approach is to replace f by an interpolating polynomial. Suppose that we have already calculated approximations $y_0, y_1, \ldots, y_n$ to $y(x)$ at the equally space points $x_r = x_0 + rh$, $r = 0, 1, \ldots, n$. We write

$$f_r = f(x_r, y_r), \qquad r = 0, 1, \ldots, n;$$

these are approximations to $f(x_r, y(x_r))$.

We construct the interpolating polynomial p_m through the $m + 1$ points $(x_n, f_n), (x_{n-1}, f_{n-1}), \ldots, (x_{n-m}, f_{n-m})$, where $m \leqslant n$. We use $p_m(x)$ as an approximation to $f(x, y(x))$ between x_n and x_{n+1} and replace (11.67) by

$$y_{n+1} = y_n + \int_{x_n}^{x_{n+1}} p_m(x) \, dx. \tag{11.68}$$

It should be remembered that $p_m(x)$, being based on the values $f_n, f_{n-1}, \ldots, f_{n-m}$, is not the exact interpolating polynomial for $f(x, y(x))$ constructed at the points $x_n, x_{n-1}, \ldots, x_{n-m}$.

By the backward difference formula (4.43),

$$p_m(x) = \sum_{j=0}^{m} (-1)^j \binom{-s}{j} \nabla^j f_n, \quad \text{where } s = (x - x_n)/h.$$

Thus, as in §6.3,

$$\int_{x_n}^{x_{n+1}} p_m(x) \, dx = h \sum_{j=0}^{m} b_j \nabla^j f_n, \tag{11.69}$$

where

$$b_j = (-1)^j \int_0^1 \binom{-s}{j} \, ds. \tag{11.70}$$

Note that the b_j are independent of m and n.

On substituting (11.69) in (11.68), we obtain the *Adams–Bashforth* algorithm for an initial value problem. To compute approximations to the solution $y(x)$ at equally spaced points $x_r = x_0 + rh$, $r = 0, 1, \ldots$, given the starting approximations $y_0, y_1, \ldots, y_m$, we use the recurrence relation

$$y_{n+1} = y_n + h[b_0 f_n + b_1 \nabla f_n + \cdots + b_m \nabla^m f_n] \tag{11.71}$$

for $n = m, m + 1, \ldots$, where $f_r = f(x_r, y_r)$.

The starting values $y_0, y_1, \ldots, y_m$ are usually found by a one-step method. Table 11.1 exhibits some of the b_j. We shall see that (11.71) is of order $m + 1$, that is, both the truncation and global errors are of order $m + 1$.

Table 11.1   The coefficients b_j.

j	0	1	2	3	4
b_j	1	$\frac{1}{2}$	$\frac{5}{12}$	$\frac{3}{8}$	$\frac{251}{720}$

In practice we usually expand the differences in (11.71) so that it takes the form

$$y_{n+1} = y_n + h[\beta_0 f_n + \beta_1 f_{n-1} + \cdots + \beta_m f_{n-m}]. \tag{11.72}$$

The β_j now depend on m. For example, when $m = 1$ we obtain the second order method

$$y_{n+1} = y_n + \frac{h}{2}(3f_n - f_{n-1}) \tag{11.73}$$

and, for $m = 3$, the fourth order method

$$y_{n+1} = y_n + \frac{h}{24}(55f_n - 59f_{n-1} + 37f_{n-2} - 9f_{n-3}). \tag{11.74}$$

To determine the truncation error, we rewrite (11.71) as

$$\frac{y_{n+1} - y_n}{h} = \sum_{j=0}^{m} b_j \nabla^j f_n \tag{11.75}$$

and consider this as an approximation to (11.1). Now from (6.65)

$$\int_{x_n}^{x_{n+1}} f(x, y(x))\, dx = \int_{x_n}^{x_{n+1}} y'(x)\, dx$$

$$= h \sum_{j=0}^{m} b_j \nabla^j y'(x_n) + h^{m+2} b_{m+1} y^{(m+2)}(\xi_n),$$

where $x_{n-m} < \xi_n < x_{n+1}$. Thus from (11.67)

$$\frac{y(x_{n+1}) - y(x_n)}{h} = \sum_{j=0}^{m} b_j \nabla^j f(x_n, y(x_n)) + h^{m+1} b_{m+1} y^{(m+2)}(\xi_n), \tag{11.76}$$

where $\nabla^j f(x_n, y(x_n)) = \nabla^{j-1} f(x_n, y(x_n)) - \nabla^{j-1} f(x_{n-1}, y(x_{n-1}))$.

The last term in (11.76) is the *truncation error*. You will notice that this is of order $m + 1$. We expand the differences in (11.76) to obtain

$$y(x_{n+1}) = y(x_n) + h[\beta_0 f(x_n, y(x_n)) + \beta_1 f(x_{n-1}, y(x_{n-1})) + \cdots$$
$$+ \beta_m f(x_{n-m}, y(x_{n-m}))] + h^{m+2} b_{m+1} y^{(m+2)}(\xi_n). \quad (11.77)$$

We assume that $y^{(m+2)}(x)$ is bounded for $x \in [x_0, b]$, so there exists M_{m+2} such that

$$\max_{x \in [x_0, b]} |y^{(m+2)}(x)| \leqslant M_{m+2}. \quad (11.78)$$

We again let

$$e_n = |y(x_n) - y_n|$$

and, by the Lipschitz condition on f,

$$|f(x_n, y(x_n)) - f(x_n, y_n)| \leqslant L|y(x_n) - y_n| = Le_n.$$

On subtracting (11.72) from (11.77), we obtain

$$e_{n+1} \leqslant e_n + hL[|\beta_0|e_n + |\beta_1|e_{n-1} + \cdots + |\beta_m|e_{n-m}]$$
$$+ h^{m+2}|b_{m+1}|M_{m+2}. \quad (11.79)$$

We suppose that the starting values $y_0, \ldots, y_m$ are such that

$$e_r \leqslant \delta \quad \text{for } r = 0, 1, \ldots, m.$$

We now apply Lemma 11.2 with

$$\alpha_r = hL|\beta_r| \quad \text{and} \quad b = h^{m+2}|b_{m+1}|M_{m+2},$$

so that

$$A = \alpha_0 + \alpha_1 + \cdots + \alpha_m = hLB_m,$$

where

$$B_m = |\beta_0| + |\beta_1| + \cdots + |\beta_m|. \quad (11.80)$$

From (11.27) we obtain, remembering that $nh = x_n - x_0$,

$$|y(x_n) - y_n| = e_n \leqslant \left(\delta + \frac{h^{m+1}|b_{m+1}|M_{m+2}}{LB_m}\right) \exp\big((x_n - x_0)LB_m\big)$$

$$- \frac{h^{m+1}|b_{m+1}|M_{m+2}}{LB_m}. \quad (11.81)$$

The inequality (11.81) provides a global error bound for the Adams–Bashforth algorithm. This bound is again more of qualitative than quantitative interest. It is usually impossible to obtain useful bounds on the derivatives of $y(x)$; however, (11.81) does provide much information about the behaviour of the error as h is decreased. From the first term on the right hand side, it can be seen that if the method is to be of order $m + 1$, we require δ, a bound for the error in the starting values, to decrease like h^{m+1} as h tends to zero. Thus we should choose a one-step method of order $m + 1$ to determine starting values.

Theorem 11.3 The Adams–Bashforth algorithm (11.71) is convergent of order $m + 1$ if $y^{(m+2)}(x)$ is bounded for $x \in [x_0, b]$ and the starting values $y_0, \ldots, y_m$ are such that

$$|y(x_r) - y_r| \leqslant Dh^{m+1}, \qquad r = 0, 1, \ldots, m,$$

for some $D \geqslant 0$ and all $h \in (0, h_0]$ where $h_0 > 0$. Furthermore, the global errors satisfy (11.81), where M_{m+2} and B_m are defined by (11.78) and (11.80) and

$$\delta = \max_{0 \leqslant r \leqslant m} |y(x_r) - y_r|. \quad \square$$

Example 11.14 As we have seen before, for simple equations it is possible to determine bounds on the derivatives of y and hence compute global error bounds. We will seek a global error bound for the second order Adams–Bashforth algorithm applied to the problem of Example 11.13. From Example 11.4, we have $L = 3\sqrt{3}/8$ and in Example 11.10 we found

$$f^{(2)}(x, y) = \frac{10y^2 - 2}{(1 + y^2)^5}.$$

Thus

$$|y'''(x)| = |f^{(2)}(x, y(x))| \leqslant \max_{-\infty < y < \infty} \left| \frac{10y^2 - 2}{(1 + y^2)^5} \right| = 2$$

and we choose $M_3 = 2$.

For the Adams–Bashforth algorithm of order 2,

$$m = 1, \qquad b_2 = \tfrac{5}{12}, \qquad B_1 = |\beta_0| + |\beta_1| = 2$$

and thus (11.81) yields

$$|y(x_n) - y_n| \leqslant \left(\delta + h^2 \frac{5}{12} \cdot \frac{8}{3\sqrt{3}} \cdot \frac{2}{2} \right) \exp(3\sqrt{3}(x_n - 0)/4)$$

$$- h^2 \frac{5}{12} \cdot \frac{8}{3\sqrt{3}} \cdot \frac{2}{2}$$

$$= \left(\delta + \frac{10h^2}{9\sqrt{3}} \right) e^{3\sqrt{3}x_n/4} - \frac{10h^2}{9\sqrt{3}}. \qquad (11.82)$$

This error bound is considerably larger than the actual error. For example, with $h = 0.1$, $n = 10$ and y_1 obtained by the simple Runge–Kutta method (which yields $\delta = 10^{-5}$), (11.82) becomes

$$|y(1.0) - y_{10}| \leqslant 0.0172.$$

In fact, to five decimal places,

$$y(1.0) = 1.40629 \quad \text{and} \quad y_{10} = 1.40584$$

so that

$$|y(1.0) - y_{10}| = 0.00045. \ \square$$

There is no reason why we should not consider the integral form of the differential equation over intervals other than x_n to x_{n+1}. For example, instead of (11.67), we may consider

$$y(x_{n+1}) = y(x_{n-1}) + \int_{x_{n-1}}^{x_{n+1}} f(x, y(x)) \, dx. \tag{11.83}$$

On replacing $f(x, y(x))$ by a polynomial as before, we derive the formula

$$y_{n+1} = y_{n-1} + h[b_0^* f_n + b_1^* \nabla f_n + \cdots + b_m^* \nabla^m f_n], \tag{11.84}$$

where

$$b_j^* = (-1)^j \int_{-1}^{1} \binom{-s}{j} \, ds. \tag{11.85}$$

This method is attributed to E. J. Nyström.

The Adams–Bashforth and Nyström algorithms are known as *explicit* or *open* methods. The recurrence relations (11.72) and (11.84) do not involve $f(x_{n+1}, y_{n+1})$; they express y_{n+1} explicitly in terms of $y_n, y_{n-1}, \ldots, y_{n-m}$. The explicit nature of these algorithms arises from the fact that we use extrapolation when obtaining an approximation to $f(x, y(x))$ between x_n and x_{n+1}. As was seen in Chapter 4, extrapolation, that is interpolation outside the range of the interpolating points, is usually less accurate than interpolation inside this range. The class of formulae considered in the next section avoid extrapolation by introducing the extra point (x_{n+1}, f_{n+1}) in the interpolatory process.

11.8 Methods based on numerical integration; implicit methods

We again approximate to the integral in (11.67) and thus obtain a recurrence relation. Suppose that $q_{m+1}(x)$ is the interpolating polynomial through the points (x_{n+1}, f_{n+1}), (x_n, f_n), $\ldots$, (x_{n-m}, f_{n-m}). We regard $q_{m+1}(x)$ as an approximation to $f(x, y(x))$ between x_n and x_{n+1} and replace (11.67) by

$$y_{n+1} = y_n + \int_{x_n}^{x_{n+1}} q_{m+1}(x) \, dx.$$

Notice that this approximation does not involve extrapolation.
By the backward difference formula (4.43),

$$q_{m+1}(x) = \sum_{j=0}^{m+1} (-1)^j \binom{-s}{j} \nabla^j f_{n+1},$$

where $s = (x - x_{n+1})/h$,
and

$$\int_{x_n}^{x_{n+1}} q_{m+1}(x)\, dx = h \sum_{j=0}^{m+1} c_j \nabla^j f_{n+1},$$

where

$$c_j = (-1)^j \int_{-1}^{0} \binom{-s}{j} ds. \tag{11.86}$$

The latter are independent of m and n and Table 11.2 gives some of the values.

Table 11.2   The coefficients c_j.

j	0	1	2	3	4
c_j	1	$-\frac{1}{2}$	$-\frac{1}{12}$	$-\frac{1}{24}$	$-\frac{19}{720}$

We thus replace (11.67) by the approximation

$$y_{n+1} = y_n + h[c_0 f_{n+1} + c_1 \nabla f_{n+1} + \cdots + c_{m+1} \nabla^{m+1} f_{n+1}]. \tag{11.87}$$

Given $y_n, y_{n-1}, \ldots, y_{n-m}$, we seek y_{n+1} satisfying (11.87). In general, (11.87) does not yield an explicit expression for y_{n+1} as each of the terms $\nabla^j f_{n+1}$ involves $f(x_{n+1}, y_{n+1})$. Usually (11.87) must be solved as an algebraic equation in y_{n+1} by an iterative method. Let us suppose that $y_{n+1}^{(i)}$ is the ith iterate in this process. Then (11.87) is already in a convenient form for iteration of the kind

$$w_{i+1} = F(w_i)$$

where F is some given function and $w_i = y_{n+1}^{(i)}$. The $y_n, y_{n-1}, \ldots, y_{n-m}$ remain unchanged throughout this iterative process to calculate y_{n+1}. We will examine the convergence of this process in the next section. To calculate a good first approximation $y_{n+1}^{(0)}$, we employ one of the explicit formulae of the last section, for example, the Adams–Bashforth formula.

This leads to the *Adams–Moulton* algorithm for an initial value problem. To compute approximations to the solution $y(x)$ at equally spaced points $x_k = x_0 + kh, k = 0, 1, \ldots$, given the starting approximations $y_0, y_1, \ldots, y_m$, we use the following for $n = m, m + 1, \ldots$.

Predictor: $y_{n+1}^{(0)} = y_n + h[b_0 f_n + b_1 \nabla f_n + \cdots + b_m \nabla^m f_n]$ (11.88a)

Corrector: $y_{n+1}^{(i+1)} = y_n + h[c_0 f_{n+1}^{(i)} + c_1 \nabla f_{n+1}^{(i)} + \cdots + c_{m+1} \nabla^{m+1} f_{n+1}^{(i)}]$, (11.88b)

$$y_{n+1} = y_{n+1}^{(I)}, \qquad\qquad i = 0, 1, \ldots, (I-1),$$

where

$$f_{n+1}^{(i)} = f(x_{n+1}, y_{n+1}^{(i)}), \qquad f_n = f(x_n, y_n)$$

and

$$\nabla^j f_{n+1}^{(i)} = \nabla^{j-1} f_{n+1}^{(i)} - \nabla^{j-1} f_n.$$

The coefficients b_j and c_j are defined by (11.70) and (11.86) respectively. Equation (11.88a) is called the *predictor* and (11.88b) the *corrector*. For each step, *I inner* iterations of the corrector (or *I* corrections) are performed and $y_{n+1}^{(I)}$ is considered a sufficiently accurate approximation to the y_{n+1} of (11.87). The flow diagram in Fig. 11.1 describes the process which stops when $n = N$. We shall prove later that (11.88b) yields a method of order $m + 2$.

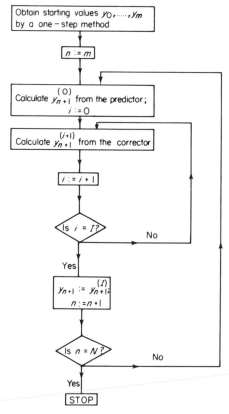

Fig. 11.1 The predictor-corrector algorithm.

We normally expand the differences so that the formulae become

Predictor: $y_{n+1}^{(0)} = y_n + h[\beta_0 f_n + \beta_1 f_{n-1} + \cdots + \beta_m f_{n-m}]$ (11.89a)

Corrector: $y_{n+1}^{(i+1)} = y_n + h[\gamma_{-1} f_{n+1}^{(i)} + \gamma_0 f_n + \cdots + \gamma_m f_{n-m}].$ (11.89b)

The β_r and γ_r are dependent on m. For $m = 0$, we obtain the second order method

$$y_{n+1}^{(0)} = y_n + h f_n,$$ (11.90a)

$$y_{n+1}^{(i+1)} = y_n + \frac{h}{2} (f_{n+1}^{(i)} + f_n)$$ (11.90b)

and for $m = 2$, the fourth order method

$$y_{n+1}^{(0)} = y_n + \frac{h}{12} (23 f_n - 16 f_{n-1} + 5 f_{n-2}),$$ (11.91a)

$$y_{n+1}^{(i+1)} = y_n + \frac{h}{24} (9 f_{n+1}^{(i)} + 19 f_n - 5 f_{n-1} + f_{n-2}).$$ (11.91b)

We may simplify the iteration procedure for the corrector by noting from (11.89b) that

$$y_{n+1}^{(i+1)} = y_{n+1}^{(i)} + h\gamma_{-1}(f_{n+1}^{(i)} - f_{n+1}^{(i-1)}), \qquad i = 1, 2, \ldots, (I - 1). \quad (11.92)$$

Therefore we only need to employ the full formula (11.89b) for the first inner iteration. For subsequent iterations we may use (11.92). Although we prove in §11.9 that the inner iteration process is convergent, we shall also show that we normally expect to make only one or two iterations.

In the rest of this section, we will discuss the truncation and global errors of the corrector and will assume that members of the sequence $y_{m+1}, y_{m+2}, \cdots$ satisfy (11.87) exactly. Of course in practice this will rarely be true, as only a finite number of inner iterations may be performed for each step. We leave the discussion of the additional errors due to this and rounding until the next section.

To derive the truncation error, we rewrite (11.87) as

$$\frac{y_{n+1} - y_n}{h} = c_0 f_{n+1} + \cdots + c_{m+1} \nabla^{m+1} f_{n+1}$$

and consider this as an approximation to (11.1). From (6.67)

$$\int_{x_n}^{x_{n+1}} f(x, y(x)) \, dx = \int_{x_n}^{x_{n+1}} y'(x) \, dx$$

$$= h \sum_{j=0}^{m+1} c_j \nabla^j y'(x_{n+1}) + h^{m+3} c_{m+2} y^{(m+3)}(\xi_n),$$

where $x_{n-m} < \xi_n < x_{n+1}$. Thus from (11.67)

$$\frac{y(x_{n+1}) - y(x_n)}{h} = \sum_{j=0}^{m+1} c_j \nabla^j f(x_{n+1}, y(x_{n+1})) + h^{m+2} c_{m+2} y^{(m+3)}(\xi_n). \quad (11.93)$$

The last term is the truncation error and is of order $m + 2$.

We expand the differences in (11.93) to obtain

$$y(x_{n+1}) = y(x_n) + h[\gamma_{-1} f(x_{n+1}, y(x_{n+1})) + \cdots$$
$$+ \gamma_m f(x_{n-m}, y(x_{n-m}))] + h^{m+3} c_{m+2} y^{(m+3)}(\xi_n). \quad (11.94)$$

As in the analysis of the Adams–Bashforth algorithm, we let

$$e_n = |y(x_n) - y_n|$$

and, on expanding and subtracting (11.87) from (11.94), we obtain

$$e_{n+1} \leqslant e_n + hL[|\gamma_{-1}|e_{n+1} + |\gamma_0|e_n + \cdots + |\gamma_m|e_{n-m}] + h^{m+3}|c_{m+2}|M_{m+3}, \quad (11.95)$$

where $|y^{(m+3)}(x)| \leqslant M_{m+3}$ for $x \in [x_0, b]$. If h is sufficiently small, so that

$$1 - hL|\gamma_{-1}| > 0, \quad (11.96)$$

then

$$e_{n+1} \leqslant \frac{1}{1 - hL|\gamma_{-1}|} \{e_n + hL[|\gamma_0|e_n + \cdots + |\gamma_m|e_{n-m}] + h^{m+3}|c_{m+2}|M_{m+3}\}$$

$$= e_n + \frac{hL}{1 - hL|\gamma_{-1}|} \{(|\gamma_0| + |\gamma_{-1}|)e_n + |\gamma_1|e_{n-1} + \cdots + |\gamma_m|e_{n-m}\}$$

$$+ \frac{h^{m+3}|c_{m+2}|M_{m+3}}{1 - hL|\gamma_{-1}|}.$$

On applying Lemma 11.2, we obtain

$$|y(x_n) - y_n| = e_n \leqslant \left(\delta + \frac{h^{m+2}|c_{m+2}|M_{m+3}}{LC_m}\right) \exp\left(\frac{(x_n - x_0)LC_m}{1 - hL|\gamma_{-1}|}\right)$$

$$- \frac{h^{m+2}|c_{m+2}|M_{m+3}}{LC_m}, \quad (11.97)$$

where

$$C_m = |\gamma_{-1}| + |\gamma_0| + \cdots + |\gamma_m| \quad (11.98)$$

and δ is the maximum starting error, so that

$$|y(x_r) - y_r| \leqslant \delta, \quad r = 0, 1, \ldots, m.$$

The inequality (11.97) provides a global error bound for the Adams–Moulton algorithm, assuming that the corrector is always satisfied exactly.

Again, this result is qualitative rather than quantitative but it may be seen that the method is of order $m + 2$ if the maximum starting error δ decreases like h^{m+2} as h is decreased. We should therefore choose a one-step method of order $m + 2$ to determine starting values.

Theorem 11.4 The Adams–Moulton algorithm (11.88), with the corrector satisfied exactly for each step, is convergent of order $m + 2$ if $y^{(m+3)}(x)$ is bounded for $x \in [x_0, b]$ and the starting values $y_0, \ldots, y_m$ are such that

$$|y(x_r) - y_r| \leqslant Dh^{m+2}, \qquad r = 0, 1, \ldots, m,$$

for some $D \geqslant 0$ and all $h \in (0, h_0]$ where $h_0 > 0$. Furthermore, the global errors satisfy (11.97) provided $h < (L|\gamma_{-1}|)^{-1}$. $\square$

You will observe that the order of the corrector in (11.88) is one higher than the order of the predictor. This is because we have chosen a pair which requires an equal number of starting values. There is no reason why we should not use other explicit formulae as predictors, as these clearly do not affect the final accuracy of the corrector, provided I is suitably adjusted.

It should be noted that, for small h and a given value of δ, the error bound for the Adams–Moulton method is smaller than that of the Adams–Bashforth method of the same order, since for $m > 0$, we can show that $|c_m| < |b_m|$ (see Problem 11.27) and $C_m < B_{m+1}$. For the second order Adams–Moulton method the relevant constants are $c_2 = -\frac{1}{12}$ and $C_0 = 1$, whereas for the second order Adams–Bashforth method we have $b_2 = \frac{5}{12}$ and $B_1 = 2$. For nearly all problems the Adams–Moulton algorithm is far superior to the Adams–Bashforth algorithm of the same order, in that global errors are much smaller. This might be expected since the Adams–Moulton method does not use extrapolation.

Again, we may obtain other corrector formulae by integrating the differential equation over two or more intervals. By integrating over two intervals, we obtain a class of formulae attributed to W. E. Milne who first suggested predictor-corrector methods. (See Problem 11.28.)

Example 11.15 Obtain a global error bound for the second order Adams–Moulton algorithm applied to the problem of Example 11.13, assuming that the corrector is satisfied exactly after each step. From Example 11.4, $L = 3\sqrt{3}/8$ and, from Example 11.14, $M_3 = 2$. Since $c_2 = -\frac{1}{12}$, $C_0 = 1$ and $\gamma_{-1} = \frac{1}{2}$, (11.97) becomes

$$|y(x_n) - y_n| \leqslant \left(\delta + h^2 \frac{1}{12} \cdot \frac{8}{3\sqrt{3}} \cdot \frac{2}{1}\right) \exp\left(\frac{x_n \cdot 3\sqrt{3}/8}{1 - h \cdot 3\sqrt{3}/16}\right) - h^2 \frac{1}{12} \cdot \frac{8}{3\sqrt{3}} \cdot \frac{2}{1}$$

$$= \left(\delta + \frac{4h^2}{9\sqrt{3}}\right) \exp\left(\frac{6\sqrt{3}x_n}{16 - 3\sqrt{3}h}\right) - \frac{4h^2}{9\sqrt{3}}$$

provided $h < 16/(3\sqrt{3})$. For example, with $h = 0.1$, $\delta = 10^{-5}$ and $x_n = 1$, this bound is

$$|y(1.0) - y_{10}| \leqslant 0.00248.$$

Notice that this is much smaller than that in Example 11.14 for the second order Adams–Bashforth algorithm.□

11.9 Iterating with the corrector

We first verify that the inner iterative process (11.88b) for solving the corrector is convergent for small h. We rearrange the expanded form (11.89b) as

$$y_{n+1}^{(i+1)} = y_n + h[\gamma_0 f_n + \cdots + \gamma_m f_{n-m}] + h\gamma_{-1}f(x_{n+1}, y_{n+1}^{(i)}), \quad (11.99)$$

which, on writing $w_i = y_{n+1}^{(i)}$, may be expressed as

$$w_{i+1} = F(w_i).$$

In Chapter 7, Theorem 7.3, it was shown that the sequence $\{w_i\}$ will converge to a solution if F satisfies closure and a suitable Lipschitz condition on some interval. In this case, we take the interval to be all the real numbers, when clearly closure is satisfied, and we seek $L_F < 1$, such that

$$|F(w) - F(z)| \leqslant L_F|w - z| \qquad (11.100)$$

for all reals w and z. From (11.99), we have

$$|F(w) - F(z)| = |h\gamma_{-1}f(x_{n+1}, w) - h\gamma_{-1}f(x_{n+1}, z)|$$
$$\leqslant hL|\gamma_{-1}| \, |w - z| \qquad (11.101)$$

where L is the usual Lipschitz constant for $f(x, y)$ with respect to y. We obtain a condition of the form (11.100) with $L_F < 1$ for

$$h < \frac{1}{|\gamma_{-1}|L}.$$

This does not impose a very severe restriction on h for most problems. You will notice that this is the same condition as (11.96) which was necessary for the validity of the error bound (11.97).

We also observe, for later use, that

$$|y_{n+1}^{(i+1)} - y_{n+1}^{(i)}| = |F(y_{n+1}^{(i)}) - F(y_{n+1}^{(i-1)})|$$
$$\leqslant hL|\gamma_{-1}| \, |y_{n+1}^{(i)} - y_{n+1}^{(i-1)}|. \qquad (11.102)$$

We now investigate the effect of performing only a finite number of iterations with the corrector and we will assume that $y_{n+1}^{(I)}$ is the last iterate which is actually computed. From (11.89b)

$$y_{n+1}^{(I)} = y_n + h[\gamma_{-1}f_{n+1}^{(I)} + \gamma_0 f_n + \cdots + \gamma_m f_{n-m}] + y_{n+1}^{(I)} - y_{n+1}^{(I+1)}.$$

We let

$$p_n = y_{n+1}^{(l+1)} - y_{n+1}^{(l)}$$

and assume that

$$|p_n| \leqslant \rho \quad \text{for all } n.$$

We will also allow for rounding errors in the process and thus, instead of computing a sequence $\{y_n\}$ which satisfies the corrector exactly for each n, we actually compute a sequence $\{z_n\}$ with

$$z_{n+1} = z_n + h[\gamma_{-1}f(x_{n+1}, z_{n+1}) + \cdots + \gamma_m f(x_{n-m}, z_{n-m})] + p_n + \varepsilon_n$$

where ε_n is due to rounding. On subtracting (11.94) and putting

$$e_n = |y(x_n) - z_n|$$

we obtain

$$e_{n+1} \leqslant e_n + hL[|\gamma_{-1}|e_{n+1} + |\gamma_0|e_n + \cdots + |\gamma_m|e_{n-m}]$$
$$+ h^{m+3}|c_{m+2}|M_{m+3} + \rho + \varepsilon, \quad (11.103)$$

where $|\varepsilon_n| \leqslant \varepsilon$ for all n. By comparing (11.103) with (11.95), it can be seen that an amount $(\rho + \varepsilon)/h$ must be added to $h^{m+2}|c_{m+2}|M_{m+3}$ in the error bound (11.97). If we have an estimate for M_{m+3}, we can then decide how small ρ and ε should be. Since $h^{m+2}|c_{m+2}|M_{m+3}$ is a bound for the truncation error of the corrector (see (11.94)), we are in fact comparing $(\rho + \varepsilon)/h$ with the truncation error and in the next section we describe how this may be estimated.

Having chosen ρ, we may use (11.102) to decide when to stop iterating with the corrector. We require

$$|y_{n+1}^{(l)} - y_{n+1}^{(l-1)}| \leqslant \frac{\rho}{h|\gamma_{-1}|L}. \quad (11.104)$$

If L is not known, as is usual, we can use the mean value theorem in (11.101) and replace L by $|\partial f/\partial y|$ at some point near $x = x_{n+1}, y = y_{n+1}$. We then use the estimate

$$L \simeq \left|\frac{\partial f}{\partial y}\right| \simeq \left|\frac{f_{n+1}^{(1)} - f_{n+1}^{(0)}}{y_{n+1}^{(1)} - y_{n+1}^{(0)}}\right|, \quad (11.105)$$

in place of L in (11.104).

11.10 A method of estimating truncation errors

If we choose the predictor to be of the same order as the corrector, it is possible to estimate the truncation error for each step. It is also only necessary to make one application of the corrector for each step, provided h is small.

We consider the following formulae.

Predictor: $y_{n+1}^{(0)} = y_n + h[b_0 f_n + b_1 \nabla f_n + \cdots + b_{m+1} \nabla^{m+1} f_n]$ (11.106a)

Corrector: $y_{n+1}^{(i+1)} = y_n + h[c_0 f_{n+1}^{(i)} + c_1 \nabla f_{n+1}^{(i)} + \cdots + c_{m+1} \nabla^{m+1} f_{n+1}^{(i)}],$

$$i = 0, 1, \ldots, (I-1), \quad (11.106b)$$

$$y_{n+1} = y_{n+1}^{(I)}.$$

Apart from the necessity of an extra starting value, these formulae are implemented in an identical manner to (11.88).

From (11.76) and (11.93),

$$y(x_{n+1}) = y(x_n) + h \sum_{j=0}^{m+1} b_j \nabla^j f(x_n, y(x_n)) + h^{m+3} b_{m+2} y^{(m+3)}(\eta_n), \quad (11.107a)$$

$$y(x_{n+1}) = y(x_n) + h \sum_{j=0}^{m+1} c_j \nabla^j f(x_{n+1}, y(x_{n+1})) + h^{m+3} c_{m+2} y^{(m+3)}(\xi_n),$$

$$(11.107b)$$

where η_n and ξ_n are intermediate points. We assume that

$$y^{(m+3)}(\eta_n) \simeq y^{(m+3)}(\xi_n)$$

and write

$$T_n = h^{m+2} c_{m+2} y^{(m+3)}(\xi_n) \tag{11.108}$$

for the truncation error of (11.107b). (An additional factor h is accounted for by writing (11.107b) in the form (11.93).) On subtracting (11.107a) from (11.107b), we obtain

$$\left(\frac{b_{m+2}}{c_{m+2}} - 1 \right) T_n \simeq \sum_{j=0}^{m+1} c_j \nabla^j f(x_{n+1}, y(x_{n+1})) - \sum_{j=0}^{m+1} b_j \nabla^j f(x_n, y(x_n)). \tag{11.109}$$

We now use approximations

$$\nabla^j f(x_n, y(x_n)) \simeq \nabla^j f_n, \qquad \nabla^j f(x_{n+1}, y(x_{n+1})) \simeq \nabla^j f_{n+1}^{(0)},$$

to obtain

$$T_n \simeq \left(\frac{c_{m+2}}{b_{m+2} - c_{m+2}} \right) \left(\sum_{j=0}^{m+1} c_j \nabla^j f_{n+1}^{(0)} - \sum_{j=0}^{m+1} b_j \nabla^j f_n \right)$$

$$= \frac{1}{h} \left(\frac{c_{m+2}}{b_{m+2} - c_{m+2}} \right) (y_{n+1}^{(1)} - y_{n+1}^{(0)}), \tag{11.110}$$

on using (11.106a) and (11.106b). We may use this last result to estimate the truncation error at each step.

If only one application of the corrector is made, $I = 1$ and we seek an estimate of

$$|\rho_n| = |y^{(2)}_{n+1} - y^{(1)}_{n+1}| \leqslant hL|\gamma_{-1}| |y^{(1)}_{n+1} - y^{(0)}_{n+1}|.$$

Now, from (11.110) and (11.108),

$$h|\gamma_{-1}|L|y^{(1)}_{n+1} - y^{(0)}_{n+1}| \simeq h^2L|\gamma_{-1}| \left| \frac{b_{m+2} - c_{m+2}}{c_{m+2}} \right| |T_n|$$

$$\leqslant h^{m+4}L|\gamma_{-1}| |b_{m+2} - c_{m+2}|M_{m+3},$$

where $M_{m+3} \geqslant |y^{(m+3)}(x)|$ for $x \in [x_0, b]$. Thus an *approximate* upper bound for the ρ_n is

$$\tilde{\rho} = h^{m+4}L|\gamma_{-1}| |b_{m+2} - c_{m+2}|M_{m+3}$$

and $\tilde{\rho}/h$ is small compared with

$$h^{m+2}|c_{m+2}|M_{m+3} > |T_n|,$$

if

$$h \ll \frac{1}{L|\gamma_{-1}|} \frac{|c_{m+2}|}{|b_{m+2} - c_{m+2}|}. \tag{11.111}$$

If this condition is satisfied, there should be no advantage in applying the corrector more than once. The effects of correcting only once are investigated more precisely by Henrici [1962]. Ralston describes how the estimate (11.110) of the truncation error may be used to improve the predicted value $y^{(0)}_{n+1}$ at each step.

Example 11.16 We consider the following fourth order pair.

Predictor:
$$y^{(0)}_{n+1} = y_n + \frac{h}{24}(55f_n - 59f_{n-1} + 37f_{n-2} - 9f_{n-3}), \tag{11.112a}$$

Corrector:
$$y^{(i+1)}_{n+1} = y_n + \frac{h}{24}(9f^{(i)}_{n+1} + 19f_n - 5f_{n-1} + f_{n-2}). \tag{11.112b}$$

Here $m = 2$, $b_{m+2} = b_4 = \frac{251}{720}$ and $c_{m+2} = c_4 = -\frac{19}{720}$. The truncation error of the predictor is

$$h^{m+2}b_{m+2}y^{(m+3)}(\eta_n) = \frac{251}{720}h^4y^{(5)}(\eta_n)$$

and that of the corrector is

$$T_n = h^{m+2}c_{m+2}y^{(m+3)}(\xi_n) = -\frac{19}{720}h^4y^{(5)}(\xi_n).$$

Thus, from (11.110),

$$T_n \simeq \frac{1}{h}\left(\frac{c_4}{b_4 - c_4}\right)(y^{(1)}_{n+1} - y^{(0)}_{n+1}) = \frac{1}{h}\left(-\frac{19}{270}\right)(y^{(1)}_{n+1} - y^{(0)}_{n+1}). \tag{11.113}$$

If only one correction is made for each step, then from (11.111) with $|\gamma_{-1}| = \frac{9}{24}$ we require

$$h \ll \frac{8}{3L} \cdot \frac{19}{270} \simeq 0.2/L. \quad \square \qquad (11.114)$$

Example 11.17   We apply the fourth order method of Example 11.16 to the problem in Example 11.13. From (11.114) with $L = 3\sqrt{3}/8$, we see that we require $h \ll 0.289$ if only one correction is made for each step. Table 11.3 gives results with $h = 0.1$; the T_n are obtained from (11.113). The starting values are calculated using the fourth order Runge–Kutta method. The last column gives the errors $y(x_n) - y_n^{(1)}$.

Table 11.3   Results for Example 11.17.

x_n	$y_n^{(0)}$	$f_n^{(0)}$	$y_n^{(1)}$	$f_n^{(1)}$	$T_{n-1} \times 10^8$	$y(x_n)$	Error $\times 10^8$
0			1	0.5		1.0	0
0.1			1.04879039	0.47619926		1.04879039	0
0.2			1.09531337	0.45460510		1.09531337	0
0.3			1.13977617	0.43495475		1.13977617	0
0.4	1.18236027	0.41701839	1.18236150	0.41701788	-87	1.18236140	-10
0.5	1.22322920	0.40059453	1.22323034	0.40059409	-80	1.22323016	-18
0.6	1.26252409	0.38550990	1.26252510	0.38550953	-71	1.26252485	-25
0.7	1.30037111	0.37161390	1.30037199	0.37161358	-62	1.30037168	-31
0.8	1.33688243	0.35877598	1.33688319	0.35877571	-54	1.33688284	-35
0.9	1.37215806	0.34688284	1.37215872	0.34688262	-46	1.37215832	-40
1.0	1.40628744	0.33583594	1.40628800	0.33583576	-39	1.40628758	-42

At first sight it might appear that for $x_n \geqslant 0.9$ the correction process makes the approximations worse, as $|y_n^{(1)} - y(x_n)| > |y_n^{(0)} - y(x_n)|$ for $n \geqslant 9$. However, omission of the corrector affects all approximations so that the fourth order Adams–Bashforth process does not give the column $y_n^{(0)}$. $\square$

11.11 Numerical stability

We have seen that the inclusion of rounding errors does not radically change the global error bounds for the methods introduced earlier in this chapter. However, the error bounds do not always provide a good description of the effects of rounding errors and starting errors.

We consider the calculation of members of a sequence $\{y_n\}_0^\infty$ by means of a recurrence relation (or difference equation) of the form

$$y_{n+1} = F(y_n, y_{n-1}, \ldots, y_{n-m}), \qquad n = m, m+1, \ldots, \qquad (11.115)$$

given starting values $y_0, y_1, \ldots, y_m$. We suppose that a single error is introduced so that, for some arbitrary r,

$$y_r^* = y_r + \varepsilon$$

and we compute the sequence

$$y_0, y_1, \ldots, y_{r-1}, y_r^*, y_{r+1}^*, \ldots.$$

We assume that this sequence satisfies (11.115) for all $n \geqslant m$ *except* $n = r - 1$. If $r \leqslant m$ the error will occur in the starting values and if $r > m$ the error will occur when (11.115) is computed for $n = r - 1$. We say that the recurrence relation (11.115) is *numerically unstable* if, for $y_n \neq 0$, the relative errors

$$\left| \frac{y_n - y_n^*}{y_n} \right| \tag{11.116}$$

are unbounded as $n \to \infty$. Clearly an unstable recurrence relation will not provide a satisfactory numerical process, as we are certain to introduce rounding errors. Of course in practice we will introduce more than one error but in the analysis it is more instructive to consider the effect of an isolated error.

Example 11.18 Suppose we wish to calculate $\{y_n\}$ from

$$y_{n+1} = 100.01 y_n - y_{n-1} \tag{11.117}$$

with $y_0 = 1$, $y_1 = 0.01$. The general solution of (11.117) is

$$y_n = c_1 \cdot 100^n + c_2 \left(\frac{1}{100} \right)^n$$

where c_1 and c_2 are arbitrary. The solution which satisfies the initial conditions is

$$y_n = \left(\frac{1}{100} \right)^n,$$

for which $c_1 = 0$ and $c_2 = 1$. The introduction of errors may be considered equivalent to changing the particular solution being computed. This will have the effect of making c_1 non-zero, so that a term $c_1 \cdot 100^n$ must be added to y_n. Clearly $c_1 \cdot 100^n$ will quickly "swamp" $c_2/100^n$. For example, if

$$y_0^* = y_0 + 10^{-6} = 1.000001,$$
$$y_1^* = y_1 = 0.01$$

then

$$y_2^* = 0.000099 \quad \text{and} \quad y_5^* \simeq -1.0001. \quad \square$$

Clearly for an approximation to an initial value differential problem, the function F in the recurrence relation (11.115) will depend on h and $f(x, y)$. In defining numerical stability we do *not* decrease h as $n \to \infty$ as was necessary when considering the convergence of y_n to $y(x_n)$. We are now interested in only the numerical stability of recurrence relations resulting from differential equations. Because the precise form of the recurrence relation depends on $f(x, y)$, it is difficult to derive general results regarding the numerical stability of such approximations. We find it instructive, however, to consider in detail the linear differential equation

$$y' = -Ay \quad \text{with } y(0) = 1, \tag{11.118}$$

where A is a positive constant, which has solution $y(x) = e^{-Ax}$. Later in this section, we show that to a large extent the lessons to be learned from this problem will apply to other initial value problems.

We start by considering the *mid-point* rule

$$y_{n+1} = y_{n-1} + 2hf_n \tag{11.119}$$

which is obtained by using

$$\frac{y_{n+1} - y_{n-1}}{2h} = f_n$$

as an approximation to (11.1) or, alternatively, from Nyström's formula (11.84) with $m = 0$. (See Problem 11.21.) For (11.118), $f(x, y) = -Ay$ and the method becomes

$$y_{n+1} = -2hAy_n + y_{n-1}. \tag{11.120}$$

This is a second order difference equation with characteristic equation

$$z^2 + 2hAz - 1 = 0,$$

whose roots are

$$\left.\begin{matrix} z_1 \\ z_2 \end{matrix}\right\} = -Ah \pm (1 + A^2h^2)^{1/2}.$$

The general solution of (11.120) is

$$y_n = c_1 z_1^n + c_2 z_2^n. \tag{11.121}$$

We find that z_1^n behaves like $y(x)$ as h is decreased with x_n fixed, but that z_2^n has a completely different behaviour. On expanding the square root, we obtain

$$z_1 = -Ah + 1 + \tfrac{1}{2}(A^2h^2) + O(h^4).$$

On comparison with

$$e^{-Ah} = 1 - Ah + \tfrac{1}{2}(A^2h^2) + O(h^3),$$

we see that

$$z_1 = e^{-Ah} + O(h^3)$$
$$= e^{-Ah}(1 + O(h^3)),$$

since

$$e^{-Ah} = 1 + O(h).$$

Thus

$$z_1{}^n = e^{-Anh}(1 + O(h^3))^n = e^{-Anh}(1 + n \cdot O(h^3))$$
$$= e^{-Ax_n}(1 + O(h^2)),$$

as $x_n = nh$. Now as $h \to 0$ with $x_n = nh$ fixed,

$$z_1{}^n \to e^{-Ax_n} = y(x_n)$$

and, for small values of h, $z_1{}^n$ behaves like an approximation to $y(x)$.
On the other hand,

$$z_2 = -Ah - (1 + A^2h^2)^{1/2}$$
$$= -Ah - 1 - \tfrac{1}{2}A^2h^2 + O(h^4)$$
$$= -e^{Ah} + O(h^3)$$
$$= -e^{Ah}(1 + O(h^3))$$

and thus

$$z_2{}^n = (-1)^n e^{Anh}(1 + O(h^3))^n$$
$$= (-1)^n e^{Ax_n}(1 + O(h^2)).$$

Clearly, as h is decreased with x_n fixed, $z_2{}^n$ does *not* approximate to $y(x_n)$.
The general solution of (11.120) is, therefore,

$$y_n = c_1 e^{-Ax_n}(1 + O(h^2)) + c_2(-1)^n e^{Ax_n}(1 + O(h^2)) \qquad (11.122)$$

and for this to be an approximation to $y(x_n)$, we require

$$c_1 = 1 \quad \text{and} \quad c_2 = 0.$$

We would therefore like to compute the particular solution

$$y_n = z_1{}^n$$

of (11.120). In practice we cannot avoid errors and we obtain, say,

$$y_n^* = (1 + \delta_1)z_1{}^n + \delta_2 z_2{}^n.$$

In this case, the relative error (11.116) is

$$\left| \frac{y_n^* - y_n}{y_n} \right| = \left| \frac{\delta_1 z_1{}^n + \delta_2 z_2{}^n}{z_1{}^n} \right| = \left| \delta_1 + \delta_2 \left(\frac{z_2}{z_1} \right)^n \right|.$$

Now

$$|z_2| = |Ah + (1 + A^2h^2)^{1/2}| > |Ah - (1 + A^2h^2)^{1/2}| = |z_1|$$

and, for a fixed step size h, the factor $(z_2/z_1)^n$ is unbounded as $n \to \infty$. Therefore the method (11.120) is numerically unstable for this differential equation.

The solution $z_2{}^n$ is called the *parasitic solution* of the difference equation (11.120). It appears because we have replaced a *first* order differential equation by a *second* order difference equation. In this case the parasitic solution "swamps" the solution we would like to determine.

Even if we could use exact arithmetic with the mid-point rule, we would not avoid numerical instability. We need starting values y_0 and y_1, and y_1 must be obtained by some other method. Let us suppose that this yields the exact solution $y(x_1)$. Clearly, with

$$y_1 = y(x_1) = e^{-Ah} \neq z_1,$$

we do not obtain $c_1 = 1$ and $c_2 = 0$ in (11.121).

Example 11.19   The initial value problem

$$y'(x) = -3y(x) \quad \text{with } y(0) = 1$$

has solution $y(x) = e^{-3x}$. We use the mid-point rule (11.119) with $h = 0.1$ and starting values

$$y_0 = y(0) = 1 \quad \text{and} \quad y_1 = 0.7408.$$

The last value is $y(h) = e^{-3h}$ correct to four decimal places. Results are tabulated in Table 11.4. The oscillatory behaviour for $x_n > 1$ is typical of unstable processes and is due to the $(-1)^n$ factor in the second term of (11.122). ☐

The mid-point rule is thus at a major disadvantage when used for problems with exponentially decreasing solutions and will not produce accurate results even though it has a smaller truncation error (see Problem 11.31) than the second order Adams–Bashforth algorithm, which we consider next.

Table 11.4   An unstable process.

x_n	y_n
0	1
0.1	0.7408
0.2	0.5555
0.3	0.4075
0.4	0.3110
0.5	0.2209
0.6	0.1785
0.7	0.1138
0.8	0.1102
0.9	0.0477
1.0	$+0.0816$
1.1	-0.0013
1.2	$+0.0824$
1.3	-0.0507
1.4	$+0.1128$

From

$$y_{n+1} = y_n + \frac{h}{2}(3f_n - f_{n-1})$$

with $f(x, y) = -Ay$ ($A > 0$, as before), we obtain the recurrence relation

$$y_{n+1} = (1 - \tfrac{3}{2}Ah)y_n + \tfrac{1}{2}Ahy_{n-1}, \qquad (11.123)$$

which has characteristic polynomial

$$z^2 - (1 - \tfrac{3}{2}Ah)z - \tfrac{1}{2}Ah.$$

The roots are

$$\left.\begin{array}{c} z_1 \\ z_2 \end{array}\right\} = \tfrac{1}{2}[(1 - \tfrac{3}{2}Ah) \pm (1 - Ah + \tfrac{9}{4}A^2h^2)^{1/2}].$$

On expanding the square root, we obtain

$$z_1 = \tfrac{1}{2}[(1 - \tfrac{3}{2}Ah) + 1 + \tfrac{1}{2}(-Ah + \tfrac{9}{4}A^2h^2) - \tfrac{1}{8}(-Ah + \tfrac{9}{4}A^2h^2)^2 + O(h^3)]$$

$$= 1 - Ah + \tfrac{1}{2}A^2h^2 + O(h^3)$$

$$= e^{-Ah}(1 + O(h^3)).$$

Thus

$$z_1^n = e^{-Anh}(1 + O(h^3))^n$$

$$= e^{-Ax_n}(1 + O(h^2)),$$

so that, for small h, z_1^n is an approximation to $y(x_n)$.

We seek the solution of (11.123) of the form

$$y_n = z_1^n$$

but, in practice, obtain

$$y_n^* = (1 + \delta_1)z_1^n + \delta_2 z_2^n.$$

Now

$$\frac{z_2}{z_1} = \frac{1 - \tfrac{3}{2}Ah - (1 - Ah + \tfrac{9}{4}A^2h^2)^{1/2}}{1 - \tfrac{3}{2}Ah + (1 - Ah + \tfrac{9}{4}A^2h^2)^{1/2}}.$$

The quantity under the square root is positive and, therefore, we have

$$\left|\frac{z_2}{z_1}\right| \leqslant 1 \quad \text{for } 1 - \tfrac{3}{2}Ah \geqslant 0.$$

Thus

$$\left| \frac{y_n^* - y_n}{y_n} \right| = \left| \delta_1 + \delta_2 \left(\frac{z_2}{z_1} \right)^n \right|$$

is bounded, and the method is numerically stable, for $h \leqslant 2/(3A)$. We will certainly also need to satisfy this inequality if the truncation error is not to be excessive and, therefore, this is not a severe restriction for the differential equation $y' = -Ay$.

Similar results may also be obtained for the equation

$$y' = -Ay + B, \tag{11.124}$$

where $A > 0$ and B are constants. The constant B will not affect the homogeneous part of the difference approximations to (11.124) and it is only necessary to add the particular solution $y_n = B/A$, to the general solution of the homogeneous equations (11.120) and (11.123). As $h \to 0$, with $nh = x_n$ fixed, this particular solution converges to $y(x) = B/A$, a particular solution of (11.124).

For both of the methods investigated in this section, we would like the solution

$$y_n = z_1^n + B/A,$$

but actually obtain

$$y_n^* = (1 + \delta_1)z_1^n + \delta_2 z_2^n + B/A.$$

Thus

$$\left| \frac{y_n^* - y_n}{y_n} \right| = \left| \frac{\delta_1 z_1^n + \delta_2 z_2^n}{z_1^n + B/A} \right|.$$

Precise stability conditions depend on A and B, but certainly if the method is stable for (11.118), it is stable for (11.124).

If we have $y' = -Ay$ with $A < 0$, so that the solution e^{-Ax} is exponentially *increasing*, the parasitic solutions introduced in a difference approximation may decay exponentially and therefore be unimportant. For example, for the mid-point rule, the second (parasitic) term of (11.122) is exponentially decreasing and we have

$$|z_2| = |Ah + (1 + A^2h^2)^{1/2}| < |Ah - (1 + A^2h^2)^{1/2}| = |z_1|,$$

so that the method is stable.

We now show that the stability behaviour of approximations to the general equation

$$y' = f(x, y) \tag{11.125}$$

will to a certain extent be similar to those for the special problem (11.118). On expanding $f(x, y)$, using Taylor series at the point $x = X$, $y = Y$, we obtain

$$f(x, y) = f(X, Y) + (y - Y)f_y(X, Y) + O(x - X) + O(y - Y)^2$$
$$\simeq yf_y(X, Y) + [f(X, Y) - Yf_y(X, Y)]$$

near the point (X, Y). Thus locally (that is, over a small range of values of x) the problem (11.125) will behave like one of the form (11.124). Alternatively, we may argue that one would need to be rather optimistic to expect success in computing a decreasing solution of (11.125) by a method which is unstable for the particular case (11.118).

There are no stability difficulties of the type described so far if first order difference equations are used, as there are no parasitic solutions in such cases. However, there is another type of stability problem which can arise even for first order difference equations. If Euler's method is applied to (11.118), we obtain

$$y_{n+1} = y_n - Ahy_n$$
$$= (1 - Ah)y_n,$$

so that

$$y_n = (1 - Ah)^n y_0.$$

For $A > 0$, $y(x) = y_0 e^{-Ax} \to 0$ as $x \to \infty$ and we shall therefore expect $y_n \to 0$ as $n \to \infty$ with *fixed h*. This last property is called *absolute stability*. For Euler's method, we can see that $y_n \to 0$ only if $|1 - Ah| < 1$, that is, $h < 2/A$. In practice, for (11.118), this restriction on h should normally be satisfied for accuracy reasons even when A is large. For a large value of A, the exact solution decreases very rapidly and a small step size would be essential. However, consider the equation

$$y' = -A(y - 1)$$

which has solution $y(x) = ce^{-Ax} + 1$ where c is arbitrary. Euler's method is

$$y_{n+1} = y_n - Ah(y_n - 1)$$

so that

$$y_n = c_1(1 - Ah)^n + 1$$

where c_1 is arbitrary. Now $y(x) \to 1$ as $x \to \infty$, but $y_n \to 1$ as $n \to \infty$ with h fixed, only if $|1 - Ah| < 1$, that is, $h < 2/A$. This condition may be severe if A is very large. If the initial value $y(0)$ is near $+1$, the solution will not vary rapidly for $x > 0$ and there should be no need to use a small step size.

There are similar restrictions for all the other methods introduced in this

book except the Adams–Moulton method of order 2. When this method is
applied to (11.118), we obtain

$$y_{n+1} = y_n + \frac{h}{2}(-Ay_n - Ay_{n+1})$$

so that

$$(1 + \tfrac{1}{2}Ah)y_{n+1} = (1 - \tfrac{1}{2}Ah)y_n.$$

Thus

$$y_n = \left(\frac{1 - \tfrac{1}{2}Ah}{1 + \tfrac{1}{2}Ah}\right)^n y_0$$

and $y_n \to 0$, as $n \to \infty$, for all $A > 0$ and fixed $h > 0$. Results for the Runge–
Kutta and Adams–Bashforth methods are included in Problem 11.34 and a
full discussion of stability is given by Lambert.

11.12 Systems and higher order equations

Suppose that

$$f_j = f_j(x, z_1, z_2, \ldots, z_k), \qquad j = 1, \ldots, k,$$

is a set of k real valued functions of $k + 1$ variables which are defined for
$a \leqslant x \leqslant b$ and all real $z_1, \ldots, z_k$. The simultaneous equations

$$\left.\begin{aligned}
y_1'(x) &= f_1(x, y_1(x), \ldots, y_k(x)) \\
y_2'(x) &= f_2(x, y_1(x), \ldots, y_k(x)) \\
&\;\vdots \\
y_k'(x) &= f_k(x, y_1(x), \ldots, y_k(x))
\end{aligned}\right\} \qquad (11.126)$$

where $x \in [a, b]$, are called a system of ordinary differential equations. Any
set of k differentiable functions†

$$y_1(x), \ldots, y_k(x)$$

satisfying (11.126) is called a solution.

In general, the solution of (11.126), if it exists, will not be unique unless we
are given k extra conditions. These usually take the form

$$y_j(x_0) = s_j, \qquad j = 1, \ldots, k, \qquad (11.127)$$

---

† Note that the subscripts denote different functions and not values or approximations
to values of a single function, as in earlier sections.

where the s_j are known and $x_0 \in [a, b]$. We are thus given the values of the $y_j(x)$ at the point $x = x_0$. The problem of solving (11.126) subject to (11.127) is again known as an initial value problem.

We write (11.126) and (11.127) in the form

$$\mathbf{y}' = \mathbf{f}(x, \mathbf{y}),$$
$$\mathbf{y}(x_0) = \mathbf{s}, \tag{11.128}$$

where $\mathbf{y}(x)$, $\mathbf{y}'(x)$, $\mathbf{f}(x, \mathbf{y})$ and $\mathbf{s}$ are k-dimensional vectors, whose ith components are $y_i(x)$, $y_i(x)$, $f_i(x, y_1, \ldots, y_k)$ and s_i respectively. The problem has a unique solution if $\mathbf{f}$ satisfies a Lipschitz condition of the form: there exists $L \geqslant 0$ such that, for all real vectors $\mathbf{y}$ and $\mathbf{z}$ of dimension k and all $x \in [a, b]$,

$$\|\mathbf{f}(x, \mathbf{y}) - \mathbf{f}(x, \mathbf{z})\|_\infty \leqslant L\|\mathbf{y} - \mathbf{z}\|_\infty. \tag{11.129}$$

The norm is defined in §9.1.

Numerical methods analogous to those for one equation may be derived in an identical manner. For the Taylor series method, we determine $\mathbf{f}^{(r)}(x, \mathbf{y})$ such that

$$\mathbf{y}(x_{n+1}) = \mathbf{y}(x_n) + h\mathbf{f}(x_n, \mathbf{y}(x_n)) + \frac{h^2}{2!} \mathbf{f}^{(1)}(x_n, \mathbf{y}(x_n)) + \cdots$$

$$+ \frac{h^p}{p!} \mathbf{f}^{(p-1)}(x_n, \mathbf{y}(x_n)) + h^{p+1}\mathbf{R}_{p+1}(x_n)$$

(cf. (11.35)), where $\mathbf{R}_{p+1}$ is a remainder term. For example,

$$\mathbf{f}^{(1)}(x, \mathbf{y}(x)) = \frac{d}{dx} \mathbf{f}(x, \mathbf{y}(x)) = \frac{\partial \mathbf{f}}{\partial x} + \sum_{j=1}^{k} \frac{dy_j(x)}{dx} \frac{\partial \mathbf{f}}{\partial y_j}$$

$$= \frac{\partial \mathbf{f}}{\partial x} + \sum_{j=1}^{k} f_j \frac{\partial \mathbf{f}}{\partial y_j}, \tag{11.130}$$

where the ith component of $\partial \mathbf{f}/\partial y_j$ is $\partial f_i/\partial y_j$. The formulae for $\mathbf{f}^{(r)}$ are very lengthy for $r \geqslant 2$. The Taylor series method of order two is thus

$$\mathbf{y}_0 = \mathbf{s},$$

$$\mathbf{y}_{n+1} = \mathbf{y}_n + h\mathbf{f}(x_n, \mathbf{y}_n) + \frac{h^2}{2!} \mathbf{f}^{(1)}(x_n, \mathbf{y}_n), \tag{11.131}$$

where $\mathbf{y}_n$ is a vector whose ith element $y_{i,n}$ is to be considered as an approximation to $y_i(x_n)$ and $x_n = x_0 + nh$.

We may derive Runge–Kutta methods by using Taylor series in several variables. Analogous to (11.46), we have the classical method

$$\mathbf{y}_{n+1} = \mathbf{y}_n + \frac{h}{6} [\mathbf{k}_1 + 2\mathbf{k}_2 + 2\mathbf{k}_3 + \mathbf{k}_4],$$

where

$$\mathbf{k}_1 = \mathbf{f}(x_n, \mathbf{y}_n), \quad \mathbf{k}_2 = \mathbf{f}(x_n + \tfrac{1}{2}h, \mathbf{y}_n + \tfrac{1}{2}h\mathbf{k}_1),$$

$$\mathbf{k}_3 = \mathbf{f}(x_n + \tfrac{1}{2}h, \mathbf{y}_n + \tfrac{1}{2}h\mathbf{k}_2), \quad \mathbf{k}_4 = \mathbf{f}(x_{n+1}, \mathbf{y}_n + h\mathbf{k}_3).$$

Predictor-corrector methods are also identical to those for single equations. We merely replace the y_n and f_n in (11.89) by vectors $\mathbf{y}_n$ and $\mathbf{f}_n$.

Global error bounds for all methods are very similar to those for a single equation. We replace the quantity $|y_n - y(x_n)|$ by $\|\mathbf{y}_n - \mathbf{y}(x_n)\|_\infty$.

Example 11.20   Consider the initial value problem

$$y_1'(x) = \quad xy_1(x) - y_2(x)$$
$$y_2'(x) = -y_1(x) + y_2(x)$$

with

$$y_1(0) = s_1, \qquad y_2(0) = s_2.$$

We have

$$\mathbf{f}(x, \mathbf{y}) = \begin{bmatrix} f_1(x, y_1, y_2) \\ f_2(x, y_1, y_2) \end{bmatrix} = \begin{bmatrix} (xy_1 - y_2) \\ (-y_1 + y_2) \end{bmatrix}$$

and

$$\mathbf{f}^{(1)}(x, \mathbf{y}) = \frac{\partial \mathbf{f}}{\partial x} + f_1 \frac{\partial \mathbf{f}}{\partial y_1} + f_2 \frac{\partial \mathbf{f}}{\partial y_2}$$

$$= \begin{bmatrix} y_1 \\ 0 \end{bmatrix} + (xy_1 - y_2)\begin{bmatrix} x \\ -1 \end{bmatrix} + (-y_1 + y_2)\begin{bmatrix} -1 \\ 1 \end{bmatrix}$$

$$= \begin{bmatrix} ((x^2 + 2)y_1 - (1 + x)y_2) \\ (-(1 + x)y_1 + 2y_2) \end{bmatrix}.$$

The Taylor series method of order 2 is

$$y_{1,0} = s_1, \qquad y_{2,0} = s_2,$$

$$y_{1,n+1} = y_{1,n} + h(x_n y_{1,n} - y_{2,n}) + \frac{h^2}{2}\{(x_n^2 + 2)y_{1,n} - (1 + x_n)y_{2,n}\},$$

$$y_{2,n+1} = y_{2,n} + h(-y_{1,n} + y_{2,n}) + \frac{h^2}{2}\{-(1 + x_n)y_{1,n} + 2y_{2,n}\},$$

for $n = 0, 1, 2, \ldots$.

The Adams–Moulton predictor-corrector method of order 2 (cf. (11.90)) is as follows.

Initial values: $y_{1,0} = s_1, \qquad y_{2,0} = s_2;$

Predictor: $\begin{cases} y_{1,n+1}^{(0)} = y_{1,n} + hf_{1,n}, \\ y_{2,n+1}^{(0)} = y_{2,n} + hf_{2,n}; \end{cases}$

Corrector: $\begin{cases} y_{1,n+1}^{(i+1)} = y_{1,n} + \dfrac{h}{2}(f_{1,n+1}^{(i)} + f_{1,n}), \\ \\ y_{2,n+1}^{(i+1)} = y_{2,n} + \dfrac{h}{2}(f_{2,n+1}^{(i)} + f_{2,n}), \end{cases}$ $\quad i = 0, 1, \ldots, (I-1);$

$$y_{1,n+1} = y_{1,n+1}^{(I)}, \qquad y_{2,n+1} = y_{2,n+1}^{(I)};$$

where $f_{j,n+1}^{(i)} = f_j(x_n, y_{1,n+1}^{(i)}, y_{2,n+1}^{(i)})$.

(The superfix i in the above is to distinguish different iterates and does not denote differentiation.) $\square$

If the real valued function

$$f = f(x, z_1, z_2, \ldots, z_m)$$

of m variables is defined for $x \in [a, b]$ and all real $z_1, \ldots, z_m$, we say that

$$y^{(m)}(x) = f(x, y(x), y'(x), \ldots, y^{(m-1)}(x)) \qquad (11.132)$$

is a differential equation of order m. If the solution $y(x)$ is to be unique we need m extra conditions. If these take the form

$$y^{(r)}(x_0) = t_r, \qquad r = 0, 1, \ldots, m-1, \qquad (11.133)$$

where $x_0 \in [a, b]$ and the t_r are given, we again say that we have an initial value problem. You will notice that we are given the values of y and its first $m-1$ derivatives at one point $x = x_0$. In Chapter 12 we will discuss problems with conditions involving more than one value of x.

We make the substitutions

$$z_j(x) = y^{(j-1)}(x), \qquad j = 1, \ldots, m,$$

so that

$$z_j'(x) = y^{(j)}(x) = z_{j+1}(x), \qquad j = 1, \ldots, m-1,$$

and thus rewrite the single equation (11.132) as the *system* of first order equations

$$\begin{aligned} z_1'(x) &= z_2(x) \\ z_2'(x) &= z_3(x) \\ &\vdots \\ z_{m-1}'(x) &= z_m(x) \\ z_m'(x) &= f(x, z_1(x), z_2(x), \ldots, z_m(x)). \end{aligned}$$

The conditions (11.133) become

$$z_1(x_0) = t_0, \qquad z_2(x_0) = t_1, \ldots, \qquad z_m(x_0) = t_{m-1},$$

and we have an initial value problem of the type considered at the beginning of this section.

Example 11.21 Consider

$$y'' = g(x, y, y')$$

with

$$y(x_0) = t_0, \qquad y'(x_0) = t_1.$$

Let

$$z_1(x) = y(x), \qquad z_2(x) = y'(x),$$

when

$$z_1'(x) = z_2(x),$$
$$z_2'(x) = g(x, z_1(x), z_2(x)).$$

Euler's method is

$$\mathbf{z}_0 = \mathbf{s},$$
$$\mathbf{z}_{n+1} = \mathbf{z}_n + h\mathbf{f}(x_n, \mathbf{z}_n).$$

For this example, we obtain

$$z_{1,0} = t_0, \qquad z_{2,0} = t_1,$$
$$z_{1,n+1} = z_{1,n} + hz_{2,n},$$
$$z_{2,n+1} = z_{2,n} + hg(x_n, z_{1,n}, z_{2,n}). \quad \square$$

There are also some special methods for higher order differential equations, particularly second order equations. See Henrici [1962].

We have already seen in §11.11 that for Euler's method applied to $y' = -Ay$ with $A > 0$, we obtain $y_n \to 0$ as $n \to \infty$ with h fixed, only if $h < 2/A$. Now suppose that we have a linear higher order equation or a linear system of equations and that one component of the general solution vector has the form

$$y(x) = c_1 e^{-A_1 x} + c_2 e^{-A_2 x} + \cdots + c_m e^{-A_m x} \qquad (11.134)$$

where $A_r > 0$, $r = 1, \ldots, m$. It can be shown that with Euler's method we need to choose

$$h < 2/A_r \quad \text{for } r = 1, \ldots, m, \qquad (11.135)$$

if the solution of the difference equation is to decrease. If $0 < A_1 \leqslant A_2 \leqslant A_3 \leqslant \cdots \leqslant A_m$ in (11.134) then, for $c_1 \neq 0$, the first term will be the dominant term for large x. The inequalities (11.135) may impose a very severe restriction on h if A_m is very much larger than A_1. We illustrate this by the following example.

Example 11.22   Given

$$y'(x) = -8y(x) + 7z(x)$$
$$z'(x) = 42y(x) - 43z(x)$$

(11.136)

with $y(0) = 1$ and $z(0) = 8$,
the solution is

$$y(x) = 2e^{-x} - e^{-50x}$$
$$z(x) = 2e^{-x} + 6e^{-50x}.$$

For large $x > 0$, the dominant term in this solution is $2e^{-x}$ but we must choose $h < 0.04$ if Euler's method is to produce a decreasing solution. $\square$

Equations such as those of Example 11.22 are called *stiff* equations. There are similar difficulties with all the methods we have discussed except the Adams-Moulton method of order 2 which, as we saw in §11.11, is always absolutely stable. The restrictions on h given in Problem 11.34 for the Runge–Kutta and second order Adams–Bashforth methods must be satisfied for each A_r in (11.134). Lambert gives a full discussion of the solution of stiff equations.

11.13 Notes comparing step-by-step methods

The types considered are as follows.

(a) Taylor series methods.
(b) Runge-Kutta methods.
(c) Explicit methods, for example, the Adams–Bashforth algorithm (11.71).
(d) Predictor-corrector methods, for example, the Adams–Moulton algorithm (11.88) and the algorithm (11.106).

Ease of use

The main consideration is how simple they are to program for a digital computer.

(a) Difficult to establish suitable formulae, except for low order methods. No special starting procedure.
(b) Very simple. No special starting procedure.
(c) Quite straightforward, but special starting procedure required.
(d) Slightly difficult to implement, due to inner iterations, unless only one correction is made per step. Special starting procedure required.

Amount of computation

We usually measure this by the number of evaluations of f that are needed for each step. If f is at all complicated, most of the calculation time will be spent on evaluations of f.

(a) Many evaluations of f and derivatives required per step, for example, six evaluations for the third order method.

(b) Comparatively many evaluations of f, for example, four for the fourth order method. For the commonly used methods, the number of evaluations equals the order of the method.

(c) One evaluation of f, regardless of the order.

(d) Two or three evaluations of f, regardless of the order.

Truncation error

The truncation error gives a measure of the accuracy of a method for a range of problems but does not necessarily provide a correct comparison of the accuracy of methods for one particular problem. We consider the magnitude of the truncation error for a *given* order.

(a) Not very good.

(b) May be quite good.

(c) Much poorer than (d).

(d) Good.

Error estimation

(a) Difficult.

(b) Difficult, though there are some methods which give some indication of truncation errors.

(c) Difficult.

(d) Fairly easy. See §11.10 for the algorithm (11.106).

Numerical stability

By this we mean stability of the type based on considering the relative errors (11.116).

(a) Stable.

(b) Stable.

(c) Some methods are unstable, but there are satisfactory stable methods, including the Adams–Bashforth algorithm.

(d) Some methods are unstable, but there are satisfactory stable methods, including the Adams–Moulton algorithm.

Stiff equations

For all the methods introduced in this chapter, except the Adams–Moulton algorithm of order 2, the step size must be restricted for absolute stability and hence there are difficulties for stiff equations.

Step size adjustment

It is sometimes desirable to adjust the step size as the solution proceeds. This is particularly necessary if there are regions in which the solution has large derivatives, in which a small step size is required, and other regions in which derivatives are small, so that a large step size may be employed.

(a) Very easy to change the step size, as these are one-step methods.
(b) As (a).
(c) Quite difficult, as special formulae are required.
(d) As (c).

Choice of method

(a) Not recommended unless the problem has some special features, for example if f and its derivatives are known from other information.

(b) Recommended for a quick calculation for which there is little concern about accuracy. Also recommended to obtain starting values for (c) and (d).

(c) Only recommended for problems for which f is particularly difficult to evaluate.

(d) Recommended (particularly (11.106)) for most problems as truncation error estimation is possible. The fourth order predictor-corrector algorithm (11.112) is probably one of the best methods, except for stiff equations. It involves comparatively few function evaluations (for one or two inner iterations) and provides truncation error estimates. There are also adaptations of the method which reduce the number of function evaluations by not computing the final $f_{n+1}^{(I)}$ at the end of each step.

Problems

Section 11.1

11.1 Show that for $0 \leqslant x \leqslant 1$

$$y' = xy \quad \text{with } y(0) = 1$$

has a unique solution and find an approximate solution by Picard iteration.

11.2   Show that

$$y' = \sin y \quad \text{with } y(x_0) = s,$$

has a unique solution on any interval containing x_0.

11.3   Given that $g(x)$ and $h(x)$ are continuous functions on some interval $[a, b]$ containing x_0, show that the linear differential equation

$$y' = g(x).y + h(x) \quad \text{with } y(x_0) = s,$$

has a unique solution on $[a, b]$.
(*Hint:* use the fact that $g(x)$ is bounded on $[a, b]$.)

Section 11.2

11.4   Find the general solution of the homogeneous difference equations:

(a) $y_{n+1} - ay_n = 0$, where a is independent of n;
(b) $y_{n+2} - 4y_{n+1} + 4y_n = 0$;
(c) $y_{n+3} - 2y_{n+2} - y_{n+1} + 2y_n = 0$;
(d) $y_{n+3} - 5y_{n+2} + 8y_{n+1} - 4y_n = 0$;
(e) $y_{n+3} + 3y_{n+2} + 3y_{n+1} + y_n = 0$.

11.5   Find the general solutions of the difference equations:

(a) $y_{n+2} - 4y_{n+1} + 4y_n = 1$;
(b) $y_{n+2} + y_{n+1} - 2y_n = 1$;
(c) $y_{n+2} - 2y_{n+1} + y_n = n + 1$.

(*Hint:* find particular solutions by trying solutions which are polynomials in n, for example, $y_n = k_0$, $y_n = k_0 + k_1 n$, $y_n = k_0 + k_1 n + k_2 n^2$, where the k_i are independent of n.)

Section 11.3

11.6   Derive an explicit formula for the Taylor series method (11.36) with $p = 3$ for the differential equation of Problem 11.2.

11.7   The initial value problem,

$$y' = ax + b \quad \text{with } y(0) = 0,$$

has solution

$$y(x) = \tfrac{1}{2}ax^2 + bx.$$

If Euler's method is applied to this problem, show that the resulting difference equation has solution

$$y_n = \tfrac{1}{2}ax_n^2 + bx_n - \tfrac{1}{2}ahx_n,$$

where $x_n = nh$, and thus that

$$y(x_n) - y_n = \tfrac{1}{2}ahx_n.$$

11.8   Show that, for any choice of v, the "Runge–Kutta" method

$$y_{n+1} = y_n + \frac{h}{2}[k_2 + k_3], \tag{11.137}$$

$$k_1 = f(x_n, y_n), \qquad k_2 = f(x_n + vh, y_n + vhk_1),$$
$$k_3 = f(x_n + [1 - v]h, y_n + [1 - v]hk_1),$$

is consistent of order 2. That is, show that the first three terms in the Taylor series expansion of (11.137) agree with the first three terms ($p = 2$) in (11.35).

11.9   Show that the "Runge–Kutta" method

$$y_{n+1} = y_n + h[\tfrac{2}{9}k_1 + \tfrac{1}{3}k_2 + \tfrac{4}{9}k_3],$$
$$k_1 = f(x_n, y_n), \qquad k_2 = f(x_n + \tfrac{1}{2}h, y_n + \tfrac{1}{2}k_1h),$$
$$k_3 = f(x_n + \tfrac{3}{4}h, y_n + \tfrac{3}{4}k_2h),$$

is consistent of order 3. (Consider four terms in the Taylor series expansion.)

11.10   Solve the equation

$$y' = x^2 + x - y \quad \text{with } y(0) = 0$$

by the simple and classical Runge–Kutta methods to find an approximation to $y(0.6)$ using a step size $h = 0.2$. Compare your solution with the exact solution

$$y(x) = -e^{-x} + x^2 - x + 1.$$

Section 11.5

11.11   Show that a suitable value of the Lipschitz constant L_ϕ, of (11.58), for the method of Problem 11.8 is

$$L_\phi = \tfrac{1}{2}L(2 + [\,|v| + |1 - v|\,]Lh_0),$$

where L is the usual Lipschitz constant for f. Deduce that the method is convergent of order 2.

11.12   Show that a suitable value of the Lipschitz constant L_ϕ, of (11.58), for the method of Problem 11.9 is

$$L_\phi = L + h_0\frac{L^2}{2!} + h_0{}^2\frac{L^3}{3!}.$$

11.13   Use Theorem 11.2 to find a global error bound for Euler's method applied to the equation given in Problem 11.10 with $0 \leqslant x \leqslant 1$.

11.14 Find a global error bound for the Taylor series method of order 2 applied to the initial value problem of Examples 11.10 and 11.13 with $0 \leqslant x \leqslant 1$. (*Hint:* show first that $|f^{(2)}(x, y)| \leqslant 2$.)

11.15 Find a global error bound for Euler's method applied to the differential equation of Problem 11.2 with $x_0 = 0$, $h = 0.1$ and $0 \leqslant x \leqslant 1$.

Section 11.7

11.16 The Taylor polynomial of degree $p - 1$ for $f(x, y(x))$ constructed at $x = x_n$ is

$$q(x) = f(x_n, y(x_n)) + \frac{(x - x_n)}{1!} f^{(1)}(x_n, y(x_n)) + \cdots$$

$$+ \frac{(x - x_n)^{p-1}}{(p - 1)!} f^{(p-1)}(x_n, y(x_n)),$$

where the $f^{(r)}$ are as defined in §11.3. Thus

$$\int_{x_n}^{x_{n+1}} q(x)\, dx = hf(x_n, y(x_n)) + \cdots + \frac{h^p}{p!} f^{(p-1)}(x_n, y(x_n)).$$

Show that, on approximating to the integral in (11.67) by the above integral of q and replacing $y(x_n)$ by y_n, we obtain the Taylor series algorithm for an initial value problem.

11.17 Determine the third order Adams–Bashforth algorithm in the form (11.72), that is, expand the differences.

11.18 Determine an approximation to $y(1.0)$, where $y(x)$ is the solution of

$$y' = 1 - y \quad \text{with } y(0) = 0,$$

using the Adams–Bashforth algorithm of order 2 with $h = 0.2$ and starting values $y_0 = 0$, $y_1 = 0.181$ ($= y(0.2)$ correct to 3 decimal places). Compare your calculated solution with

$$y(x) = 1 - e^{-x}.$$

11.19 Evaluate the error bound (11.81) for the initial value problem and method of Problem 11.18. Compare the global error bound with the actual error for $x_n = 1.0$. (Initial error bound, $\delta = 0.0005$.)

11.20 Obtain a global error bound for the Adams–Bashforth algorithm of order 2 applied to the equation in Problem 11.2. (*Hint:* show that $f^{(2)}(x, y) = \sin y \cos 2y$ has maximum modulus $+1$.)

11.21 Derive explicit forms of the Nyström method (11.84) for $m = 0, 1, 2$. (Expand the differences.)

Section 11.8

11.22 Determine the third order Adams–Moulton algorithm ($m = 1$) in the form (11.89).

11.23 Show that if a predictor-corrector method is used to solve the linear differential equation of Problem 11.3, then y_{n+1} may be found *explicitly* from the corrector formula.

11.24 Use the Adams–Moulton algorithm of order 2 with $h = 0.2$ to determine an approximation to $y(1.0)$, where $y(x)$ is the solution of the differential equation of Problem 11.18. (Note the result of Problem 11.23.)

11.25 Evaluate the global error bound (11.97) at $x_n = 1.0$ for the calculation of Problem 11.24. Compare the actual error with this error bound.

11.26 Obtain a global error bound for the Adams–Moulton method of order 2 applied to the equation in Problem 11.2. (Compare with the result of Problem 11.20 which also contains a hint.)

11.27 Show that

$$\left| \int_0^1 \binom{-s}{j} ds \right| > \left| \int_0^1 \binom{s}{j} ds \right|, \quad \text{for } j \geq 2,$$

and thus that $|b_j| > |c_j|$ for $j \geq 2$, where b_j and c_j are defined by (11.70) and (11.86) respectively.

11.28 Corrector formulae may be obtained by approximating to (11.83), using q_{m+1} as in §11.8. Show that, with both $m = 1$ and $m = 2$, there results the formula

$$y_{n+1} = y_{n-1} + \frac{h}{3} [f_{n-1} + 4f_n + f_{n+1}].$$

This is called the Milne–Simpson formula and it reduces to Simpson's rule if f is independent of y.

Section 11.10

11.29 The predictor-corrector formulae

$$y_{n+1}^{(0)} = y_n + \tfrac{1}{2}h(3f_n - f_{n-1}),$$
$$y_{n+1}^{(i+1)} = y_n + \tfrac{1}{2}h(f_n + f_{n+1}^{(i)}),$$

obtained from (11.106) with $m = 0$, are both of order 2. Show that the truncation error of the corrector as given by (11.108) is

$$T_n = -\frac{1}{12} h^2 y^{(3)}(\xi_n)$$

and that

$$T_n \simeq -\frac{1}{6h}(y^{(1)}_{n+1} - y^{(0)}_{n+1}).$$

Discuss the choice of h if only one correction is to be made for each step.

Section 11.11

11.30 Investigate the numerical stability of the following methods for the differential equation $y' = -Ay$ with $A > 0$.

(i) $y_{n+1} = y_n + \dfrac{h}{12}[5f_{n+1} + 8f_n - f_{n-1}]$

    (Adams–Moulton corrector of order 3.)

(ii) $y_{n+1} = y_{n-1} + \dfrac{h}{3}[f_{n+1} + 4f_n + f_{n-1}]$

    (Milne–Simpson formula. See Problem 11.28.)

(iii) $y_{n+1} = y_{n-1} + \dfrac{h}{2}[f_n + 3f_{n-1}].$

Show that for (iii) the absolute errors, $|y_n - y^*_n|$, (notation as in §11.11) will decrease for sufficiently small h, even though the method is unstable.

11.31 Show that the truncation error of the mid-point rule (11.119) is $\frac{1}{6}h^2 y'''(\xi_n)$.

11.32 Show that the method

$$y_{n+1} = y_{n-1} + \frac{h}{2}(f_{n+1} + 2f_n + f_{n-1})$$

is numerically unstable for the equation $y' = -Ay$ and stable for $y' = -Ay + B$, where $A > 0$ and $B \neq 0$.

11.33 The equation $y' = -Ay + B$ has general solution $y(x) = ce^{-Ax} + B/A$ where c is arbitrary and thus $y(x) \to B/A$ as $x \to \infty$. If Euler's method is applied to this equation, show that $y_n \to B/A$ as $n \to \infty$ with the step size h fixed, only if $h < 2/A$.

11.34 The equation $y' = -Ay$ with $A > 0$ is to be solved. Show that $y_n \to 0$ as $n \to \infty$ with the step size h fixed (and thus there is absolute stability), only if

(a) $|1 - Ah + \frac{1}{2}A^2 h^2| < 1$, for the simple Runge–Kutta method;

(b) $\left| 1 - Ah + \dfrac{1}{2!} A^2h^2 - \dfrac{1}{3!} A^3h^3 + \dfrac{1}{4!} A^4h^4 \right| < 1$, for the classical Runge–Kutta method;

(c) $h < 1/A$, for the Adams–Bashforth method of order 2.

Section 11.12

11.35 Show that, for the function **f** of the equations in Example 11.20 with $x \in [0, b]$ where $b > 1$, a suitable choice of L, the Lipschitz constant in (11.129), is $L = 1 + b$.

11.36 Determine, as in Example 11.20, the difference equations for the Taylor series method of order 2 applied to the two simultaneous equations

$$y_1'(x) = x^2 . y_1(x) - y_2(x)$$
$$y_2'(x) = - y_1(x) + xy_2(x)$$

with $y_1(0) = s_1$ and $y_2(0) = s_2$.

11.37 Solve the equation

$$y'' + 4xy'y + 2y^2 = 0$$

with $y(0) = 1$, $y'(0) = 0$, using Euler's method with $h = 0.1$, to obtain approximations to $y(0.5)$ and $y'(0.5)$.

11.38 Determine the difference equations for the classical Runge–Kutta method applied to the differential equation of Problem 11.37.

11.39 Derive the difference equations for the Taylor series method of order 2 applied to

$$y''' = 2y'' + x^2y + 1 + x,$$

assuming that $y(x)$, $y'(x)$ and $y''(x)$ are known for $x = x_0$.

11.40 Determine the difference equations for the Adams–Moulton predictor-corrector method of order 2 applied to the differential equation in Problem 11.39.

Chapter 12

Boundary Value and Other Methods for Ordinary Differential Equations

12.1 Shooting method for boundary value problems

Two associated conditions are required for the second order equation

$$y'' = f(x, y, y') \tag{12.1}$$

if the solution y is to be unique. If these two conditions are imposed at two distinct values of x, we call the differential equation and conditions a *boundary value* problem. The conditions are called *boundary* conditions. The simplest type are

$$y(a) = \alpha \quad \text{and} \quad y(b) = \beta, \tag{12.2}$$

where $a < b$ and α and β are known. Thus $y(x)$ is given at the distinct points $x = a$ and $x = b$. For such problems, we seek $y(x)$ for $x \in [a, b]$.

We may adapt the step-by-step methods employed for initial value problems to boundary value problems. Suppose that we have an estimate w of $y'(a)$, where $y(x)$ is the solution of (12.1) subject to (12.2). We now solve the initial value problem (12.1) with

$$y(a) = \alpha$$
$$y'(a) = w$$

using a step-by-step method and suppose that β_w is the resulting approximation to $y(b)$. It is most likely that $\beta_w \neq \beta$. We adjust w and repeat the calculation to try to produce $\beta_w = \beta$. Fig. 12.1 describes a typical problem. In the upper curve, w is too large and in the lower too small. The broken line is the required solution. We call the process of starting from one end, $x = a$, to try to produce the correct value at the other end, $x = b$, a "shooting" method.

Of course we should adjust w in a systematic fashion. It is clear that

$$\beta_w = F(w)$$

for some function $F(w)$ and we are therefore trying to find w such that

$$F(w) = \beta.$$

This is an algebraic equation and the bisection, regula falsi and Muller methods are suitable for its solution. (See Chapter 7.)

More general boundary conditions than (12.2) are

$$p_1 y(a) + q_1 y'(a) = \alpha_1, \tag{12.3a}$$
$$p_2 y(b) + q_2 y'(b) = \alpha_2, \tag{12.3b}$$

where p_i, q_i and α_i, $i = 1, 2$, are constants. The shooting method may be adapted to these conditions by choosing approximations to $y(a)$ and $y'(a)$

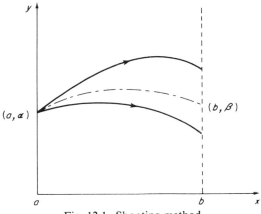

Fig. 12.1 Shooting method.

which satisfy (12.3a) exactly and calculating approximations to $y(b)$ and $y'(b)$ by a step-by-step method. These approximations will probably not satisfy (12.3b), so we adjust the approximations to $y(a)$ and $y'(a)$ and repeat the calculation.

12.2 Boundary value method

An alternative to step-by-step methods, when solving a boundary value problem, is to make difference approximations to derivatives and impose boundary conditions on the solution of the resulting difference equation. We illustrate the method first by considering the linear second order equation on $[a, b]$,

$$u(x)y'' + v(x)y' + w(x)y = f(x), \tag{12.4}$$

where u, v, w and f are continuous on $[a, b]$. We are also given that $u(x) > 0$ and $w(x) \leqslant 0$ on $[a, b]$. (Not all of these restrictions on u, v, w and f are necessary but they do simplify our analysis.) The given boundary conditions are again taken to be

$$y(a) = \alpha, \qquad y(b) = \beta \tag{12.5}$$

and we will assume that (12.4) and (12.5) have a unique solution $y(x)$ on $[a, b]$.

We divide the interval $[a, b]$ into $N + 1$ intervals each of length h and seek approximations y_n to the solution $y(x_n)$ at the points $x_n = a + nh$, $n = 1$, $2, \ldots, N$. We choose the mesh length $h = (b - a)/(N + 1)$ so that $x_{N+1} = b$. We approximate to (12.4) at the point x_n by replacing derivatives by differences as in (6.14) and (6.17) and so obtain

$$u_n\left(\frac{y_{n+1} - 2y_n + y_{n-1}}{h^2}\right) + v_n\left(\frac{y_{n+1} - y_{n-1}}{2h}\right) + w_n y_n = f_n, \tag{12.6}$$

$$n = 1, 2, \ldots, N,$$

where $u_n = u(x_n)$, $v_n = v(x_n)$, $w_n = w(x_n)$ and $f_n = f(x_n)$. Corresponding to (12.5) we also have

$$y_0 = \alpha, \qquad y_{N+1} = \beta. \tag{12.7}$$

Equations (12.6) are N linear equations in the N unknowns $y_1, y_2, \ldots, y_N$ and may be written in the form

$$\mathbf{Ay} = \mathbf{f}, \tag{12.8}$$

where $\mathbf{A}$ is the $N \times N$ tridiagonal matrix

$$\mathbf{A} = \begin{bmatrix} \left(\dfrac{-2u_1}{h^2} + w_1\right) & \left(\dfrac{u_1}{h^2} + \dfrac{v_1}{2h}\right) & & & 0 \\ \left(\dfrac{u_2}{h^2} - \dfrac{v_2}{2h}\right) & \left(\dfrac{-2u_2}{h^2} + w_2\right) & \left(\dfrac{u_2}{h^2} + \dfrac{v_2}{2h}\right) & & \\ & \ddots & \ddots & \ddots & \\ & & & \ddots & \left(\dfrac{u_{N-1}}{h^2} + \dfrac{v_{N-1}}{2h}\right) \\ 0 & & & \left(\dfrac{u_N}{h^2} - \dfrac{v_N}{2h}\right) & \left(\dfrac{-2u_N}{h^2} + w_N\right) \end{bmatrix},$$

$$\mathbf{y} = \begin{bmatrix} y_1 \\ y_2 \\ \cdot \\ \cdot \\ \cdot \\ y_N \end{bmatrix} \quad \text{and} \quad \mathbf{f} = \begin{bmatrix} f_1 - \left(\dfrac{u_1}{h^2} - \dfrac{v_1}{2h}\right)\alpha \\ f_2 \\ f_3 \\ \vdots \\ f_{N-1} \\ f_N - \left(\dfrac{u_N}{h^2} + \dfrac{v_N}{2h}\right)\beta \end{bmatrix}.$$

Because of the tridiagonal nature of **A**, the equations (12.8) may easily be solved by an elimination method, as was seen in §8.12. It can be shown that if h is sufficiently small the equations are non-singular. Problem 12.5 gives a proof of this for the case $v(x) \equiv 0$.

If the boundary conditions are of the form (12.3) with $q_1 \neq 0$ and $q_2 \neq 0$ so that derivatives are involved, we must make an approximation to (12.3) using (6.14). At the end $x = a$ we introduce the point $x_{-1} = a - h$ and a "fictitious" value y_{-1} at this point.† We replace (12.3a) by

$$p_1 y_0 + q_1\left(\frac{y_1 - y_{-1}}{2h}\right) = \alpha_1, \qquad (12.9)$$

which on rearrangement yields

$$y_{-1} = \frac{2h(p_1 y_0 - \alpha_1)}{q_1} + y_1. \qquad (12.10)$$

Since y_0 is unknown we must also use (12.6) with $n = 0$, the difference approximation to (12.4) at $x = a$. We use (12.10) to eliminate the y_{-1} introduced and obtain

$$\left(\frac{-2u_0}{h^2} + w_0 + \frac{2hp_1}{q_1}\left(\frac{u_0}{h^2} - \frac{v_0}{2h}\right)\right)y_0 + \frac{2}{h^2}u_0 y_1 = f_0 + \frac{2h\alpha_1}{q_1}\left(\frac{u_0}{h^2} - \frac{v_0}{2h}\right),$$

$$(12.11)$$

where again $u_0 = u(x_0)$, etc. We can obtain a similar equation at x_{N+1} relating y_N and y_{N+1} by approximating to (12.3b) and using (12.6) with $n = N + 1$. We now have linear equations of the form

$$\mathbf{Bz} = \mathbf{g},$$

where

$$\mathbf{z} = \begin{bmatrix} y_0 \\ y_1 \\ \vdots \\ y_{N+1} \end{bmatrix}.$$

It can be seen from (12.6) and (12.11) that the $(N + 2) \times (N + 2)$ coefficient matrix **B** is again tridiagonal and therefore the equations are easily solved.

---

† We cannot say that y_{-1} is an approximation to $y(a - h)$ since the latter is not defined.

Example 12.1 We use the above method for the linear boundary value problem on $[0, 1]$,

$$(1 + x^2)y'' - xy' - 3y = 6x - 3 \tag{12.12}$$

with

$$y(0) - y'(0) = 1 \tag{12.13a}$$
$$y(1) = 2. \tag{12.13b}$$

We take $h = 0.2$ and seek y_n, $n = 0, 1, 2, 3, 4$, approximations to $y(x_n)$ where $x_n = 0.2n$.

We replace the boundary conditions (12.13) by

$$y_0 - \left(\frac{y_1 - y_{-1}}{0.4} \right) = 1 \tag{12.14}$$

and

$$y_5 = 2.$$

At $x = 0$ the approximation to (12.12) is

$$\frac{y_1 - 2y_0 + y_{-1}}{(0.2)^2} - 3y_0 = -3$$

and, on multiplying by $(0.2)^2$ and substituting for y_{-1} using (12.14) we obtain

$$-2.52y_0 + 2y_1 = -0.52.$$

The difference approximation to (12.12) is

$$(1 + 0.04n^2)\left(\frac{y_{n+1} - 2y_n + y_{n-1}}{(0.2)^2} \right) - 0.2n\left(\frac{y_{n+1} - y_{n-1}}{0.4} \right) - 3y_n$$

$$= 1.2n - 3, \quad n = 1, 2, 3, 4.$$

We multiply these equations by $(0.2)^2$ and rearrange them to obtain the tridiagonal set

$$\begin{bmatrix} -2.52 & 2 & 0 & 0 & 0 \\ 1.06 & -2.20 & 1.02 & 0 & 0 \\ 0 & 1.20 & -2.44 & 1.12 & 0 \\ 0 & 0 & 1.42 & -2.84 & 1.30 \\ 0 & 0 & 0 & 1.72 & -3.40 \end{bmatrix} \begin{bmatrix} y_0 \\ y_1 \\ y_2 \\ y_3 \\ y_4 \end{bmatrix} = \begin{bmatrix} -0.52 \\ -0.072 \\ -0.024 \\ +0.024 \\ -3.048 \end{bmatrix}.$$

The solution, correct to five significant decimal digits, is $y_0 = 1.0132$, $y_1 = 1.0167$, $y_2 = 1.0693$, $y_3 = 1.2188$, $y_4 = 1.5130$. The required solution of the differential equation is $y(x) = 1 + x^3$, so that

$$y(x_0) = 1, y(x_1) = 1.008, y(x_2) = 1.064, y(x_3) = 1.216, y(x_4) = 1.512. \quad \square$$

We now return to the non-linear problem

$$y'' = f(x, y, y') \tag{12.15}$$

with $y(a) = \alpha$, $y(b) = \beta$. We again seek approximations y_n to $y(x_n)$ at the N equally spaced points $x_n = a + nh$, $n = 1, 2, \ldots, N$, with $h = (b - a)/(N + 1)$. The usual approximation to (12.15) is

$$\frac{y_{n+1} - 2y_n + y_{n-1}}{h^2} = f\left(x_n, y_n, \frac{y_{n+1} - y_{n-1}}{2h}\right), \qquad n = 1, 2, \ldots, N, \tag{12.16}$$

with $y_0 = \alpha$ and $y_{N+1} = \beta$. If the boundary conditions are of the form (12.3), we introduce extra equations as in the linear case. Equations (12.16) form N *non-linear* equations in the unknowns $y_1, y_2, \ldots, y_N$. The solution must usually be found by an iterative method. For example, we could try keeping the right side in (12.16) fixed and solving the resulting linear equations. We use this solution to compute new values for the right side and then solve the linear equations again. However, the resulting iterative process will not always converge. A much more satisfactory approach is to use Newton's method, especially as each of the equations (12.16) involves at most three unknowns so that the Jacobian matrix is tridiagonal. (See Problem 12.7.) Henrici [1962] discusses in detail the convergence of Newton's process when f is independent of y'.

So far we have not considered the global errors $|y_n - y(x_n)|$ and the possible convergence of these to zero as the mesh length h is decreased. In this context, we also will assume that f is independent of y' and will investigate only problems of the form

$$y'' = f(x, y) \tag{12.17}$$

with

$$y(a) = \alpha \quad \text{and} \quad y(b) = \beta. \tag{12.18}$$

The simplest difference approximation is

$$\frac{y_{n+1} - 2y_n + y_{n-1}}{h^2} = f(x_n, y_n), \qquad n = 1, 2, \ldots, N, \tag{12.19}$$

with

$$y_0 = \alpha, \qquad y_{N+1} = \beta. \tag{12.20}$$

If $y^{(4)}(x)$ is continuous on $[a, b]$ we note from (6.25) that

$$\frac{y(x_{n+1}) - 2y(x_n) + y(x_{n-1})}{h^2} = f(x_n, y(x_n)) + \frac{h^2}{12} y^{(4)}(\xi_n) \tag{12.21}$$

where $\xi_n \in (x_{n-1}, x_{n+1})$. The last term in (12.21) is called the truncation error of the method (12.19). To obtain a bound on global errors, we need the following lemma, which is based on the so-called "maximum principle".

Lemma 12.1 If w_n, $n = 0, 1, \ldots, N + 1$, satisfy the difference equation

$$w_{n+1} - (2 + c_n)w_n + w_{n-1} = d_n, \qquad n = 1, 2, \ldots, N,$$

where $c_n \geqslant 0$ and $d_n \geqslant 0$, $n = 1, 2, \ldots, N$, and $w_0 \leqslant 0$ and $w_{N+1} \leqslant 0$ are given, then

$$w_n \leqslant 0, \qquad n = 1, 2, \ldots, N.$$

Proof For $n = 1, 2, \ldots, N$,

$$w_n = \frac{w_{n-1} + w_{n+1}}{2 + c_n} - \frac{d_n}{2 + c_n}$$

$$\leqslant \frac{w_{n-1} + w_{n+1}}{2}.$$

Hence w_n cannot exceed both of its neighbours, w_{n-1} and w_{n+1}, and, therefore, the maximum w_n must occur at one or both of the boundaries $n = 0$ and $n = N + 1$. Thus

$$w_n \leqslant w_0 \leqslant 0 \quad \text{or} \quad w_n \leqslant w_{N+1} \leqslant 0, \qquad 0 \leqslant n \leqslant N + 1. \quad \square$$

We will make the additional assumption that in (12.17) $f_y(x, y)$ exists and is continuous with $f_y(x, y) \geqslant 0$ for $x \in [a, b]$ and $-\infty < y < \infty$. Henrici [1962] shows that this condition is sufficient to ensure the existence of a solution of the boundary value problem.

Theorem 12.1 Given the boundary value problem (12.17) and (12.18), let y_n, $n = 0, 1, \ldots, N + 1$, satisfy (12.19) and (12.20). If

$$M = \max_{x \in [a,b]} |y^{(4)}(x)|$$

exists and $\partial f / \partial y \geqslant 0$ for $x \in [a, b]$, $-\infty < y < \infty$, the global error satisfies

$$|y(x_n) - y_n| \leqslant \frac{Mh^2}{24} (x_n - a)(b - x_n). \tag{12.22}$$

Proof It is easily verified by direct substitution that the solution of

$$z_{n+1} - 2z_n + z_{n-1} = \frac{h^4 M}{12}, \qquad n = 1, 2, \ldots, N, \tag{12.23}$$

with

$$z_0 = z_{N+1} = 0,$$

is

$$z_n = \frac{-h^2 M}{24} (x_n - a)(b - x_n). \tag{12.24}$$

From (12.21) we have

$$y(x_{n+1}) - 2y(x_n) + y(x_{n-1}) = h^2 f(x_n, y(x_n)) + \frac{h^4}{12} y^{(4)}(\xi_n).$$

On subtracting $h^2 \times$ (12.19), putting $e_n = y(x_n) - y_n$ and using the mean value theorem, we obtain

$$e_{n+1} - 2e_n + e_{n-1} = h^2 g_n e_n + \frac{h^4}{12} y^{(4)}(\xi_n), \qquad (12.25)$$

where $g_n = f_y(x_n, \eta_n) \geqslant 0$, with η_n lying between y_n and $y(x_n)$. On adding (12.23) and (12.25), we see that

$$(z_{n+1} + e_{n+1}) - (2 + h^2 g_n)(z_n + e_n) + (z_{n-1} + e_{n-1})$$

$$= \frac{h^4}{12} (M + y^{(4)}(\xi_n)) - h^2 g_n z_n. \quad (12.26)$$

The right side of (12.26) is non-negative, as can be seen from $z_n \leqslant 0$, $g_n \geqslant 0$ and the definition of M. Also $z_0 + e_0 = 0$ and $z_{N+1} + e_{N+1} = 0$ and, on applying Lemma 12.1 to (12.26) with $w_n = z_n + e_n$, we deduce that

$$z_n + e_n \leqslant 0,$$

that is,

$$e_n \leqslant -z_n. \qquad (12.27)$$

Similarly, by subtracting (12.25) from (12.23), we deduce that

$$z_n - e_n \leqslant 0$$

so that

$$e_n \geqslant z_n. \qquad (12.28)$$

Combining (12.27), (12.28) and (12.24) provides the result (12.22). □

For many problems the bound in Theorem 12.1 is very much larger than the actual error. The bound also requires an upper bound of $|y^{(4)}(x)|$, which may not be readily available. The theorem does show, however, that the global error is $O(h^2)$, as h is decreased, and therefore we say that the method is of order 2.

As in the discussion of errors for initial value problems, we can include the effects of rounding errors. Suppose that, instead of finding $y_1, \ldots, y_N$, we actually calculate $w_1, \ldots, w_N$ satisfying

$$w_{n+1} - 2w_n + w_{n-1} = h^2 f(x_n, w_n) + \varepsilon_n,$$

where ε_n is the error introduced in solving the simultaneous algebraic equations. If for some ε, $|\varepsilon_n| \leqslant h^4 \varepsilon$, $n = 1, \ldots, N$, we must add ε to $M/12$ in the bound (12.22) to obtain a bound of $|w_n - y(x_n)|$.

There is a more accurate three-point difference approximation than (12.19) to (12.17), the special case of (12.1) when f is independent of y'. The formula

$$\frac{y_{n+1} - 2y_n + y_{n-1}}{h^2} = \tfrac{1}{12}(f(x_{n+1}, y_{n+1}) + 10f(x_n, y_n) + f(x_{n-1}, y_{n-1}))$$

$$(12.29)$$

has a truncation error $-h^4 y^{(6)}(\xi_n)/240$, $\xi_n \in (x_{n-1}, x_{n+1})$. See Problem 12.2 for a proof that the error is $O(h^4)$. We can obtain a global error bound which is of $O(h^4)$ but is otherwise similar to that in Theorem 12.1. Since (12.29) again involves at most three unknowns in each equation, it is no more diffi-cult to implement than (12.19) and is, therefore, to be preferred for problems of the form (12.17).

Example 12.2 We consider the simple linear equation on [0, 1]

$$y'' = y$$

with

$$y(0) = 0, \; y(1) = 1.$$

The solution is $y(x) = \sinh x/\sinh 1$.†
  With $h = 0.2$, the difference approximation (12.19) becomes

$$y_{n+1} - 2.04y_n + y_{n-1} = 0, \qquad n = 1, 2, 3, 4,$$

with $y_0 = 0$, $y_5 = 1$. Table 12.1 shows results including a comparison of the solutions of the difference equation and the differential equation. The global error bound is calculated from (12.22) with

$$M = \max_{0 \leqslant x \leqslant 1} |y^{(4)}(x)| = \max_{0 \leqslant x \leqslant 1} \left|\frac{\sinh x}{\sinh 1}\right| = 1.$$

The difference approximation (12.29) is in this case

$$y_{n+1} - 2y_n + y_{n-1} = \frac{0.04}{12}(y_{n+1} + 10y_n + y_{n-1}), \qquad n = 1, 2, 3, 4,$$

Table 12.1   Solution of Example 12.2.

n	y_n	$y(x_n)$	$\lvert y(x_n) - y_n \rvert \times 10^5$	Error bound (12.22) $\times 10^5$
1	0.17140	0.17132	8	27
2	0.34967	0.34952	15	40
3	0.54192	0.54174	18	40
4	0.75584	0.75571	13	27

† $\sinh x \equiv \tfrac{1}{2}(e^x - e^{-x})$.

that is,

$$0.996667y_{n+1} - 2.033333y_n + 0.996667y_{n-1} = 0, \qquad n = 1, 2, 3, 4,$$

with $y_0 = 0$, $y_5 = 1$. The solution to five significant decimal digits is identical to $y(x_n)$, $n = 1, 2, 3, 4$. $\square$

12.3 Extrapolation to the limit

In solving a differential equation we can often apply an extrapolation to the limit process to results using different step sizes as in Chapter 6 for numerical differentiation and integration. We consider, for example, the solution of the first order initial value problem (11.1) using Euler's method (11.11). We write $Y(x_n; h)$ to denote the approximation y_n to $y(x_n)$ when a step size h is used. We will assume that the global error may be expressed as a power series in h of the form

$$y(x) - Y(x; h) = A_1(x)h + A_2(x)h^2 + A_3(x)h^3 + \cdots, \qquad (12.30)$$

where $A_1(x)$, $A_2(x)$, ... are unknown functions of x but are independent of h. The existence of an expansion of the form (12.30) is dependent on there being no singularities in the differential equation and the existence of "sufficient" partial derivatives of $f(x, y)$. Henrici [1962] proves the existence of $A_1(x)$ under these conditions. As in §6.2, we usually halve the step size and eliminate the first term on the right of (12.30). By repeating this process we can eliminate terms in h^2, h^3, ... Analogous to (6.44), we compute†

$$Y^{(m)}(x; h) = \frac{2^m Y^{(m-1)}(x; h/2) - Y^{(m-1)}(x; h)}{2^m - 1}, \qquad (12.31)$$

where $Y^{(0)}(x; h) = Y(x; h)$.

Example 12.3 We use Euler's method and repeated extrapolation to the limit to find approximations to $y(0.8)$, where $y(x)$ is the solution of

$$y' = x^2 - y$$

with $\qquad\qquad\qquad y(0) = 1.$

The results are tabulated in Table 12.2. The layout is similar to that of Table 6.2. The solution of the differential equation is

$$y(x) = 2 - 2x + x^2 - e^{-x}$$

and, therefore, to five decimal places $y(0.8) = 0.59067$. $\square$

---

† The superscripts m do not denote derivatives.

In general, if a method is of order p, the power series for the global error will take the form

$$y(x) - Y(x;h) = A_p(x)h^p + A_{p+1}(x)h^{p+1} + A_{p+2}(x)h^{p+2} + \cdots, \quad (12.32)$$

if such a power series exists. We need to modify (12.31) by replacing 2^m by 2^{m+p-1} in both the denominator and numerator. Some methods will yield a power series in only even powers of h, so that

$$y(x) - Y(x;h) = E_1(x)h^2 + E_2(x)h^4 + E_3(x)h^6 + \cdots$$

for some functions $E_1, E_2, \ldots$. We must then use the relation (6.44) (with G replaced by Y). The Adams–Moulton corrector (11.90b) and the difference approximation (12.16) are both examples of such methods.

Table 12.2   Extrapolation to the limit, Example 12.3.

h	$Y^{(0)}$	$Y^{(1)}$	$Y^{(2)}$	$Y^{(3)}$
0.8	0.20000			
		0.64800		
0.4	0.42400		0.58485	
		0.60064		0.59101
0.2	0.51232		0.59024	
		0.59284		0.59069
0.1	0.55258		0.59063	
		0.59118		
0.05	0.57188			

Finally, we must add a warning that extrapolation will produce misleading results if a series like (12.32) does not exist. This can be caused by singularities in the differential equation, that is, one of the coefficients in the equation or one or more of the solutions of the equation is singular at some point. It is also important that derivatives of the solutions should exist. There may be trouble even if the solution which misbehaves is not the particular solution we are seeking. We have already seen in numerical integration, §6.4, that singular derivatives at one point may have an adverse effect on extrapolation.

12.4 Deferred correction

It is often possible to estimate the truncation error of difference approximations to differential equations by using differences of the computed solution. We can then use these estimates to improve (or "correct") the computed solution. This technique can be used for both initial value and boundary value problems and we will illustrate it by considering the solution of the boundary value problem (12.17) and (12.18) using the difference approximation (12.29). From Problem 12.2 it follows that, if $y(x)$ is the solution of the differential equation and $y(x)$ has a continuous eighth derivative,

$$\frac{y(x_{n+1}) - 2y(x_n) + y(x_{n-1})}{h^2}$$

$$= \tfrac{1}{12}(f(x_{n+1}, y(x_{n+1})) + 10f(x_n, y(x_n)) + f(x_{n-1}, y(x_{n-1})))$$

$$- \frac{h^4}{240} y^{(6)}(x_n) + O(h^6). \quad (12.33)$$

Now from (6.20), using Taylor series, we see that

$$y^{(6)}(x_n) = \frac{1}{h^6} \Delta^6 y(x_{n-3}) + O(h^2). \quad (12.34)$$

Having computed the y_n, the solution of (12.29), we approximate to the truncation error in (12.33), using $-\Delta^6 y_{n-3}/(240h^2)$, and solve

$$\frac{z_{n+1} - 2z_n + z_{n-1}}{h^2} = \tfrac{1}{12}(f(x_{n+1}, z_{n+1}) + 10f(x_n, z_n) + f(x_{n-1}, z_{n-1}))$$

$$- \frac{1}{240h^2} \Delta^6 y_{n-3}. \quad (12.35)$$

We expect z_n to be a better approximation to $y(x_n)$. In general, (12.35) gives a non-linear set of equations in the z_n, which are therefore difficult to compute (although no more difficult than the y_n from (12.29)). We can replace (12.35) by the linearization

$$\frac{e_{n+1} - 2e_n + e_{n-1}}{h^2} =$$

$$\tfrac{1}{12}(f_y(x_{n+1}, y_{n+1}) \cdot e_{n+1} + 10f_y(x_n, y_n) \cdot e_n + f_y(x_{n-1}, y_{n-1}) \cdot e_{n-1})$$

$$- \frac{1}{240h^2} \Delta^6 y_{n-3}, \quad (12.36)$$

where $f_y = \partial f/\partial y$ and e_n is to be the correction to y_n, that is we take $y_n + e_n$ as an improved approximation to $y(x_n)$. We derive (12.36) by subtracting (12.29) from (12.35) and using

$$f(x_n, z_n) - f(x_n, y_n) \simeq (z_n - y_n)f_y(x_n, y_n)$$

from the mean value theorem. We also replace $z_n - y_n$ by e_n.

It can be shown that, for both (12.35) and (12.36), the global error of the resulting process is $O(h^6)$, that is $|y(x_n) - z_n| = O(h^6)$ and $|y(x_n) - (y_n + e_n)| = O(h^6)$, compared with $O(h^4)$ for (12.29). It is possible to use a more accurate approximation to the truncation error in (12.33) and so increase the order of the method even more. To achieve this we also need to iterate. We compute a succession of approximations, each of which is used to estimate the truncation error from which the next approximation can be calculated. Fox [1957] gives many details of deferred correction.

With boundary conditions (12.18), we must apply (12.29) and (12.35) for $n = 1, 2, \ldots, N$ where $N = (b - a)/h - 1$. To use (12.35) we need values $y_{-2}, y_{-1}, y_{N+2}, y_{N+3}$ and these are found by using (12.29) in a step-by-step procedure. For example, we use (12.29) with $n = 0$ to compute y_{-1} from y_0 and y_1, and then with $n = -1$, we compute y_{-2} from y_{-1} and y_0.

Example 12.4 Compute an approximation to the solution of

$$y'' = y + (2 - x^2)$$

with

$$y(0) = 0, \qquad y(4) = 17.$$

Use (12.29) with $h = 1$ and the deferred correction process (12.36). The linearization (12.36) is identical to (12.35) with $e_n = z_n - y_n$ because f is linear in y. The equations (12.29), $n = 1, 2, 3$, become

$$(-2 - \tfrac{10}{12})y_1 + (1 - \tfrac{1}{12})y_2 \qquad\qquad = \tfrac{10}{12}$$
$$(1 - \tfrac{1}{12})y_1 + (-2 - \tfrac{10}{12})y_2 + (1 - \tfrac{1}{12})y_3 = -\tfrac{26}{12}$$
$$(1 - \tfrac{1}{12})y_2 + (-2 - \tfrac{10}{12})y_3 = -\tfrac{86}{12} - (1 - \tfrac{1}{12}).17.$$

Solving these gives the values y_1, y_2, y_3 shown correct to five decimal places

Table 12.3   Solution of Example 12.4 before correction.

n	y	Δy	$\Delta^2 y$	$\Delta^3 y$	$\Delta^4 y$	$\Delta^5 y$	$\Delta^6 y$
-2	3.86762						
		-2.91045					
-1	0.95717		1.95328				
		-0.95717		0.04672			
0	0		2.00000		0.00001		
		1.04283		0.04673		0.05094	
1	1.04283		2.04673		0.05095		0.05569
		3.08956		0.09768		0.10663	
2	4.13239		2.14441		0.15758		0.17177
		5.23397		0.25526		0.27840	
3	9.36636		2.39967		0.43598		0.47571
		7.63364		0.69124		0.75411	
4	17.00000		3.09091		1.19009		
		10.72455		1.88133			
5	27.72455		4.97224				
		15.69679					
6	43.42134						

in Table 12.3. We now find y_{-1}, y_{-2}, y_5 and y_6 using (12.29) with $n = 0, -1$, 4 and 5 respectively. Thus, for example, with $n = 4$,

$$\tfrac{11}{12}y_5 = (2 + \tfrac{10}{12})y_4 - \tfrac{11}{12}y_3 - \tfrac{170}{12}.$$

From (12.36), the corrections satisfy

$$(1 - \tfrac{1}{12})e_{n-1} + (-2 - \tfrac{10}{12})e_n + (1 - \tfrac{1}{12})e_{n+1} = -\tfrac{1}{240}\Delta^6 y_{n-3},$$

$n = 1, 2, 3$, with $e_0 = e_4 = 0$. The solution of these equations and the corrected solution $z_n = y_n + e_n$ are shown in Table 12.4, which also gives a comparison with the exact solution $y(x) = x^2 + \sinh x/\sinh 4$. $\square$

Table 12.4   Solution of Example 12.4 with correction.

n	y_n	$e_n \times 10^5$	z_n	$y(x_n)$	$(y(x_n) - y_n) \times 10^5$	$(y(x_n) - z_n) \times 10^5$
1	1.04283	29	1.04312	1.04306	23	−6
2	4.13239	64	4.13303	4.13290	51	−13
3	9.36636	91	9.36727	9.36709	73	−18

12.5 Chebyshev series method

We will show how to obtain a Chebyshev series approximation to the solution of a differential equation with polynomial coefficients. We shall obtain approximations which are valid over an interval and not just approximations to the solution at particular points. We need the following identities (see Problem 12.13) concerning products and indefinite integrals of Chebyshev polynomials:

$$x^s T_r(x) = 2^{-s} \sum_{j=0}^{s} \binom{s}{j} T_{r-s+2j}(x) \tag{12.37}$$

$$T_r(x)T_s(x) = \tfrac{1}{2}[T_{r+s}(x) + T_{r-s}(x)] \tag{12.38}$$

$$\int T_r(x)\, dx = \begin{cases} \dfrac{1}{2}\left[\dfrac{T_{r+1}(x)}{r+1} - \dfrac{T_{r-1}(x)}{r-1}\right], & r \neq 1 \\[2mm] \tfrac{1}{4}T_2(x), & r = 1 \end{cases} \tag{12.39}$$

$$T_r'(x) = \begin{cases} 0, & r = 0 \\ 2r(\tfrac{1}{2}T_0 + T_2 + \cdots + T_{r-1}), & r \text{ odd} \\ 2r(T_1 + T_3 + \cdots + T_{r-1}), & r \text{ even} \end{cases} \tag{12.40}$$

for $r, s = 0, 1, 2, \ldots$, where we define $T_{-r}(x) = T_r(x)$.

We consider first the initial value problem on $[-1, 1]$,

$$q_1(x)y'(x) + q_2(x)y(x) = q_3(x) \tag{12.41}$$

with

$$y(-1) = s, \tag{12.42}$$

where q_1, q_2 and q_3 are polynomials. On integrating (12.41) we obtain

$$q_1(x) \cdot y(x) + \int (q_2(x) - q_1'(x))y(x)\, dx = \int q_3(x)\, dx + C, \tag{12.43}$$

where C is an arbitrary constant. We now express q_1, q_2 and q_3 in terms of Chebyshev polynomials and seek a series†

$$y(x) = \sum_{r=0}^{\infty}{}' a_r T_r(x)$$

which satisfies (12.43). On using (12.37)–(12.40) we can express (12.43) as

$$\sum_{r=0}^{\infty}{}' b_r T_r(x) = \sum_{r=0}^{m}{}' c_r T_r(x), \tag{12.44}$$

if $q_3 \in P_{m-1}$. Each coefficient b_r is some linear combination of the unknown a_r; the coefficients c_r are known, being determined by q_3, with the exception of c_0 which also depends on the arbitrary constant C. On equating coefficients b_r and c_r, we obtain an infinite set of linear equations in the a_r. We usually solve the first few of these equations, except for $r = 0$ as this involves the arbitrary constant, and so find an approximation to $y(x)$. We also impose the initial condition (12.42) on our approximation and so obtain a further linear equation in the a_r. The following example illustrates the process.

Example 12.5 Consider

$$y' + xy = x, \qquad -1 \leqslant x \leqslant 1, \tag{12.45}$$

with

$$y(-1) = 0. \tag{12.46}$$

On integrating, (12.45) becomes

$$y + \int xy\, dy = \tfrac{1}{2}x^2 + C. \tag{12.47}$$

If

$$y = \sum_{r=0}^{\infty}{}' a_r T_r$$

---

† Σ' denotes summation with the first term halved.

then, from (12.37) and (12.39),

$$\int xy\,dx = \tfrac{1}{4}(a_1 - a_3)T_1 + \tfrac{1}{4}\sum_{r=2}^{\infty}\left(\frac{a_{r-2} - a_{r+2}}{r}\right)T_r.$$

Thus

$$y + \int xy\,dx = \tfrac{1}{2}a_0 + \tfrac{1}{4}(5a_1 - a_3)T_1 + \tfrac{1}{4}\sum_{r=2}^{\infty}\left(\frac{a_{r-2}}{r} + 4a_r - \frac{a_{r+2}}{r}\right)T_r.$$

$$(12.48)$$

Also,

$$\tfrac{1}{2}x^2 + C = (\tfrac{1}{4} + C) + \tfrac{1}{4}T_2. \qquad (12.49)$$

On equating coefficients (except constant terms) in (12.48) and (12.49), we obtain

$$\left.\begin{array}{c} \tfrac{5}{4}a_1 - \tfrac{1}{4}a_3 = 0 \\[4pt] \tfrac{1}{8}a_0 + a_2 - \tfrac{1}{8}a_4 = \tfrac{1}{4} \\[4pt] \tfrac{1}{12}a_1 + a_3 - \tfrac{1}{12}a_5 = 0 \\[4pt] \tfrac{1}{16}a_2 + a_4 - \tfrac{1}{16}a_6 = 0 \end{array}\right\} \qquad (12.50)$$

and

$$\left.\frac{1}{4r}a_{r-2} + a_r - \frac{1}{4r}a_{r+2} = 0, \qquad r = 5, 6, 7, \ldots\right\}$$

We look for an approximation $p_4 \in P_4$ to y of the form

$$p_4(x) = \sum_{r=0}^{4}{}' \tilde{a}_r T_r(x),$$

where the $\tilde{a}_r$ satisfy the first four of the equations (12.50) with $\tilde{a}_5 = \tilde{a}_6 = \cdots$ $= 0$. If we solve these equations, keeping $\tilde{a}_0$ as a free parameter, we find that

$$\tilde{a}_1 = \tilde{a}_3 = 0$$

$$\tilde{a}_4 = -\tfrac{1}{16}\tilde{a}_2 = \frac{\tilde{a}_0 - 2}{129}.$$

Finally, p_4 satisfies the initial condition (12.46) if

$$\tfrac{1}{2}\tilde{a}_0 - \tilde{a}_1 + \tilde{a}_2 - \tilde{a}_3 + \tilde{a}_4 = 0,$$

that is,

$$\tfrac{1}{2}\tilde{a}_0 - \tfrac{16}{129}(\tilde{a}_0 - 2) + \tfrac{1}{129}(\tilde{a}_0 - 2) = 0,$$

whence

$$\tilde{a}_0 = -\tfrac{20}{33}.$$

Thus the approximation is

$$p_4(x) = -\tfrac{10}{33} + \tfrac{32}{99}T_2(x) - \tfrac{2}{99}T_4(x)$$
$$= -\tfrac{16}{99}(4 - 5x^2 + x^4).$$

The solution of the problem is

$$y(x) = 1 - e^{1/2(1-x^2)}$$

and on $[-1, 1]$ the maximum error in using p_4 as an approximation to y occurs at $x = 0$, when

$$y(0) - p_4(0) \simeq 22 \times 10^{-4}.$$

One interesting feature of the method is that we can make a backward error analysis. The coefficients $\tilde{a}_r$ satisfy all the equations (12.50) with the exception of that with $r = 6$, which must be replaced by

$$\frac{1}{4.6}\tilde{a}_4 + \tilde{a}_6 - \frac{1}{4.6}\tilde{a}_8 = \frac{1}{4.6}\tilde{a}_4.$$

Now this equation would be obtained if a term $\tilde{a}_4 T_6/24$ were added to the right side of (12.47). We thus find that p_4 satisfies an equation of the form

$$p_4(x) + \int xp_4(x)\, dx = \tfrac{1}{2}x^2 - \tfrac{1}{1188}T_6(x) + C.$$

We have perturbed (12.47) by an amount $T_6/1188$. □

The method is easily extended to second order equations of the form

$$q_1 y'' + q_2 y' + q_3 y = q_4, \tag{12.51}$$

where q_r, $r = 1, 2, 3, 4$, are polynomials. We integrate (12.51) twice to obtain

$$q_1 y + \int (q_2 - 2q_1')y\, dx + \int\int (q_3 - q_2' + q_1'')y\, dx = \int\int q_4\, dx + C_1 x + C_2,$$

where C_1 and C_2 are arbitrary constants. We now proceed as before and obtain a system of linear equations in the Chebyshev coefficients. This time we omit the equations equating coefficients for both T_0 and T_1, as these involve the arbitrary constants. These equations are replaced by the two boundary conditions which are needed with the second order equation (12.51).

The method can also be used to deal with boundary conditions of a more general form than (12.42). For example, we could deal with the unlikely condition

$$\alpha y(-1) + \beta y(1) = s,$$

where $\alpha \neq 0$, $\beta \neq 0$ are constants. This will just give a linear equation in the

a_r. For a second order problem there is no difficulty in dealing with conditions of the form

$$\alpha y(a) + \beta y'(a) = s.$$

If the interval on which we require a solution is not $[-1, 1]$ we must use a simple transformation of variable. (See Problem 4.28.) Polynomial terms in the differential equations will remain polynomials after such a transformation. Fox and Parker discuss the use of Chebyshev methods for differential equations and similar problems more fully. Much of the pioneering work on these methods is due to C. Lanczos.

Problems

Section 12.1

12.1 Draw a flow diagram showing the principal steps in the numerical solution of the boundary value problem (12.1) and (12.2) by a predictor-corrector method. Note that you should have three "nested" iteration loops corresponding to:

(i) choice of starting conditions

(ii) step-by-step advancement

(iii) use of the corrector.

Section 12.2

12.2 Suppose that y satisfies

$$y'' = f(x, y)$$

and let

$$t(x_n) = \frac{1}{h^2}\left(y(x_{n+1}) - 2y(x_n) + y(x_{n-1})\right)$$

$$- \tfrac{1}{12}(f(x_{n+1}, y(x_{n+1})) + 10f(x_n, y(x_n)) + f(x_{n-1}, y(x_{n-1}))),$$

where $x_{n\pm1} = x_n \pm h$. If the eighth derivative of y exists and is continuous on $[x_{n-1}, x_{n+1}]$, show that

$$t(x_n) = -\frac{h^4}{240} y^{(6)}(x_n) + O(h^6).$$

(*Hint:* use the Taylor series for both y and y'' at $x = x_n$.)

12.3 Use the difference approximation (12.19) with $h = 0.2$ to solve

$$y'' = 1$$

with

$$y(0) = y(1) = 0.$$

Show that the calculated solution is identical to the exact solution

$$y(x) = (x^2 - x)/2.$$

12.4 Repeat Problem 12.3 with the boundary conditions replaced by

$$y'(0) = -\tfrac{1}{2}, \qquad y(1) = 0.$$

Use (12.9) to approximate to the first of these conditions.

12.5 If $v(x) \equiv 0$ in (12.4) and we scale the equation so that $u(x) \equiv 1$, the $N \times N$ coefficient matrix $\mathbf{A}$ in (12.8) is the tridiagonal matrix

$$\mathbf{A} = -\frac{1}{h^2}
\begin{bmatrix}
(2 - h^2 w_1) & -1 & & 0 \\
-1 & (2 - h^2 w_2) & -1 & \\
& \ddots & \ddots & \ddots \\
0 & & &
\end{bmatrix}
= -\frac{1}{h^2}\mathbf{B},$$

say. Show that if $\mathbf{z}$ is any N-dimensional (real) vector with components z_n then

$$\mathbf{z}^T \mathbf{B} \mathbf{z} = z_1{}^2 + (z_1 - z_2)^2 + (z_2 - z_3)^2 + \cdots + (z_{N-1} - z_N)^2 + z_N{}^2$$
$$- h^2(w_1 z_1{}^2 + w_2 z_2{}^2 + \cdots + w_N z_N{}^2).$$

Hence show that, if $w(x) \leqslant 0$ on $[a, b]$, the matrix $\mathbf{B}$ is symmetric and positive definite (see §8.10) and thus $\mathbf{A}$ is non-singular.

12.6 Given the non-linear boundary value problem

$$y'' = -2yy'$$

with

$$y(0) + y'(0) = 0,$$
$$y(1) = \tfrac{1}{2},$$

derive a system of non-linear equations for approximations y_r to $y(0.2r)$, $r = 0, 1, 2, 3, 4$, using a difference approximation of the form (12.16) with $h = 0.2$.

12.7 The equations (12.16) can be written in the form

$$F_n(y_{n-1}, y_n, y_{n+1}) = 0, \qquad n = 1, 2, \ldots, N,$$

with $y_0 = \alpha$, $y_{N+1} = \beta$. We can write these as

$$\mathbf{F}(\mathbf{y}) = \mathbf{0},$$

where $\mathbf{F}$ is an N-dimensional vector whose nth component is F_n. Show that the Jacobian matrix (see §10.2) of $\mathbf{F}$ is tridiagonal.

12.8   The boundary value problem on $[0, 1]$,

$$y'' = 4(y + x)$$

with

$$y(0) = 1, \qquad y(1) = e^2 - 1,$$

is to be solved using the difference approximation (12.19). Use Theorem 12.1 to determine what size of interval should be used if the global error is to be less than 10^{-2} at all grid points. (Note that the required solution of the differential equation is $y(x) = e^{2x} - x$.)

Section 12.3

12.9   A one-step method for first order initial value problems may be formed as follows. Suppose y_n is known.

(a) Advance one step using Euler's method (11.11) with step size h; let $y(x_{n+1}; h)$ be the resulting approximation to $y(x_{n+1})$.

(b) Advance from y_n again, using two steps of Euler's method with step size $h/2$. Let $y(x_{n+1}; h/2)$ be the resulting approximation to $y(x_{n+1})$.

(c) Extrapolate using

$$y_{n+1} = 2y(x_{n+1}; h/2) - y(x_{n+1}; h).$$

Show that the resulting method is precisely the modified Euler method (11.44).

12.10   Use Euler's method with extrapolation to the limit to obtain approximations to $y(0.8)$, where $y(x)$ is the solution of

$$y' = 1 - y$$

with

$$y(0) = 0.$$

Take $h = 0.4, 0.2, 0.1$ and 0.05. Compare your computed solutions with the exact solution $y(x) = 1 - e^{-x}$.

Section 12.4

12.11   Consider deferred correction for Euler's method. Show that, if

$$y' = f(x, y)$$

and $y^{(3)}(x)$ exists for $x_n < x < x_{n+1}$, then

$$\frac{y(x_{n+1}) - y(x_n)}{h} = f(x_n, y(x_n)) + \frac{h}{2} y''(x_n) + \frac{h^2}{6} y^{(3)}(\xi_n),$$

where $\xi_n \in (x_n, x_{n+1})$. Thus we can estimate the truncation error using (see §6.1)

$$\frac{h}{2} y''(x_n) \simeq \frac{h}{2} \frac{\Delta^2 y_{n-1}}{h^2} = \frac{1}{2h} \Delta^2 y_{n-1}.$$

Having computed approximations y_n, we seek new approximations z_n satisfying (cf. (12.35))

$$\frac{1}{h} (z_{n+1} - z_n) = f(x_n, z_n) + \frac{1}{2h} \Delta^2 y_{n-1}.$$

Apply this method to the initial value problem given in Problem 12.10, taking $h = 0.2$ and $0 \leqslant x \leqslant 0.8$. Note that the method requires y_{-1}, which must be calculated from

$$y_0 = y_{-1} + hf(x_{-1}, y_{-1}),$$

where $x_{-1} = x_0 - h = -h$.

12.12   Derive the formulae needed for applying the deferred correction process to the method (12.19) for boundary value problems of the form (12.17).

Section 12.5

12.13   Verify the formulae (12.37)–(12.40). (*Hints:* (12.37) may be proved by mathematical induction and the recurrence relation (4.64) written in the form

$$xT_r(x) = \tfrac{1}{2}(T_{r-1}(x) + T_{r+1}(x));$$

the other formulae may be proved using $T_r(x) = \cos r\theta$ where $\theta = \cos^{-1} x$. See also §6.6.)

12.14   Use the method of §12.5 to obtain an approximation $p_2 \in P_2$ to the solution of the initial value problem given in Example 12.5. Compare the accuracy of p_2 with that of the p_4 given in the example.

12.15   Use the method of §12.5 to determine an approximation $p_3 \in P_3$ to the solution of the initial value problem

$$y' + y = x^2$$

with

$$y(0) = 1.$$

(See Example 12.3 for the exact solution.)

12.16   Show that we obtain the same approximation p_4 as in Example 12.5 if we replace the initial condition (12.46) by

$$y(-1) + y(1) = 0.$$

12.17   Use the method of §12.5 to determine an approximation $p_4 \in P_4$ to the solution of the boundary value problem on $[-1, 1]$

$$y'' = y$$

with $y(-1) = -1$, $y(1) = 1$. (The solution of the differential equation is $y(x) = \sinh x/\sinh 1$.)

Appendix

Computer Arithmetic

A.1 Binary numbers

For reasons of economy in design, computers use *binary* representation of numbers instead of the *decimal* system adopted by humans. In the decimal system we use ten as a *base* and there are ten different digits, 0, 1, 2, 3, 4, 5, 6, 7, 8, 9. Numbers are represented as sequences of these digits. For example,

$$739 = 7 \times 10^2 + 3 \times 10^1 + 9 \times 10^0$$
$$= 7 \times 100 + 3 \times 10 + 9 \times 1$$

and

$$910.68 = 9 \times 10^2 + 1 \times 10^1 + 0 \times 10^0 + 6 \times 10^{-1} + 8 \times 10^{-2}$$
$$= 9 \times 100 + 1 \times 10 + 0 \times 1 + 6 \times \tfrac{1}{10} + 8 \times \tfrac{1}{100}.$$

In the binary system we use two as the base and there are two different digits, 0 and 1. For example,

$$11011 = 1 \times 2^4 + 1 \times 2^3 + 0 \times 2^2 + 1 \times 2^2 + 1 \times 2^0$$

and

$$11.101 = 1 \times 2^1 + 1 \times 2^0 + 1 \times 2^{-1} + 0 \times 2^{-2} + 1 \times 2^{-3}.$$

It is shown in number theory that every real number can be expressed as a finite or infinite sequence of digits, regardless of the base used in the representation (except for base one). It is also shown that if the representation of a rational number is an infinite sequence of digits then these must include a pattern which repeats after a certain point. It is interesting to note that a number which can be represented by a finite sequence of digits in one representation may need an infinite sequence in another representation. (See Problem A.1.)

In this book we normally discuss decimal representations because of their familiarity. Further details of computer arithmetic are given in Phillips and Taylor [1969].

358

A.2 Integers and fixed point fractions

All computers can deal with integers but, because a computer store must be finite, only a restricted range is available. Usually this is more than adequate and typically a machine will store integers of up to 16 binary digits so that the range is $-65,535 \leqslant n \leqslant 65,535$. Arithmetic with such integers is exact provided the result is an integer in the permitted range. If the result is too large most computers give some kind of "overflow" indication.

The earliest types of computer dealt only with *fixed point* fractions as an alternative to integers. These fractions are restricted to the range $-1 < x < 1$ and a finite number of digits, usually about 30–40 binary digits (approximately 9–12 decimal digits). If we think in terms of decimal numbers, a fixed point fraction consists of only a finite number t (say) of decimal digits after the decimal point. In general, we have to terminate the decimal representation of a number after t digits. To do this we *round off* the number by chopping the decimal representation after t digits and then adding 10^{-t} if the amount neglected exceeds $\frac{1}{2}10^{-t}$. (If the amount neglected is exactly $\frac{1}{2}10^{-t}$ we can avoid statistical bias by forcing the last digit in the rounded number to be even.) In binary fixed point representation with t digits, we replace 10^{-t} by 2^{-t}.

Addition and subtraction of two fixed point fractions does not introduce any error unless the result lies outside the permitted range. Multiplication of two t-digit numbers yields a $2t$-digit product which is said to be *double-length*, as it consists of twice as many digits as the operands, which are said to be *single-length*. We must usually round the product to single-length (or t digits) and thus, provided the result lies in the permitted range, an error of at most $\frac{1}{2}10^{-t}$ is introduced. Similarly, division gives an error of at most $\frac{1}{2}10^{-t}$. If fix(a) represents the fixed point representation of a real number a, we have the following results.

For any a,

(i) $|\text{fix}(a) - a| \leqslant \frac{1}{2}10^{-t}$.

If a and b are any two fixed point fractions with t digits:

(ii) $\text{fix}(a \pm b) = a \pm b$,

(iii) $|\text{fix}(ab) - ab| \leqslant \frac{1}{2}10^{-t}$, (A.1)

(iv) $|\text{fix}(a/b) - a/b| \leqslant \frac{1}{2}10^{-t}$. (A.2)

You will notice that multiplication and division will in general introduce rounding errors. These results are only valid if all quantities lie in the range†
$-1 < x < 1$. If working is to t *binary* digits, 10^{-t} must be replaced by 2^{-t}.

† Strictly, this range should be $-1 + \frac{1}{2}10^{-t} \leq x \leq 1 - \frac{1}{2}10^{-t}$ but some computers allow $x = +1$ and/or $x = -1$ as "fixed point fractions".

If "chopping" is used, without rounding up if the error exceeds $\frac{1}{2}10^{-t}$, the bounds must be made twice as large.

On some computers, integers are stored in exactly the same way as fixed point fractions. The main differences lie in the way that the binary digits are interpreted. For integers, the binary point is considered to be at the least significant end and for fractions it is at the most significant end. There is also a difference in the way that the double-length product is interpreted.

In dealing with decimal numbers (not necessarily fractions) in fixed point form, we sometimes refer to the number of *decimal places*, by which we mean the number of decimal digits appearing after the decimal point. Thus 3.460 is a number given to three decimal places. If this is an approximation to some value, the least significant zero indicates that it is correct to three decimal places whereas 3.46 would suggest it is correct to only two decimal places. In general, we say that two numbers agree to t decimal places if their absolute difference does not exceed $\frac{1}{2}10^{-t}$.

A.3 Floating point arithmetic

One difficulty with fixed point arithmetic is that we usually have to worry about scaling so that results of arithmetical operations lie in the permitted range. This adds considerable complexity to computer programs. We can avoid nearly all of such problems with *floating point* representation of numbers. Each non-zero number is expressed in the form $a \times 10^p$ where a is a fraction in the range $1 > |a| \geqslant 0.1$ and p is an integer called the *exponent*.† We now represent a number by a pair (a, p). For example,

$$
\begin{aligned}
27.3 &= & 0.273 \times 10^2 &= (0.273, 2) \\
-832,000 &= & -0.832 \times 10^6 &= (-0.832, 6) \\
0.000641 &= & 0.641 \times 10^{-3} &= (0.641, -3).
\end{aligned}
$$

We represent zero by the pair $(0, 0)$. On a computer we work with powers of two and numbers are usually expressed in the form $a \times 2^p$ where $1 > |a| \geqslant \frac{1}{2}$. Some computers use powers of eight for the exponent even though numbers are kept in binary form.

In practice a, the main part of the number, must consist of a finite number of digits and we will assume that only t decimal digits are permitted. The exponent must also be restricted in range and we assume that $-N \leqslant p \leqslant N$ so that numbers expressed in floating point form must lie within $10^N > |x| \geqslant 10^{-N-1}$. We say that the floating point number consists of t digits, irre-

---

† There is no special name for a although some people misname this the "mantissa".

spective of the number of digits in the exponent. Typically $t \simeq 10$ and $N \simeq 77$† so that a very wide range of numbers is available.

Suppose $x \neq 0$ is any real number which we wish to express in the nearest floating point form of t decimal digits. It is always possible to find p such that $x = b.10^p$ and $0.1 \leqslant |b| < 1$. We assume that p is in the permitted range $-N \leqslant p \leqslant N$. We now choose a to be a t-digit fraction obtained by rounding b, so that‡

$$|b - a| \leqslant \tfrac{1}{2}10^{-t}$$

and

$$0.1 \leqslant |a| < 1.$$

The floating point number we choose in place of x is (a, p). Since $|b| \geqslant 0.1$,

$$|x| = |b|.10^p \geqslant 0.1 \times 10^p = 10^{p-1} \tag{A.3}$$

and we find that

$$
\begin{aligned}
|x - a.10^p| &= |b.10^p - a.10^p| \\
&= |b - a|.10^p \\
&\leqslant \tfrac{1}{2}10^{-t}.10|x|.
\end{aligned}
$$

Hence

$$\left| \frac{x - a.10^p}{x} \right| \leqslant \tfrac{1}{2}10^{-t+1} \tag{A.4}$$

and, therefore, the relative error in replacing x by (a, p) is at most $\tfrac{1}{2}10^{-t+1}$. We shall write $\mathrm{fl}(x)$ for the floating point approximation to x, so that $\mathrm{fl}(x) = (a, p)$.

All the basic arithmetic operations on floating point numbers may introduce a rounding error and, from (A.4), the following statements are valid for any non-zero real numbers x and y, provided all quantities lie in the permitted range.

(i) $\mathrm{fl}(x) = x(1 + \varepsilon)$,

(ii) $\mathrm{fl}(x \pm y) = (x \pm y)(1 + \varepsilon)$,

(iii) $\mathrm{fl}(xy) = xy(1 + \varepsilon)$,

(iv) $\mathrm{fl}(x/y) = (x/y)(1 + \varepsilon)$,

where in each case $|\varepsilon| \leqslant \tfrac{1}{2}10^{-t+1}$. The results (ii)–(iv) cannot be improved even if x and y are floating point numbers. If binary arithmetic is used,

---

† This corresponds to binary exponents in the range $-255 \leqslant p \leqslant 255$, as $2^{255} \simeq 10^{77}$.
‡ If $1 - \tfrac{1}{2}10^{-t} \leqslant |b| < 1$ it is necessary to increase p by $+1$ and take $a = +0.1$ or -0.1.

10^{-t+1} must be replaced by 2^{-t+1}. The arithmetic in most computers is arranged so that rounding is optimal and thus the above conditions are satisfied. Sometimes "chopping" without rounding is used, so that errors can be up to twice as large. Usually the computer will give an "overflow" indication if the exponent of a result is too large, that is $p > N$. "Underflow", when $p < -N$, usually means that the result is treated as zero.

We often refer to the first r *significant digits* of a number and by this we mean the first r digits in the main part of the floating point representation of the number. Thus the first three significant digits of 730,000 are 7, 3 and 0 and the first two significant digits of 0.000681 are 6 and 8. Two numbers are said to be the same (or agree) to r significant digits if, after rounding to r-digit floating point form, the numbers are identical. Thus 730,423 and 730,000 are the same to three significant decimal digits. We are concerned here with relative accuracy rather than absolute accuracy, which is usually expressed in terms of decimal places. For example, although 0.00126 and 0.0013 agree to four decimal places, they agree to only two significant digits.

Problems

Section A.1

A.1   In each of the following the number on the left is in decimal and that on the right is in binary. Verify that they are both the same number.

(i) $39.25 = 100111.01$

(ii) $0.2 = 0.0011\ 0011\ 0011\ldots$

(iii) $1 = 0.1111111\ldots$

Use (ii) and (iii) to verify that $0.2 \times 5 = 1$ by working in binary arithmetic.

Section A.2

A.2   Give examples of fixed point fractions a and b such that equality holds in (A.1) and (A.2).

A.3   Show by means of examples that, if x, y and z are any real numbers, then in general

$$\text{fix}(x + y) \neq \text{fix}(x) + \text{fix}(y)$$
$$\text{fix}(\text{fix}(x.y).z) \neq \text{fix}(x.\text{fix}(y.z)).$$

Show that the last relation can hold even if x, y and z are already fixed point fractions with t digits.

Section A.3

A.4   We do not obtain equality in (A.3) unless $|b| = 0.1$ in which case $\text{fl}(x) = x$. Use this result to show that we must have strict inequality in (A.4).

A.5   As in Problem A.3, show that if x, y and z are t-digit floating point numbers, then in general

$$\text{fl}(\text{fl}(x + y) + z) \neq \text{fl}(x + \text{fl}(y + z))$$

$$\text{fl}(\text{fl}(x.y).z) \neq \text{fl}(x.\text{fl}(y.z)).$$

References

Davis, P. J. (1963). "Interpolation and Approximation" Blaisdell, New York.

Davis, P. J. and Rabinowitz, P. (1967). "Numerical Integration" Blaisdell, New York.

Forsythe, G. E. and Moler, C. B. (1967). "Computer Solution of Linear Algebraic Systems" Prentice-Hall, Englewood Cliffs, New Jersey.

Fox, L. (1957). "The Numerical Solution of Two-point Boundary Problems in Ordinary Differential Equations" Oxford University Press, Oxford.

Fox, L. and Parker, I. B. (1968). "Chebyshev Polynomials in Numerical Analysis" Oxford University Press, Oxford.

Henrici, P. (1962). "Discrete Variable Methods in Ordinary Differential Equations" John Wiley and Sons, New York.

Henrici, P. (1964). "Elements of Numerical Analysis" John Wiley and Sons, New York.

Hildebrand, F. B. (1956). "Introduction to Numerical Analysis" McGraw-Hill, New York.

Isaacson, E. and Keller, H. B. (1966). "Analysis of Numerical Methods" John Wiley and Sons, New York.

Lambert, J. D. (1973). "Computational Methods in Ordinary Differential Equations" John Wiley and Sons, New York.

Noble, B. (1969). "Applied Linear Algebra" Prentice-Hall, Englewood Cliffs, New Jersey.

Phillips, G. M. and Taylor, P. J. (1969). "Computers" Methuen and Co., London.

Ralston, A. (1965). "A First Course in Numerical Analysis" McGraw-Hill, New York.

Rivlin, T. J. (1969). "An Introduction to the Approximation of Functions" Blaisdell, New York.

Solutions to Selected Problems

Chapter 1

1.1 (ii) Constructive proof: $\cos(2\cos^{-1} x) = \frac{1}{2} \Rightarrow 2\cos^{-1} x = (2k \pm \frac{1}{3})\pi$, where k is an integer. Thus $x = \cos\frac{1}{6}\pi$, $\cos\frac{7}{6}\pi$ are two real roots.

   Non-constructive proof: let $f(x) = 2\cos(2\cos^{-1} x) - 1$, when $f(0) = -3$ and $f(1) = +1$. Thus as the function is continuous (see §2.2) and has opposite signs at the end points of $[0, 1]$, it has at least one zero in $[0, 1]$. Similarly, there is at least one zero in $[-1, 0]$.

1.4 Condition (i) on p. 6 is not satisfied. There is no solution if we change any one of the coefficients in the first two equations by an arbitrary small amount, keeping the others fixed.

Chapter 2

2.1 (i) $[\frac{3}{4}, \infty)$, (ii) $[0, \frac{2}{3}]$, (iii) $[1, \infty)$. Inverses exist for (ii) and (iii) only.

2.10 All exist; only (iii) and (iv) are attained.

2.13 Only (i) and (ii) are continuous. At $x = -1$, (iii) is not even defined; similarly for (iv) at $x = 0$.

2.15 If $|f(x) - f(x')| \leqslant L|x - x'|$ for all $x, x' \in [a, b]$ then Definition 2.10 holds if we choose $\delta = \varepsilon/L$.

2.17 (ii) We have

$$\frac{f(x + h) - f(x)}{h} = \frac{1}{h}\left[\frac{1}{x + h + 1} - \frac{1}{x + 1}\right] = \frac{-1}{(x + h + 1)(x + 1)}$$

and thus $f'(x) = -1/(x + 1)^2$.

2.19 $|f(x) - f(x')| = \big| |x| - |x'| \big| \leqslant |x - x'|$.

2.21 Since $|u_n - \frac{1}{2}| = \frac{1}{2n}$, $\{u_n\}$ converges to $\frac{1}{2}$.

Chapter 3

3.2 $\cos x$: $p_6(x) = 1 - \dfrac{1}{2!} x^2 + \dfrac{1}{4!} x^4 - \dfrac{1}{6!} x^6.$

$(1 + x)^{1/2}$: $p_6(x) = 1 + \dfrac{1}{2} x - \dfrac{1}{2.4} x^2 + \dfrac{1.3}{2.4.6} x^3 - \dfrac{1.3.5}{2.4.6.8} x^4$

$$+ \dfrac{1.3.5.7}{2.4.6.8.10} x^5 - \dfrac{1.3.5.7.9}{2.4.6.8.10.12} x^6.$$

$\ln (1 - x)$: $p_6(x) = -x - \tfrac{1}{2}x^2 - \tfrac{1}{3}x^3 - \tfrac{1}{4}x^4 - \tfrac{1}{5}x^5 - \tfrac{1}{6}x^6.$

3.4 $n = 9$.

3.6 $R_n(x) = (-1)^{n+1}(n + 1)x\left(\dfrac{x - \xi}{1 + \xi}\right)^n \dfrac{1}{(1 + \xi)^2}$, where $-1 < a \leqslant x < \xi < 0$. Thus

$$|R_n(x)| < \dfrac{(n + 1)|a|^n}{(1 + a)^2} \quad \text{for } -1 < a \leqslant x < 0.$$

Since $(n + 1)|a|^n \to 0$ as $n \to \infty$, we have uniform convergence.

3.9 The nth term in the series is $(-1)^{n+1}x^n/n$. For $|x| > 1$, this term increases in magnitude as $n \to \infty$, showing that the series cannot be convergent. Combine this last result with that of Problem 3.8 to show that the radius of convergence is $+1$.

3.11 For $x > 0$ use $|R_1(x)| = \tfrac{1}{8}x^2(1 + \xi)^{-3/2} < \tfrac{1}{8}x^2 = O(x^2)$, where $0 < \xi < x$. For $x < 0$ use $|R_1(x)| = |\tfrac{1}{4}x(x - \xi)(1 + \xi)^{-3/2}| < \tfrac{1}{4}|x^2/(1 + x)^{-3/2}| = O(x^2)$, where $-1 < x < \xi < 0$.

3.13 By Taylor's theorem

$$e^x = 1 + x + \tfrac{1}{2}x^2 e^{\xi_1}$$
$$e^{nx} = 1 + nx + \tfrac{1}{2}n^2 x^2 e^{\xi_2}$$
$$(1 + x)^n = 1 + nx + \tfrac{1}{2}x^2 n(n - 1)(1 + \xi_3)^{n-2}.$$

Thus $e^x \geqslant 1 + x$, $e^x = 1 + x + O(x^2)$, $e^{nx} = (e^x)^n \geqslant (1 + x)^n$ and $e^{nx} - (1 + x)^n = O(x^2)$.

Chapter 4

4.1 $p_1(x) = 4x + 1$.

4.4 0.6381.

4.5 $p_1(x) = \tfrac{1}{3}(x + 2)$, so $2^{1/2} \simeq 1\tfrac{1}{3}$, $3^{1/2} \simeq 1\tfrac{2}{3}$.

4.6 $p_2(x) = 2x^2 - 1$.

4.7 $p_3(x) = x^2 + 1$. (See Theorem 4.1.)

4.8 $p(x) = 4x^3 - 3x$.

4.9 Error lies between 0.003 and 0.006.

4.11 $h = 0.002$.

4.14 Estimate is $x = 1.11$.

4.15 $p_3(x) = -9 + (x + 3).7 + (x + 3)(x + 1).(-3) + (x + 3)(x + 1)x.1$

$$= x^3 + x^2 - 2x + 3.$$

4.18 Polynomial is $2x^2 + 3x + 1$.

4.19 $S(r) = 1 + (r - 1).4 + \frac{1}{2}(r - 1)(r - 2).5 + \frac{1}{6}(r - 1)(r - 2)(r - 3).2$
$= \frac{1}{6}(2r^3 + 3r^2 + r)$.

4.22 Degree three.

4.24 $T_6(x) = 32x^6 - 48x^4 + 18x^2 - 1$.

4.27 $n = 7$.

Chapter 5

5.1 Normal equations are $a_0 + \frac{1}{2}a_1 = \frac{2}{3}$, $\frac{1}{2}a_0 + \frac{1}{3}a_1 = \frac{2}{5}$, with solution $a_0 = \frac{4}{15}$, $a_1 = \frac{4}{5}$.

5.2 $2.97 - 2x$.

5.3 Show that if the ψ_r are linearly *dependent*, they do not form a Chebyshev set.

5.4 $4x - 5x^3 + x^5$ is zero for every $x \in \{-2, -1, 0, 1, 2\}$.

5.7 $1.06 + 1.05x + 0.95y$.

5.9 The smoothed values are 2.9, 2.8, 2.6, 2.37, 2.17, 1.97, 1.77.

5.13 $|x| \sim \frac{1}{2}\pi - \frac{4}{\pi} \sum_{r=0}^{\infty} \frac{\cos (2r + 1)x}{(2r + 1)^2}$.

5.14 $x^2 \sim \frac{\pi^2}{3} + 4 \sum_{r=1}^{\infty} (-1)^r \frac{\cos rx}{r^2}$.

5.16 Make the substitution $x = \cos \theta$.

5.21 $\cos^{-1} x \sim \frac{1}{2}\pi - \frac{4}{\pi} \sum_{r=0}^{\infty} \frac{T_{2r+1}(x)}{(2r + 1)^2}$.

5.22 $(1 - x^2)^{1/2} \sim -\dfrac{4}{\pi} \sum\limits_{r=0}^{\infty}{}' \dfrac{T_{2r}(x)}{4r^2 - 1}.$

5.23 The orthogonal polynomials are $1, x, x^2 - \frac{5}{2}, x^3 - \frac{17}{5}x.$

5.25 $x + \frac{1}{8}.$

5.26 $\frac{1}{4} + \frac{1}{2}\sqrt{2} - \frac{1}{2}x.$

5.28 Error is not greater than $\frac{1}{48}.2^{-k}$. Approximations to $\sqrt{2}, \sqrt{3}, \sqrt{10}$ are $1\frac{3}{8}, 1\frac{17}{24}, 3\frac{1}{12}$ with errors less than $\frac{1}{24}, \frac{1}{24}, \frac{1}{12}$ respectively.

5.34 Degree 5 or 6; further analysis is necessary to determine which is smallest required.

5.35 The last two terms can be "economized".

Chapter 6

6.2 Error is $\frac{1}{3}h^2 f^{(3)}(\xi_0)$, where $\xi_0 \in (x_0, x_2)$.

6.3 We obtain the differentiation rule of Problem 6.1.

6.8 $f'(x_0) \simeq (f(x_0 + h) - f(x_0 - h))/2h.$

6.9 We need to differentiate $E(h) = 2.10^{-k}/h^2 + \frac{1}{12}h^2 M_4.$

6.10 For $h = 0.1, 0.2, 0.3$ we obtain $0.9600, 0.9675, 0.9656$ respectively, with $h = 0.2$ yielding the best value. This agrees with the result of Problem 6.9, which predicts $h \simeq 0.22$.

6.14 We obtain $a_0 = a_3 = \frac{3}{8}, a_1 = a_2 = \frac{9}{8}$. (This is called the three-eighth's rule.)

6.15 Follow the method of Problem 6.14 and obtain $b_1 = b_2 = \frac{3}{2}$.

6.16 It is sufficient to put $x_0 = 0, h = 1$ and show that the rule is exact for $f = 1, x, x^2$ and x^3.

6.26 Use the fact that $\sum w_i = b - a$.

6.27 The required interpolating polynomial for e^{-t^2} is $1 + 10t(e^{-0.01} - 1)$. Since

$$\left| \frac{d^2}{dt^2} e^{-t^2} \right| \leqslant 2$$

the error of the interpolating polynomial is less than 0.0025. Thus, on integrating, $F(x) \simeq x + 5x^2(e^{-0.01} - 1)$ with error less than 0.00017 on $[0, 0.1]$.

6.29 Choose $M = 3$. Then Simpson's rule on $[0, 3]$ with $h < 0.2$ will give adequate accuracy.

6.30 Error is bounded by $(b - a)^2 h^2 (M_x + M_y)/12$, where M_x is a bound for $|\partial^2 f/\partial x^2|$ and M_y is defined similarly.

Chapter 7

7.1 If the equation is $f(x) = 0$, $f(0) > 0$ and $f(1) < 0$ and since $f'(x) < 0$ on $[0, 1]$ there is exactly one root. Fourteen iterations are required if we use the mid-point of the final subinterval $[x_0, x_1]$ to estimate the root.

7.2 Root is $x \simeq 1.6$.

7.4 One step of Muller's method gives $x = 0.3822$ (compared with $\cos(3\pi/8) = 0.3827$).

7.5 The root is 0.091 to three decimal places; (i) requires ten iterations and (ii) only three iterations.

7.7 We need to choose λ near to $\cos(0.5) \simeq 0.9$. (See (7.20).) With $\lambda = 0.9$ we obtain 0.510828, 0.510971, 0.510973 for x_1, x_2, x_3. The last iterate is correct to six decimal places.

7.10 We have $x_{10} = 0.64563$, $a_{11} = 0.09091$, $a_{12} = -0.08333$ so that $x_{12}^* = 0.69306$ which agrees with ln 2 to four decimal places.

7.12 The equation is $ax^2 + bx + c = 0$. If $c = 0$ the roots are $x = 0$ and $-b/a$ and the iterative method finds the second root in one iteration if $x_0 \neq 0$.

7.14 The pth root process is $x_{r+1} = \dfrac{1}{p}\left[(p - 1)x_r + \dfrac{c}{x_r^{p-1}}\right]$.

7.15 We find (see Theorem 7.2):

  (i) $f(a) < 0$ and $f(b) > \frac{1}{4}(a + c/a)^2 - c = \frac{1}{4}(a - c/a)^2 > 0$.

  (ii) $f''(x) = 2$.

  (iii) We need to show that $\frac{1}{2}(x + c/x) \in [a, b]$ for $x = a$ (which is easy) and $x = b$. For $x = b$, we find $\frac{1}{2}(b + c/b) > c^{1/2} > a$ and, since $b > \frac{1}{2}(a + c/a) > c^{1/2}$, we have $b^2 > c$ and thus $\frac{1}{2}(b + c/b) < b$.

7.16 Show that $x_{r+1} \pm c^{1/2} = \dfrac{1}{2x_r}(x_r \pm c^{1/2})^2$.

7.17 We obtain a sequence which converges to $-c^{1/2}$.

7.19 If $g(x) = x - \lambda f(x)/f'(x)$, then

$$g'(x) = 1 - \lambda + \lambda \frac{f(x)f''(x)}{[f'(x)]^2}$$

and if $f(x) = (x - \alpha)^2 h(x)$, the numerator and denominator of the last term each has a factor $(x - \alpha)^2$. We find that

$$g'(\alpha) = 1 - \lambda + \frac{1}{2}\lambda = 0 \quad \text{if } \lambda = 2.$$

7.20 Iterative process for finding $1/c$ is $x_{r+1} = 2x_r - cx_r^2$.

7.27 The closure condition of Theorem 7.3 is easily verified, using the given inequality. For the Lipschitz condition, if $g(x) = 2 - x^{-k}$, $g'(x) = kx^{-k-1}$ and on $[2 - 2^{1-k}, 2]$

$$x^k \geqslant (2 - 2^{1-k})^k = 2^k(1 - 2^{-k})^k \geqslant 2^{k-1} \geqslant k$$

for $k = 2, 3, \ldots$. Thus

$$|g'(x)| \leqslant \frac{1}{|x|} = 1/(2 - 2^{1-k}) < 1$$

and g satisfies a Lipschitz condition with $L < 1$. With $k = 6$ and $x_0 = 2$ we have $x_1 = 1.984$ which is correct to three decimal places.

7.28 By continuity of g' we can find $\delta > 0$ such that $|g'(x)| \leqslant L < 1$ for all $x \in [\alpha - \delta, \alpha + \delta] \subset I$. For closure,

$$|g(x) - \alpha| = |g(x) - g(\alpha)| \leqslant |x - \alpha| \cdot |g'(\xi)| \leqslant \delta.$$

7.29 We have $\psi'(y) = \dfrac{dx}{dy} = 1 \bigg/ \dfrac{dy}{dx} = 1/g'(x)$. Thus $|\psi'(y)| \geqslant 1/L > 1$.

7.30 If $g(x) = c/x^{p-1}$, $g'(x) = (1 - p)cx^{-p}$ and $|g'(c^{1/p})| = |1 - p| \geqslant 1$.

Chapter 8

8.2 $\mathbf{A} = [1 \quad 0]$, $\mathbf{B} = [0 \quad 1]^T$.

8.3 $\mathbf{A} = [1 \quad 0]$, $\mathbf{B} = [0 \quad 1]$, $\mathbf{C} = [1 \quad 1]^T$.

8.4 (a) Let $\mathbf{C} = \mathbf{AB}$, where $\mathbf{A}$ and $\mathbf{B}$ are lower triangular.

$$c_{ij} = \sum_{k=1}^{n} a_{ik}b_{kj} = \sum_{k=1}^{i} a_{ik}b_{kj}$$

as $a_{ik} = 0$ for $k > i$. Thus $c_{ij} = 0$ for $j > i$ as $b_{kj} = 0$ for $j > i \geqslant k$.

8.6 The ith column of $\mathbf{AD}$ is the ith column of $\mathbf{A}$ multiplied by d_i.

8.11 Need to interchange rows (or columns) e.g. rows 2 and 3.

$$\mathbf{x} = [-3 \quad 2 \quad -1 \quad 4]^T.$$

8.16 $\lambda = 0$, $\mathbf{x} = [\alpha \quad 0 \quad -\alpha]^T$; $\lambda = 2$, $\mathbf{x} = [3\alpha \quad 2\alpha \quad \alpha]$; $\lambda = -2$, $\mathbf{x} = [3\alpha \quad -2\alpha \quad \alpha]$ where $\alpha \neq 0$ is arbitrary.

8.18 $3\mathbf{u}_1 - \mathbf{u}_2 - \mathbf{u}_3 = 0$.

8.20

$$\int \left[\sum_i c_i \psi_i \right]^2 dx = \int \sum_j c_j \psi_j \left[\sum_i c_i \psi_i \right] dx$$

$$= \sum_j c_j \int \psi_j \sum_i c_i \psi_i \, dx = \sum_j c_j \left[\sum_i c_i \int \psi_i \psi_j \, dx \right] = 0.$$

Finally use the argument at the end of §8.3.

8.25

$$\mathbf{A} = \begin{bmatrix} 1 & 0 & 0 & 0 \\ 2 & 1 & 0 & 0 \\ -2 & 2 & 1 & 0 \\ 1 & 1 & -1 & 1 \end{bmatrix} \begin{bmatrix} 1 & 0 & 0 & 0 \\ 0 & 1 & 0 & 0 \\ 0 & 0 & -3 & 0 \\ 0 & 0 & 0 & 3 \end{bmatrix} \begin{bmatrix} 1 & 2 & -2 & 1 \\ 0 & 1 & 2 & 1 \\ 0 & 0 & 1 & -1 \\ 0 & 0 & 0 & 1 \end{bmatrix}.$$

8.30 $(\mathbf{ABC})^T = ([\mathbf{AB}]\mathbf{C})^T = \mathbf{C}^T[\mathbf{AB}]^T = \mathbf{C}^T\mathbf{B}^T\mathbf{A}^T.$

8.31 (b)

$$\begin{bmatrix} 1 & 1 \\ 1 & 1 \end{bmatrix}\begin{bmatrix} 1 & 0 \\ 0 & 0 \end{bmatrix} = \begin{bmatrix} 1 & 0 \\ 1 & 0 \end{bmatrix}.$$

8.33

$$\mathbf{M} = \begin{bmatrix} 2 & 0 & 0 \\ -1 & 4 & 0 \\ -2 & 2 & 1 \end{bmatrix}, \qquad \mathbf{x} = \begin{bmatrix} 2 \\ 1 \\ -1 \end{bmatrix}.$$

8.34

$$\mathbf{x}^T\mathbf{A}^2\mathbf{x} = \mathbf{x}^T\mathbf{A}^T\mathbf{A}\mathbf{x} = (\mathbf{A}\mathbf{x})^T\mathbf{A}\mathbf{x} = \mathbf{y}^T\mathbf{y},$$

where $\mathbf{y} = \mathbf{A}\mathbf{x}$. As $\mathbf{A}$ is non-singular, $\mathbf{x} \neq \mathbf{0} \Rightarrow \mathbf{y} \neq \mathbf{0}$ and $\mathbf{y}^T\mathbf{y} = y_1^2 + \cdots + y_n^2 > 0$ for $\mathbf{y} \neq \mathbf{0}$.

8.37 Suppose $\mathbf{b} \neq \mathbf{0}$ and $\mathbf{\Psi}^T\mathbf{b} = \mathbf{0}$. The jth component of this last equation is

$$b_0\psi_0(x_{j-1}) + b_1\psi_1(x_{j-1}) + \cdots + b_n\psi_n(x_{j-1}) = 0$$

for $0 \leq j \leq N$. Thus $b_0\psi_0 + \cdots + b_n\psi_n$ has $N+1$ zeros, which is not possible for a Chebyshev set.

8.39 From $\mathbf{A}\mathbf{A}^{-1} = \mathbf{I}$ we have

$$[i\text{th row of } \mathbf{A}] \begin{bmatrix} j\text{th} \\ \text{column} \\ \text{of } \mathbf{A}^{-1} \end{bmatrix} = \begin{cases} 1, & i = j \\ 0, & i \neq j \end{cases}$$

so that if rows of $\mathbf{A}$ are interchanged corresponding columns of $\mathbf{A}^{-1}$ must be interchanged.

8.40 The equations (8.46) are $\mathbf{L}\mathbf{x}_j = \mathbf{e}_j$, where $\mathbf{x}_j$ is the jth column of $\mathbf{L}^{-1}$. Consider the solution of these equations by forward substitution.

8.43 $\mathbf{x} = [2 \quad 3 \quad 4 \quad 5 \quad 6]^T.$

Chapter 9

9.1 $\|x\| = \|(x - y) + y\| \leqslant \|x - y\| + \|y\|$ by (9.1). Thus $\|x\| - \|y\| \leqslant \|x - y\|$. Similarly $\|y\| - \|x\| \leqslant \|x - y\|$.

9.4

$$x = \begin{bmatrix} 1 \\ 0 \end{bmatrix}, \qquad y = \begin{bmatrix} 0 \\ 1 \end{bmatrix}.$$

9.6 $AA^{-1} = I \Rightarrow \|AA^{-1}\| = \|I\| = 1$. By Lemma 9.2(v), $\|A\| \cdot \|A^{-1}\| \geqslant \|A^{-1}A\| = 1$. For second part use $A^{-1} - B^{-1} = A^{-1}(B - A)B^{-1}$ and Lemma 9.2(v).

9.7 Let $B = I + A$, when $\|A\|_1 = 0.9$. By Lemma 9.3, B is non-singular and $\|B^{-1}\|_1 \leqslant 1/(1 - \|A\|_1) = 10$.

9.9 $\|A^r\| \leqslant \|A\|^r$ so that by Lemma 9.2

$$\|p(A)\| \leqslant a_0\|I\| + a_1\|A\| + \cdots + a_n\|A^n\|$$
$$\leqslant a_0 + a_1\|A\| + \cdots + a_n\|A\|^n = p(\|A\|).$$

9.11 $\left\|\dfrac{1}{n!} A^n\right\| \leqslant \dfrac{1}{n!} \|A\|^n \to 0$ as $n \to \infty$, regardless of the value of $\|A\|$. Thus

$$\lim_{n \to \infty} \left\|\frac{1}{n!} A^n - 0\right\| = 0.$$

9.13 $k_1(A) = k_\infty(A) \simeq 12011$.

9.17 $A^{-1} = \dfrac{1}{\lambda}\begin{bmatrix} 1 & -\lambda \\ -1 & 2\lambda \end{bmatrix}$. For $|\lambda| \leqslant \frac{2}{3}$, $\|A\|_\infty = 2$, $\|A^{-1}\|_\infty = 2 + 1/|\lambda|$ and $k_\infty(A) = 4 + 2/|\lambda|$. For $|\lambda| > \frac{2}{3}$, $\|A\|_\infty = 3|\lambda|$, $\|A^{-1}\|_\infty = 2 + 1/|\lambda|$ and $k_\infty(A) = 3 + 6|\lambda|$.

9.19 Use (9.39) with $\delta b = r_0$.

9.20 $0.498 \times$ first equation is subtracted from the second equation.

$$r_0 = -[0.000360 \quad 0.00100]^T, \qquad \delta x_0 = [0.139 \quad -0.276]^T$$

so that $x_1 = [0.973 \quad -1.95]^T$. A further application of the process yields $x_2 = [0.996 \quad -1.99]^T$.

9.22 $\|I - \alpha A\| < 1 \Rightarrow \alpha \neq 0$. With $W_0 = (A + \delta A)^{-1} = \alpha I$,

$$\delta A(A + \delta A)^{-1} = \left(\frac{1}{\alpha} I - A\right)\alpha I = I - \alpha A.$$

Choose $\alpha = \frac{1}{4}$, when $\|\mathbf{I} - \alpha\mathbf{A}\|_\infty = \frac{1}{2}$.

$$\mathbf{W}_0 = \begin{bmatrix} 0.25 & 0 \\ 0 & 0.25 \end{bmatrix}, \quad \mathbf{W}_1 = \begin{bmatrix} 0.3125 & -0.0625 \\ -0.125 & 0.25 \end{bmatrix},$$

$$\mathbf{W}_3 = \begin{bmatrix} 0.399 & -0.099 \\ -0.199 & 0.299 \end{bmatrix}.$$

9.23 (a) $\mathbf{x}_3 = [-0.895 \quad -3.865 \quad -2.830]^T$

(b) $\mathbf{x}_3 = [-0.978 \quad -3.983 \quad -2.990]^T$.

Chapter 10

10.1 Closure condition is straightforward. Second, from (10.9),

$$|g_1(\mathbf{x}) - g_1(\mathbf{x}')| \leqslant \frac{1}{12}[|x_1 - x_1'|.2 + |x_3 - x_3'|.2] \leqslant \frac{1}{6}\|\mathbf{x} - \mathbf{x}'\|_1.$$

Similarly $|g_2(\mathbf{x}) - g_2(\mathbf{x}')| \leqslant \frac{1}{6}\|\mathbf{x} - \mathbf{x}'\|_1$ and $|g_3(\mathbf{x}) - g_3(\mathbf{x}')| \leqslant \frac{1}{3}\|\mathbf{x} - \mathbf{x}'\|_1$ and so $\|\mathbf{g}(\mathbf{x}) - \mathbf{g}(\mathbf{x}')\|_1 \leqslant \frac{2}{3}\|\mathbf{x} - \mathbf{x}'\|_1$, giving a contraction mapping.

10.2 If there exists a solution (x, y), then we see that $-\frac{1}{2} \leqslant x, y \leqslant \frac{1}{2}$. The closure and contraction mapping properties are easily verified, showing that the solution exists and is unique. With $x_0 = y_0 = 0$, we obtain $x = 0.49$, $y = 0.23$ in four iterations.

10.5

$$\mathbf{G} = \begin{bmatrix} \frac{1}{3} & \frac{1}{3} & \frac{1}{3} \\ \frac{1}{3} & \frac{1}{6} & \frac{1}{6} \\ \frac{1}{3} & \frac{1}{6} & \frac{1}{6} \end{bmatrix},$$

where $\mathbf{G}$ has elements g_{ij} given by (10.10). Thus from (10.12) $L = \sqrt{(2/3)}$, showing there is a contraction mapping. Note that in this case we cannot infer that a contraction mapping exists with either of the other two common norms.

10.7 Two iterations of Newton's method gives $x_1 = -0.82$, $x_2 = 1.91$.

10.10 The first two iterations give $x = \frac{1}{2}, y = \frac{1}{4}$ and $x = 0.4865, y = 0.2338$.

$$\mathbf{J}^{-1} = \frac{1}{1 + \frac{1}{4}\sin y \cos x} \begin{bmatrix} 1 & -\frac{1}{2}\sin y \\ \frac{1}{2}\cos x & 1 \end{bmatrix},$$

which exists for all x and y.

10.11 In this case $\mathbf{J} = \mathbf{A}$ and Newton's method is

$$\mathbf{x}_{r+1} = \mathbf{x}_r - \mathbf{A}^{-1}(\mathbf{A}\mathbf{x}_r - \mathbf{b})$$

which gives $\mathbf{x}_{r+1} = \mathbf{A}^{-1}\mathbf{b}$ and thus we do not obtain an iterative method.

Chapter 11

11.1 $L = 1$. With $y_0(x) = 1$, $y_n(x) = 1 + \dfrac{x^2}{2} + \dfrac{x^4}{2.4} + \cdots + \dfrac{x^{2n}}{2.4.\cdots.2n}$.

11.2 $L = 1$.

11.4 (a) $c_1 a^n$. (d) $c_1 + c_2 2^n + c_3 n 2^n$.

11.5(a) $c_1 2^n + c_2 n 2^n + 1$. (c) $c_1 + c_2 n + \frac{1}{6} n^3$.

11.7   Difference equation: $y_{n+1} = y_n + anh^2 + bh$ with $y_0 = 0$. Solution (verify by induction): $y_n = \frac{1}{2} an^2 h^2 + bnh - \frac{1}{2} anh^2$.

11.10   Simple: $y_1 = 0.024$, $y_2 = 0.09488$, $y_3 = 0.2186$.

Classical: $y_1 = 0.02127$, $y_2 = 0.08969$, $y_3 = 0.2112$.

$y(0.6) = y(x_3) = 0.2112$.

11.13 $L_\phi = L = 1$. $|y''| = |2 - e^{-x}| \leqslant 2 - e^{-1} \simeq 1.6321$. Thus $|t(x; h)| = \frac{1}{2} h |y''(\xi)| \leqslant 0.8161h$ and (11.59) becomes $|y(x_n) - y_n| \leqslant (e^{x_n} - 1)0.8161h$.

11.14 $p = 2$, $L_0 = L = 3\sqrt{3}/8$, $L_1 = 2$, $L_\phi = 3\sqrt{3}/8 + h_0$, $N = \frac{1}{3}$.

$$|y(x_n) - y_n| \leqslant \frac{1}{3} \left(\frac{e^{L_\phi x_n} - 1}{L_\phi} \right) h^2.$$

11.17 $y_{n+1} = y_n + \dfrac{h}{12} (23 f_n - 16 f_{n-1} + 5 f_{n-2})$.

11.18 $y_5 = 0.6265$, $y(1.0) = 0.6321$.

11.19 $B_1 = 2$, $b_2 = \frac{5}{12}$, $L = 1$, $M_3 = 1$, $e_5 \leqslant 0.0569$.

11.21 $m = 0, 1$: $y_{n+1} = y_{n-1} + 2hf_n$.

$m = 2$: $y_{n+1} = y_{n-1} + \dfrac{h}{3} (7 f_n - 2 f_{n-1} + f_{n-2})$.

11.22   See Problem 11.30(i).

11.24 $y_5 = 0.6334$.

11.25 $\delta = 0$, $|\gamma_{-1}| = \frac{1}{2}$, $|c_2| = \frac{1}{12}$, $M_3 = 1$, $C_0 = 1$, $L = 1$, $e_5 \leqslant 0.00679$.

11.30 (i)   Stable for $h \leqslant 3/(2A)$. (ii) Unstable. (iii) Unstable.

11.34   (a) $y_{n+1} = (1 - Ah + \frac{1}{2} A^2 h^2) y_n \Rightarrow y_n = (1 - Ah + \frac{1}{2} A^2 h^2)^n y_0 \to 0$ only if $|1 - Ah + \frac{1}{2} A^2 h^2| < 1$.

(c) From §11.11, $y_n = c_1 z_1^n + c_2 z_2^n$ and it can be shown that $|z_1| < 1$ for all $Ah > 0$, $|z_2| < 1$ for $1 > Ah > 0$ and $|z_2| \geqslant 1$ for $Ah \geqslant 1$.

11.36 $y_{1,n+1} = y_{1,n} + h(x_n{}^2 y_{1,n} - y_{2,n})$

$$+ \frac{h^2}{2}[(1 + 2x_n + x_n{}^4)y_{1,n} - x_n(1 + x_n)y_{2,n}]$$

$y_{2,n+1} = y_{2,n} + h(-y_{1,n} + x_n y_{2,n})$

$$+ \frac{h^2}{2}[-x_n(1 + x_n)y_{1,n} + (2 + x_n{}^2)y_{2,n}].$$

11.37 $y_5 = 0.8187$, $y_5' = -0.7301$. Exact solution $y(x) = (1 + x^2)^{-1}$.

11.40 Let $z = y'$, $w = z' = y''$. Corrector is

$y_{n+1}^{(r+1)} = y_n + \frac{1}{2}h(z_n + z_{n+1}^{(r)})$

$z_{n+1}^{(r+1)} = z_n + \frac{1}{2}h(w_n + w_{n+1}^{(r)})$

$w_{n+1}^{(r+1)} = w_n + \frac{1}{2}h(2w_n + 2w_{n+1}^{(r+1)} + x_n^2 y_n + x_{n+1}^2 y_{n+1}^{(r+1)} + 2 + x_n + x_{n+1}).$

Chapter 12

12.6
$$-1.6y_0 + 2y_1 - 0.08y_0{}^2 = 0$$
$$y_0 - 2y_1 + y_2 + 0.2y_1(y_2 - y_0) = 0$$
$$y_1 - 2y_2 + y_3 + 0.2y_2(y_3 - y_1) = 0$$
$$y_2 - 2y_3 + y_4 + 0.2y_3(y_4 - y_2) = 0$$
$$y_3 - 1.9y_4 - 0.2y_3 y_4 + 0.5 = 0.$$

12.8 $h < \sqrt{6}/(10e) \simeq 0.0901$.

12.10 0.6400
   0.5408
0.5904   0.5512
   0.5486   0.5509
0.5695   0.5509
   0.5503
0.5599

$y(0.8) = 0.55067$.

12.11 $y_{-1} = -0.2500$, $y_0 = 0$, $y_1 = 0.2000$, $y_2 = 0.3600$, $y_3 = 0.4880$, $y_4 = 0.5904$; $z_1 = 0.1750$, $z_2 = 0.3200$, $z_3 = 0.4400$, $z_4 = 0.5392$.

12.12 $(z_{n+1} - 2z_n + z_{n-1})/h^2 = f(x_n, z_n) + \Delta^4 y_{n-2}/(12h^2)$.

12.15 $p_3(x) = (27T_0 - 19T_1 + 5T_2 + T_3)/22 = (2x^3 + 5x^2 - 11x + 11)/11$.

12.17 $p_4(x) = (51T_1 + 2T_3)/53 = (8x^3 + 45x)/53$.

Index